Current Developments in Biological Nitrogen Fixation

Current Developments in Biological Nitrogen Fixation

Edited by
N.S. SUBBA RAO

CAMBRIDGE UNIVERSITY PRESS
Cambridge, New York, Melbourne, Madrid, Cape Town, Singapore, São Paulo, Delhi

Cambridge University Press
The Edinburgh Building, Cambridge CB2 8RU, UK

Published in the United States of America by Cambridge University Press, New York

www.cambridge.org
Information on this title: www.cambridge.org/9780521105750

First published by Edward Arnold (Publishers) Ltd 1984
This digitally printed version by Cambridge University Press 2009

A catalogue record for this publication is available from the British Library

ISBN 978-0-521-41753-2 hardback
ISBN 978-0-521-10575-0 paperback

Foreword

From time immemorial legumes have been of prime importance in Indian agriculture and together with millets and rice (associated with nitrogen-fixing blue-green algae), they form the staple of primary production. With this background of history and continuing recognition of the value of legumes, a book from New Delhi on nitrogen fixation is specially welcome.

The editor is exceptionally well qualified to organise and assemble the material of this volume having worked throughout his career on many aspects of nitrogen fixation in nodulated legumes and by free-living microorganisms. He has also been personally involved in a number of projects for legume improvement, notably the 'All India Coordinated Pulse Improvement Programme', the 'International Biological Programme on Nitrogen Fixation' and the 'Coordinated Project on Biological Nitrogen Fixation' of the Indian Council of Agricultural Research.

Over the last two decades there has been a remarkable resurgence of interest and research in biological nitrogen fixation caused by mounting worldwide concern over the depletion of energy resources. This was sharply accentuated by the 1973 and subsequent oil price rises and the consequent several-fold increases in the cost of nitrogen fertilizer, which has had a particularly serious impact upon the agriculture of developing countries. The research on nitrogen fixation stimulated by the energy crisis has led to a better assessment of its potential, and an appreciation of the need to promote its efficient use in agriculture by all possible means: by extending the areas of leguminous and non-leguminous nitrogen-fixing plants especially those with special relationships with free-living nitrogen-fixing microorganisms. Equally important is the need to increase the amounts of nitrogen fixed in these crops by the application of this newly gained knowledge as well as by ensuring their optimal nutrition and pest and disease control by better agronomy.

Agricultural statistics show that most crop legumes and forage plants fall far short of their potential yields and that in general those of the tropics fix appreciably less nitrogen than those of temperate agriculture. This disparity may reflect the larger environmental stresses experienced in the tropics, especially respecting temperature and water relation. It may also be partly a consequence of problems caused by symbiotic specificity. Because many tropical legumes belong to the 'cowpea miscellany', they are readily cross-infected and the consequent symbioses are not always the most effective. The widespread occurrence of naturally existing strains of *Rhizobium* of less than elite status present considerable problems in their replacement with better ones by inoculation.

Recent research has brought us to the verge of complete understanding of some of the most intractable problems of symbiosis and nitrogen fixation: the mechanism and biochemistry of the infection processes, the nature of specificity, the economy of nitrogen transportation and use within the plant and the interrelation of bacterial and host physiology and its genetic control. Some of these advances have come from the use of new techniques, notably the acetylene reduction assay for nitrogenase, the use of isotopes and the relatively new and powerful tools of microbial genetics. Some also have come from originality in outlook and interpretation.

New ideas and methodologies will doubtless promote new advances, but if the past is any guide, progress is most likely to be fostered by the multidisciplinary approach either by research teams or by individuals expert in more than one field.

This volume gives an up-to-date assessment of the present knowledge in some of these aspects of special interest, viz. the genetics, ecology, physiology and metabolism of certain nitrogen fixing systems; necessarily a single volume can cover only a part of so large a subject. As appropriate to a work of this scope the authors are cosmopolitan and do not restrict their interest to the tropical scene. Nevertheless students, research workers and agriculturalists attracted by the challenging problems of biological nitrogen fixation will find this volume valuable and stimulating.

P.S. NUTMAN, FRS
Great Hackworthy Cottage,
Tedburn, St. Mary, Exeter,
Ex 66 DIN
England

Preface

This volume on 'Current Developments in Biological Nitrogen Fixation' is an extension of the objectives with which the earlier volume on 'Recent Advances in Biological Nitrogen Fixation' was brought out a few years ago. The objective was to focus attention on the many research and developmental aspects of the unique process of microbiologically mediated reduction of dinitrogen of the atmosphere into ammoniacal form of nitrogen. In recent years, while the developing countries of the world are eagerly attempting to harness the practical aspects of the biological nitrogen-fixing processes on the farm front by field oriented technologies as a specific measure to relieve the stress on chemical fertilizers, the affluent nations are equally engaged in understanding the genetic control and the mechanism of fixation in biological nitrogen-fixing systems in general so as to pave the way for a new bio-technology to render the crop plants self-sufficient with regard to their nitrogen nutrition. As a result of these increased research activities, it was felt desirable and appropriate to take stock of the current situation again by summarizing the present state of art on some aspects of biological nitrogen fixation in the form of a book.

The nitrogen nutrition of nodulated legumes is dependent upon the effective rhizobial symbiosis with the appropriate host plant. In rain-fed tropical soils where most of the grain legumes are nodulated by *Rhizobium* sp. (cowpea miscellany), the competition with native ineffective rhizobia together with the stress factors operating on both the symbionts by way of high soil temperature and drought often tend to limit the formation of effective root nodules. These aspects of rhizobial survival in tropical soils form the basis of the first chapter written by Eaglesham and Ayanaba. A related subject is the interaction of nitrogen-fixing microorganisms in soil with other soil microorganisms. The associative and antagonistic factors operating in *Rhizobium* survival and performance in soil is governed by the activity of other microorganisms which perform other functions such as the solubilization of bound phosphates, provision of auxiliary hormones and growth factors or secretion of antibacterial substances. Among the beneficial associates, the vesicular-arbuscular mycorrhizal (VAM) fungi appear to be significant. The non-symbiotic nitrogen-fixing species, *Azotobacter chroococcum* is known to secrete an antifungal antibiotic substance active against plant pathogens. Many viral infections of leguminous plants diminish the benefits of root nodulation. Therefore, an integrated view of these interactions is taken by Subba Rao who has written the second chapter on microbial interactions with reference to biological nitrogen fixation.

The meeting point between a nitrogen-fixing microorganism and the host plant benefiting from root associations is the root surface. There are biochemical events which are very significant in the pre- as well as post-penetration stages of interactions between roots of plants and nitrogen-fixing microorganisms. One of the widely discussed factors operating in the pre-penetration stages is the specific protein called lectin which binds the surface of two symbionts of legume-*Rhizobium* symbiosis and determines the attachment of nitrogen-fixing bacteria to the root surface. Recently, however, an intense controversy has emerged on the involvement of lectins in legume-*Rhizobium* infection, a process which still remains highly enigmatic. Nevertheless Dazzo and his associate, Truchet who uphold the lectin mediated recognition phenomenon explain the recent developments on this aspect in the third chapter.

Nitrogen-fixing nodules on stems of plants would indeed be a convenient way to increase the surface area for plant-*Rhizobium* interaction and stem nodulation has indeed been recorded in the genera *Aeschynomene* and *Sesbania*. The fourth chapter on stem nodules by Subba Rao and Yatazawa highlights the current status of work on this aspect and points out the importance of extending research in this area.

Ironically, nitrogen fixation is an anaerobic process even in highly aerobic systems such as *Azotobacter chroococcum*. While this is acknowledged universally, precise explanations for oxygen control mechanisms have not come forth in any of the free-living systems. Needless to say, an understanding of the oxygen control mechanisms is a necessary prelude to achieving the transplantation of nitrogen-fixing bacteria into highly aerobic plant protoplasts. A review of the current knowledge on oxygen control mechanisms has been given by Shaw of New Zealand in the fifth chapter.

Genes responsible for nodulation and nitrogen fixation situated in megaplasmids have been demonstrated in *Rhizobium meliloti*. Being extrachromosomal in occurrence, plasmids offer tremendous potentiality as vehicles for *nif* gene transfer to higher plants. Although there is yet a long way to go in this exercise, Kondorosi and his colleagues explain in the sixth chapter how plasmids work in the expression of symbiosis in legumes.

Apart from cultivated legumes, forest trees such as alder (*Alnus* spp.) and *Casuarina* spp. benefit from nitrogen-fixing symbiosis through an actinomycetous microsymbiont designated as *Frankia*. The microsymbiont has been isolated successfully in *Alnus* and *Comptonia* and many of the physiological aspects of symbiosis have now been well understood. The seventh chapter by Wheeler summarizes the recent advances in the knowledge on *Frankia* symbiosis.

Lichens are composite structures consisting of blue-green algae and fungal symbionts, the nitrogen-fixing blue-green alga helping the fungal partner in nitrogen nutrition while the fungal partner provides protection for the survival of the alga even under desiccated condition. Lichens have been recognized as primary colonizers of rocks and exhibit an extreme capacity to survive under environmental hazards. They also serve as an excellent example of symbiosis. The intricate aspects of nitrogen fixation in lichens have been covered by Millbank in the eighth chapter.

In Brazil, sugar cane is grown on virgin soil requiring very little mineral nitrogen fertilizer. Biological nitrogen fixation has been demonstrated by finite methods in this crop and several non-symbiotic nitrogen-fixing bacteria have been implicated in this process. The role of bacteria in nitrogen fixation in sugar cane has been dealt with by Ruschel and Vose in the ninth chapter.

Wetland rice fields provide an ideal microhabitat for photosynthetic as well as heterotrophic nitrogen-fixing microorganisms which live and fix nitrogen under submerged conditions. The International Rice Research Institute, Manila, Philippines has carried out intensive work on this system and Watanabe and Roger provide an integrated information on this subject in the tenth chapter.

Graminaceous plants have nitrogen-fixing bacteria in their root system among which *Azospirillum* has been highlighted in recent years. The role of *Azospirillum* and other diazotrophic nitrogen-fixing bacteria in the nitrogen nutrition of grasses such as sorghum and millets has been substantiated from more than one laboratory in recent years. Many of the recent developments in this exciting area have been elucidated by Boddey and Döbereiner in the eleventh chapter.

The genetics of *Azotobacter* and *Azospirillum* is receiving increasing attention in recent years, more particularly on plasmid controlled molecular biology. From the world famous Pasteur Institute, Claudine Elmerich explains the methodology and results connected with this area of research in the twelfth chapter.

Admittedly, it is difficult to bring home all the developments in biological nitrogen fixation in a single handy volume but nevertheless, the present exercise highlights potential areas where significant advances are being currently made. In this task, I owe a great deal to the various contributors for their prompt compliance with my request in providing the manuscript. I wish to express my indebtedness to Dr. O.P. Gautam, Director-General, Indian Council of Agricultural Research, Dr. N.S. Randhawa, DDG(SAE), and Dr. H.K. Jain, Director, Indian Agricultural Research Institute, New Delhi for their kind encouragement. Finally, I will be failing in my duty if I do not say a word of thanks to my wife Gowri Subba Rao and my daughters

Shambhavi Subba Rao and Shalini Subba Rao who have helped me in so many ways in the preparation of this volume.

N.S. SUBBA RAO
Microbiology Division,
Indian Agricultural
Research Institute,
New Delhi 110012

Contents

Contributors

A. Ayanaba, International Institute of Tropical Agriculture, Ibadan, Nigeria

R.M. Boddey, Programma Fixacao Biologica de Nitrogenio, EMBRAPA/SALCS Seropedica, 23460, Rio de Janerio, Brazil

F.B. Dazzo, Department of Microbiology and Public Health, Michigan State University, East Lansing, Michigan 48824, USA

J. Döbereiner, Programma Fixacao Biologica de Nitrogenio, EMBRAPA/SALCS Seropedica, 23460, Rio de Janerio, Brazil

I. Dusha, Institute of Genetics, Biological Research Centre, Hungarian Academy of Sciences, H-6701, Szeged, P.O. Box 521, Hungary

A.R.J. Eaglesham, Boyce Thompson Institute, Tower Road, Ithaca, NY, 14853, USA

C. Elmerich, Physiology and Genetics Department, Pasteur Institute, 28, Rue Du D'Roux, 75724, Paris Cedex 15

G.B. Kiss, Institute of Genetics, Biological Research Centre, Hungarian Academy of Sciences, H-6701, Szeged, P.O. Box 521, Hungary

A. Kondorosi, Institute of Genetics, Biological Research Centre, Hungarian Academy of Sciences, H-6701, Szeged, P.O. Box 521, Hungary

J.W. Millbank, Department of Pure and Applied Biology, Imperial College of Science and Technology, London, SW7 2BB, UK

P.A. Roger, Office de la Recherche Scientifique et Technique Ontre Mer, France

A.P. Ruschel, CENA, Piracicaba, Sao Paulo, Brazil

B.D. Shaw, Plant Physiology Division, DSIR, Private Bag, Palmerston North, New Zealand

N.S. Subba Rao, Microbiology Division, Indian Agricultural Research Institute, New Delhi 110012, India

G.L. Truchet, Institut de Cytologie et Biologie Cellulaire, Faculte des Sciences Marseille—Lumimy L.A./C.N.R.S. 179, Marseille Cedex 2, 13288, France

P.B. Vose, CENA, Piracicaba, Sao Paulo, Brazil

I. Watanabe, Soil Microbiology Department, The International Rice Research Institute, P.O. Box 933, Manila, Philippines

C.T. Wheeler, Department of Botany, University of Glasgow, Glasgow, G12 8QQ, UK

M. Yatazawa, Faculty of Agriculture, Nagoya University, Chikusa, Nagoya, 464 Japan

1. Tropical Stress Ecology of Rhizobia, Root Nodulation and Legume Fixation

A.R.J. Eaglesham and A. Ayanaba

INTRODUCTION

To adequately feed the world's increasing population it is essential that food production be increased in the Third World where it will be most needed. This aim is the mandate of the International Institutes funded by the Consultative Group on International Agricultural Research and the many national agricultural institutes throughout the tropical world. Moreover, an increasing number of scientists in the developed countries are becoming involved in solving problems which relate directly to tropical agriculture.

The potentially important role of legumes in maintaining soil fertility is well established. Grain legumes provide valuable nutritious seed and, when effectively nodulated, can yield in nitrogen-deficient soils where cereals and other non-leguminous crops would barely survive. It is no coincidence that legumes are a component of many of the traditional farming systems throughout the tropics. If food production is to be increased with more productive farming systems, the package of improvement practices is likely to include an increased input of biologically fixed nitrogen to complement the use of fertilizer nitrogen if it is available. It is important to bear in mind, however, that contrary to popular opinion, the growing of a legume crop does not necessarily result in a nitrogen gain for a farming system, except where an effectively nodulated forage legume is ploughed under as a green manure. If more nitrogen is removed in the harvested grain than was fixed in the nodules, a net depletion results, even if all vegetative residues are ploughed under. The greater the amount of soil nitrogen that is available to inhibit nodulation and fixation, and the higher the harvest index for nitrogen, the more is the likelihood that a nitrogen depletion would occur [1]. The correct grain legume should be selected for a specific use, be it high grain yield at the possible expense of some soil nitrogen or less grain yield with the expectation of nitrogen accretion to the soil.

Those involved with rhizobia and legumes should not assume that "improving" nitrogen fixation is necessarily the main priority in increasing legume production [2]. With judicious use of fertilizers, water management, insect control, disease control, superior cultivars etc., the inputs of biologically-fixed nitrogen may increase concomitantly and spontaneously. On the other hand there is no room for complacency: the natural environment is constantly changing and field crops are subjected to stresses throughout the growth cycle. There exist too many gaps in the understanding, particularly in terms of the responses of biological nitrogen-fixing systems to these stresses. The objective of this review is to focus on the main environmental stresses commonplace in tropical agriculture and on how they may affect survival of rhizobia in soil, the root-nodulation process, and the functioning of the effectively nodulated legume.

For background reading and to gain access to related aspects of *Rhizobium* and legume research the reader is referred to the excellent reviews on *Rhizobium* ecology [3], nodule initiation and development [4, 5], functioning of legume nodules [6], environmental effects [7], legumes in acid soils [8, 9], legume nutrition [10–13], and tropical agricultural legumes [14].

ACIDITY

In wet equatorial zones rainfall exceeds evapotranspiration for much of the year and as a result soils become thoroughly leached of calcium and magnesium leaving them markedly acid. Acid soils are characterized by high concentrations of hydrogen ions and free aluminium, and low concentrations of calcium and available phosphate. Some acid soils contain manganese at phytotoxic levels and in some molybdenum is unavailable.

It appears that acidity is less of a constraint to cowpea rhizobial survival in soil than is desiccation or high temperature [15]. Thus, in an acid soil at Onne in Nigeria (pH 4.6, annual rainfall 2,500 mm) the cowpea rhizobial count was 4.3×10^4/g soil, whereas at Maradi in the sahel-savannah zone in Niger Republic (pH 6.1, annual rainfall 600 mm) the count was 4.9×10^2/g soil. Laboratory studies of the effects of acidity on rhizobia from soils such as these have been based on growth in synthetic media. However, because rhizobia vary in the ability to withstand conditions associated with acid soils, the acid tolerance of rhizobia cannot be predicted from the growth rate or acid production characteristics in liquid media at higher pH [16]. Moreover, because slow-growing rhizobia produce an alkaline reaction in most growth media [9] caution is needed when testing their ability to tolerate low pH-using conventional techniques. Growth media may be modified by changing the carbon source to arabinose from mannitol which is customarily recommended [17], so that pH is stabilized and remains at the initial low value throughout much of the growth cycle [18]. Alternatively, studies may be made during the early phase of growth up to visible turbidity before significant pH changes occur [19]. Although slow-growing rhizobia are in general more tolerant of low pH than the fast-growers, strain to strain differences exist [20]. *Rhizobium meliloti* is particularly sensitive to acid conditions [21]. Some slow-growing rhizobia native to acid soils are acid-requiring and grow only at approximately pH 4.5 [18]. On the other hand in a survey of 65 strains of slow-growing rhizobia of mixed origin, in liquid media acidity (pH 4.5 or 4.8) prevented the growth of 29 per cent of the strains and slowed the growth of most of the rest. Low phosphate levels limited the growth of some strains but with less severity than did acid. Aluminium (50 μM) was the most severe stress factor, stopping growth of 40 per cent of the strains. Tolerance of acidity was not necessarily correlated with tolerance of aluminium, since aluminium increased the lag time or slowed the growth rate of almost all of the strains which were tolerant of low pH [22]. A complementary study showed that while high manganese (200 μM) and low calcium (50 μM) had adverse effects on slow-growing rhizobia which varied from strain to strain in severity, neither was as severe a stress as aluminium, and strains which were tolerant of aluminium were also tolerant of manganese and low calcium [23]. The adverse effects of acid and aluminium on rhizobial growth appear to be bacteriostatic rather than bacteriocidal [24]. Unfortunately because of the lack of information in these surveys of rhizobial responses to acid-associated stresses, no correlation can be drawn between patterns of resistance or susceptibility and the type of soil from which the rhizobia originated.

The ability to grow in liquid media which mimic acid soil conditions may indicate an ability to survive in such soils with the potential to later colonize the host rhizosphere. An agar plate method for the rapid screening of rhizobia for tolerance to acidity and aluminium has recently been developed and is amenable for screening large numbers of rhizobial isolates [25]. This rapid screening technique may be used to reduce numbers for subsequent more critical examination in liquid media. Rhizobia which had been identified as tolerant of acid (pH 4.5) and aluminium (50 μM) in liquid media nodulated better and were more effective on cowpeas in an acid (pH 4.6) cum high aluminium soil than strains which had been identified as sensitive [26]. However, the pre-screening of rhizobia in acid liquid media was less useful when mung bean was used as the host in a soil of pH 5.0. The main cause of symbiotic failure in mung bean was sparsity of nodules which occurred even with some of the strains which grew in acid-defined media containing aluminium. These strains nodulated cowpea reasonably well in the same soil adjusted to pH 4.6, indicating a greater acid sensitivity in the nodulation process of mung bean [16]. Clearly, when testing the nodulating ability of rhizobia in acid conditions, an acid tolerant host is required.

When growing on mineral nitrogen, most legume species are only slightly adversely affected by acidity down to pH 4.0. Indeed some species actually grow better at pH 4.0 than in less acidic conditions, e.g. *Stylosanthes humilis* [27]. Legumes dependent on the root nodule symbiosis for nitrogen showed a range of responses to low pH, but in general nodulation was reduced or eliminated at pH values below 5 [27, 28]. In a survey of the effects of liming on eight soils of pH 3.4–4.25 the critical pH for nodule initiation and development in soybean was in the range 4.5–4.8 [29]. The inhibition of nodulation appears to result from a combination of low calcium and low pH since it was alleviated by increasing either calcium or pH [9]. Nodule initiation was more restricted when pea plants were exposed to pH 4.5 at two or three days after inoculation than at one or four days [30]. The lesion in the infection process which is induced by calcium deficiency and acidity has not been identified [9]. The nitrogen fixing activity of nodules is also adversely affected by acidity in many species [27, 28].

The presence of available aluminium in acid soils inhibits nodulation directly [31] and indirectly by stunting root growth, and also tends to compound the effects of low levels of calcium by inhibiting its uptake [8]. The mean nodule number of twelve soybean cultivars was highly correlated with primary root calcium content and inversely correlated with level of available aluminium in the soil. Exposure of nodulated roots of *Phaseolus vulgaris* to aluminium, however, had no effect on nodule development or function [31]. The inhibitory effect of the aluminium-calcium interaction on soybean nodulation varied with soil type. In two soils of higher Ca : Al ratio, mean growth of thirteen soybean cultivars at pH 4.5, although reduced in comparison with plants at pH 6, was the same whether they were relying on mineral nitrogen or nodule-fixed nitrogen. With 2×10^6 rhizobia/seed as inoculum nodule number and weight were the same at pH 4.5 as at pH 6 [33]. These findings indicate that at least for soils of this type, improvement of aluminium tolerance is more likely to be achieved by manipulating the plant rather than the *Rhizobium*. However, taking pains to use rhizobia that are stress tolerant has proven to be a wise precaution with soybean. In the acid, high aluminium soil at Onne in Nigeria mentioned earlier, soybean supplied with 150 kg N/ha gave a 74 per cent increase in grain yield in response to a lime application of 1 t/ha. A prior screening of effectiveness of rhizobia in the same soil led to the identification of two superior strains of *R. japonicum*. Plants inoculated with either of these strains did not respond to liming and yielded 1.9–2.1 t/ha of seed, approximately 43 per cent higher than the N + lime treatment [34].

Phosphorus deficiency is common in the acid soils of the tropics and in clay soils of high iron and aluminium content, phosphorus may be strongly adsorbed making the use of fertilizers uneconomic [12]. In some soils liming alleviates phosphorus deficiency but in others it may exacerbate it [35]. Although species and cultivars differ in their nutritional needs, legumes have a relatively high phosphorus requirement for optimum growth [36]. Some require significantly more phosphate to reach optimum yields when relying on symbiotically fixed nitrogen in comparison to when supplied with fertilizer nitrogen [37].

Little work has been done on root nodulation at low phosphorus levels. Indications are that phosphorus deficiency limits nodulation indirectly by limiting legume growth rather than the infection process *per se* [8, 38], although there is evidence to imply that some rhizobia are more able to nodulate at lower phosphate levels than others [11]. Nodule development requires adequate phosphorus [39, 40] and nodules accumulate a higher phosphorus content than roots [41]. A number of experiments in which sterile soils were inoculated with vesicular-arbuscular mycorrhizal fungi have established their important role in the phosphorus nutrition of plants, particularly in phosphorus-deficient soils [41, 42–46]. In comparison to the gramineae the legume root system is typically restricted in its geometry, making it particularly dependent on mycorrhizal infection [11]. For example, with a phosphate-deficient Brazilian soil (sterilized, pH 5.3) applications of rock phosphate or mycorrhiza increased the vegetative yield of *Stylosanthes guyanensis* by factors of 2.2 and 8 respectively and increased whole-plant per cent phosphorus by factors of 1.6 and 3 respectively; increased nodulation and nitrogen fixation were concomitant with improved phosphorus nutrition [41]. Very little mycorrhizal work has been done in very acid soils and although their efficiency is known to be influenced by pH [47] at least some mycorrhizas do function below pH 5. Inoculation of a sterile soil (pH 4.5) with mycorrhiza increased the growth of *Pueraria phaseoloides* by 12-fold. Addition of rock phosphate further improved yield by only 10–25 per cent, showing that without amendments phosphate was unavailable rather than grossly deficient [46].

The scant evidence available indicates that high manganese levels are unlikely to inhibit the growth or survival of rhizobia in acid soils [9]. Manganese toxicity mainly affects legume growth *per se* rather than nodulation in particular [8, 9, 48], and tolerance of manganese varies considerably between and within legume species [9]. However, variations among strains of *R. phaseoli* were found in their ability to nodulate and fix nitrogen in conditions where manganese was marginally phytotoxic [49]. The adverse effects of manganese were alleviated by liming [49].

Legume species also vary in the capacity to tolerate molybdenum deficiency in acid soils. In a Brazilian acid soil *Stylosanthes* grew well without molybdenum fertilization, whereas siratro and *Centrosema* responded positively to molybdenum fertilization. In the same soil six cultivars of *Phaseolus vulgaris* responded to molybdenum fertilization only in conjunction with liming to a pH greater than 5.4. Further liming in excess of pH 5.8 caused a sufficient desorption of molybdenum in the soil that a response to molybdenum fertilization was no longer obtained [13, 50]. The lack of response to added molybdenum was later shown to be the result of the presence in the soil of an inhibitory factor for nodulation which was removed by liming to pH 5.9, rather than an inability of *P. vulgaris* to take up molybdenum or translocate it at the lower pH [51].

Depending on the degree of tolerance to acid conditions, some species respond more strongly than others to lime application [52]. Legumes in acid soils dependent on nodule fixed nitrogen generally derive greater benefit from liming than when there is sufficient mineral nitrogen available [27]. Two notable exceptions to this rule have recently been reported with soybean, as already outlined [16, 34] although available phosphorus level may be critical in this regard since basal applications of phosphorus were made in both of these cases. High levels of lime application, where pH is raised to 6–7, can have deleterious effects on plant growth [35]. The reasons for this are not well understood, particularly since the adverse effects may be transient [53], and in fact the explanation may vary with different soils and different species. Phosphate deficiency has been implicated [35, 54], and so also have magnesium and zinc deficiencies [54]. Neither plant symptoms nor leaf analyses for eight elements gave an indication of the reason for growth depression by liming in eight legumes growing in an oxisol. In the same trial seven other species showed no adverse effect of liming [53]. It appears unlikely that yield depression of legumes at high lime rates results from an adverse effect on the root nodule symbiosis since non-legumes may be similarly limited in growth [35]. Clearly, if economic considerations alone do not dictate judicious use of lime to mitigate the effects of acid soil conditions, the possibility of compounding the nutritional problems should do so.

SALINITY AND ALKALINITY

In reviewing the literature on salinity a basic problem emerges. The many different notations used to quantify salinity often make it difficult to readily compare data from different studies. For convenience here, salinity levels in soils will be quoted with the S.I. unit mS/cm, and *in vitro* as per cent concentration on a w/v basis so that comparison may be made between soil and *in vitro* data and some of the latter will also be quoted as mS/cm (for general reference, 1 mS/cm = 1 mmho/cm, 1 per cent NaCl has an electrical conductivity of 16 mS/cm).

Saline soils are common in regions of arid or semi-arid climate where transport of soluble salts to the ocean does not occur because of low rainfall [55, 56]. They are characterized by the presence of high levels of neutral salts in the surface layers resulting from the capillary rise of water when evaporation exceeds precipitation. In the flood plains of rivers, low-lying lake margins and coastal plains, saline groundwater within a few metres of the soil surface can be a major contributory factor. The predominant salts are usually sulphates and chlorides of sodium, calcium and sometimes magnesium, and small quantities of carbonates and bicarbonates are often present. These soils are only moderately alkaline, with pH about 8, and their agricultural use usually demands irrigation. However, this may actually exacerbate the problem since river water may contain significant levels of dissolved salts [55]. Artificial saline soils have resulted from irrigation [56].

Rhizobia exhibit a large range of sensitivities to salinity. Three of eleven diverse rhizobial types failed to grow at 0.6 per cent sodium chloride (an electrical conductivity of 10 mS/cm) and the others showed a 3.7-fold range in sensitivity in terms of fractional reduction in growth [57]. *R. trifolii* appears to be a saline sensitive species; eight strains isolated from *Trifolium alexandrinum* were completely growth-inhibited at 0.7 per cent sodium chloride [58] and another six *T. alexandrinum* isolates had their growth severely reduced or eliminated at 1 per cent levels of chlorides and sulphates of sodium, potassium and magnesium [59]. In contrast five rhizobial strains from *Sesbania cannabina, Crotolaria juncea* and soybean were uninhibited by 1 per cent levels of these same salts and four of them grew at 3 per cent levels [60]. The growth of six strains isolated from *Dolichos lablab* was unaffected or actually stimulated at sodium chloride levels up to 1.6 per cent (25 mS/cm) but was severely inhibited although not eliminated at 3.2 per cent [61]. Four strains of *R. meliloti* grew well at 3 per cent levels of chloride and sulphate salts of sodium, potassium and magnesium. A fifth strain of *R. meliloti* showed differential tolerance to these salts. It was 17, 49 and 80 per cent inhibited in growth at 3 per cent levels of sodium chloride, magnesium chloride and magnesium sulphate respectively, but failed to grow at only 0.4 per cent potassium chloride and 0.6 per cent potassium or sodium sulphate [59].

No correlation was found between the ability of a *Rhizobium* strain to grow at 0.6 per cent sodium chloride and whether it originated from a saline or non-saline soil [57]. Considering this in conjunction with the fact that rhizobia are differentially tolerant to high levels of different salts leads to the conclusion that rhizobial growth studies in saline media should be interpreted with caution. Moreover, rhizobia may survive in soils at salinity levels much higher than those at which growth is restricted or eliminated in media [57].

The addition of salt to three soils of different textures greatly reduced the ability of four strains of *R. japonicum* to withstand air-drying. Survival was best in the more organic, higher clay soil and there were significant differences in the numbers of the four strains after four weeks [62].

The mechanism of salt tolerance of a *Rhizobium* strain isolated from mesquite in the Sonoran Desert has been explained in terms of its ability to accumulate L-glutamate intracellularly at high concentrations, i.e. at a 34-fold higher level than normal at 2.9 per cent sodium chloride, constituting 88 per cent of the amino acid pool [63]. *R. japonicum* strains forming non-slimy small colonies were found to be more salt labile (at 0.26 per cent sodium chloride) than those forming slimy large colonies [64]. A similar correlation between slimy colony type and tolerance of 0.5 per cent sodium chloride was observed in a survey of 139 rhizobial isolates from cowpeas grown at three West African locations (Table 1). However, it is noteworthy that the mesquite *Rhizobium* described above has a non-slimy colony type [65].

Table 1. Correlation of cowpea rhizobial colony type (on yeast extract mannitol agar) with salt tolerance

Colony type	No. of isolates tested	Fraction (%) growing at NaCl levels of			
		0%	0.5%	1.0%	2.0%
Slimy	73	100	100	12.3	0
Non-slimy	66	100	4.6	0	0

Source: Stowers, Kormendy and Eaglesham [66].

The degree of salinity tolerance of a *Rhizobium* strain appears to be of only limited use in predicting the effects of salinity on its ability to form root nodules. The nodulation potential of eleven strains of *R. meliloti* was significantly reduced at 0.4 per cent sodium chloride (6.6 mS/cm) and eliminated at 0.7 per cent sodium chloride (11 mS/cm), although none of the strains was significantly growth-inhibited by these salt levels in liquid media [67]. Extracts of a saline soil which eliminated infection thread formation in alfalfa did not affect survival of *R. meliloti* [68].

Very large differences in salt tolerance were found among twelve pasture and forage legumes [69]. A survey of the salt tolerance of 31 plant species included temperate legumes, tropical legumes and tropical grasses [70]. The legumes were inoculated with rhizobia, but the excellent growth of the grasses indicated that the soil was abundant in nitrogen. The mean salinity values (mS/cm) which reduced plant yields by 50 per cent were estimated by a mathematical model to be 7.6±0.8, 6.4±1.9 and 14.2±5.6 (in sodium chloride equivalents: 0.5, 0.4 and 0.9%) for the temperate and tropical legumes and grasses respectively; mean values of soil salinity which reduced plant yields to zero were 11.9±4.1, 12.6±3.0 and 21.6±8.3 (0.75, 0.8 and 1.4%). The most salt-tolerant legume was alfalfa (10.2 mS/cm for half yield and 18.8 mS/cm for zero yield), but *Macroptilium lathyroides* and siratro were almost equally tolerant. These levels of salinity which had serious effects on legume growth would not be excessive for many rhizobial strains, as detailed above.

Few studies have been reported on the effects of salt stress on legumes grown with nitrogen supplied in comparison with those dependent on nodule-fixed nitrogen. It is usually difficult, therefore, to appraise to what extent the adverse effects of salinity are on nodulation and fixation rather than on plant growth *per se*. The indications are that root-nodule dependent legumes are more saline-labile than when nitrogen-fed. This has been clearly shown in the case of *Cicer arietinum*, although significant variation in effectiveness among rhizobial strains existed at 0.46 per cent sodium chloride [71]. The growth of nodulated *Glycine wightii* was more adversely affected by a two-week salinity treatment of 8 mS/cm than was that of nitrogen fertilized plants [72]. The nodulation process, nodule development and the amount of nitrogen fixed were all inhibited, but were quickly regained after removal of the stress. Tissue injury resulting in leaf abscission occurred only in the nodulated plants, probably the result of nitrogen stress coupled with saline stress, but no nodule loss was noted indicating adaptability of nodules to increases in substrate salinity [72]. A comparison of salinity effects was made with alfalfa and soybean, using both nodulated and nitrogen-fertilized plants [73]. The two species responded differently to sodium chloride applications up to 8 mS/cm. At the highest salinity, alfalfa nodule weight was reduced by only 18 per cent whereas the whole plant weight was reduced by 55 per cent in both the nodulated and non-nodulated plants. At the same salt level soybean nodule weight and nodulated plant weight were both reduced by about 80 per cent whereas nitrogen-fed soybean weight was reduced by only 60 per cent. Clearly the root nodule symbiosis in alfalfa was relatively resistant to saline conditions, while adverse effects on the symbiosis *per se* contributed to poor growth in soybean. However alfalfa's salinity tolerance varied with the salt. Potassium chloride had a more strongly adverse effect on alfalfa nodulation than did magnesium chloride. At 0.3 per cent, both salts significantly reduced nodulation but had no effect on seed germination or early seedling growth [74].

In a comparison of salt tolerance of nodule-dependent cowpeas and mung beans, nodule initiation was severely reduced at 0.3 per cent sodium chloride [75]. Although nodule development was not affected in either species at this level of salinity, nitrogen fixing efficiency (nitrogen accumulated per unit of nodule tissue) was 43 per cent reduced in mung bean and unaffected in cowpea. This reduction in symbiotic efficiency may constitute the degree to which mung bean growth potential was limited by the saline conditions. The most saline-susceptible aspect of cowpea growth was nodule initiation.

Nodulation of the soybean cultivar Amsoy was inhibited at sodium chloride concentrations above 0.6 per cent (9.8 mS/cm) and eliminated at 1.2 per cent. Microscopic examination of the roots showed that at 1 per cent sodium chloride very few rhizobia were attached to root hairs and very few root hairs were deformed, indicating a salinity-induced lesion in the very early stages of the infection process [76]. Detached soybean nodules were sensitive to relatively low levels of salinity, their acetylene reducing activity being reduced by approximately 20 per cent by contact with a sodium chloride solution of 0.25 per cent [77].

Large variations in sensitivity to salt stress may exist between cultivars of a single species. When salt was applied in irrigation water from three weeks after planting to soybean, two cultivars, Lee and N53-509 were unaffected in their seed yields at a soil salinity of 9.6 mS/cm, whereas four other cultivars including Jackson and Improved Pelican failed to yield. Nitrogen was supplied at 112 kgN/ha so no conclusion may be drawn regarding relative tolerances of the symbioses. There was no correlation between the effects of salinity on seed germination and its effects on later vegetative and reproductive growth, e.g. at 10.1 mS/cm the germination rates of Lee, Jackson and Improved Pelican were equally reduced, by 15-20 per cent [78].

Alkaline soils can develop from saline soils with low calcium reserves. After a drop in the water table, soluble salts are washed down the profile and exchangeable calcium is replaced by sodium. Soil carbon dioxide forms carbonate and bicarbonate ions and these react with sodium to raise the pH [56]. Soil conductivity is usually less than 4 mS/cm with the pH in the range 8.5-10 [55]. Less intensive leaching of saline soils with higher calcium content can produce saline-alkaline soils with a conductivity of more than 4 mS/cm and a pH usually less than 8.5 [55, 56].

Clearly, the constraints to rhizobial survival, nodulation and legume growth pertaining to saline soils also apply to saline-alkaline soils. Very few data have been reported on the effects of high pH on rhizobial growth, nodulation or legume growth. Six legume species, *Melilotus parviflora, Sesbania aculeata, Trifolium alexandrinum, Cyamopsis tetragonoloba,* cowpea, lentil and pea, were planted in a highly saline-alkaline soil (36.5 mS/cm, pH 10.3) which had been uncultivated for 65 years and were examined for nodules after 60 days [79]. They were found on all species (although not on every plant examined) except pea, demonstrating that at least some rhizobial types can survive extreme saline-alkaline conditions for long periods. Soil dilution and plant infection tests indicated that rhizobia capable of nodulating *M. parviflora* and *S. aculeata* were in preponderance at $2.8-3.0\times 10^{3}$/g soil. In view of the saline lability of the infection process noted earlier, the observation of nodulation in this soil is surprising. However it may be significant that nodule scoring was delayed until 60 days; rhizosphere effects may have gradually ameliorated conditions sufficiently for the sparse nodulation observed. Beneficial effects of legume roots on both high and low extremes of soil pH have been reported [68].

None of 17 strains of *R. japonicum* showed significant growth in liquid media at pH 8.5 [80]. By contrast, in two surveys of 23 rhizobial isolates from eight diverse legume species (not including soybean) all were found to grow well in non-saline conditions at pH values up to 10 [81]. An examination of six *Rhizobium* species, however, demonstrated their markedly greater lability to carbonates and bicarbonates than to chloride salts [82]. These findings were supported by a survey of eleven rhizobial strains from five species; rhizobia which grew well at 3 per cent sodium chloride failed to grow at 0.2 per cent sodium bicarbonate [81].

Nodulation of alfalfa was reduced to zero in an agar medium containing 0.2 per cent sodium bicarbonate or 0.3 per cent sodium carbonate, but occurred, although delayed, at 0.6 per cent sodium chloride [82]. Whether the adverse effect of these alkaline salts was mediated by rhizobial death, nodulation failure or poor plant growth is not clear. It does appear that the carbonates and bicarbonates in alkaline soils constitute a significant stress factor in addition to high salt content and/or high pH.

MOISTURE DEFICIENCY

Climates with alternate wet and dry periods are experienced in vast areas of the tropics. At one extreme is a climate with high total precipitation falling in two peak rainy seasons separated by two short drier spells of reduced rainfall. At the other extreme is a climate of low total rainfall in one short season and a long dry season. In general the seasonal variation in temperature as well as the length and severity of the dry season are greater as one moves from the equator [56]. Distribution of the rains during the wet season can sometimes be uneven resulting in the surface layers of the soil becoming desiccated during crop growth.

Bacterial populations are reduced in size when soils become desiccated [83]. In a silt loam soil two distinct phases of decline were observed with seven diverse types of slow- and fast-growing rhizobia. The first phase occurred during the loss of water from the soil, when the bacterial numbers fell rapidly and extensively at an exponential rate. The second phase was a much slower linear decline in numbers after the soil had reached a constant moisture level or the dry state. On wetting and re-drying the biphasic cycle was repeated. Whether moisture loss was rapid or slow, the same two phases were observed and the final population size was not significantly affected [84].

Slow-growing rhizobia survived desiccation in a sandy soil better than did fast-growing types. After overnight drying nine slow-growing rhizobia showed a 0.6–6 per cent survival and nine strains of fast-growers showed only a 0.02–2.5 per cent survival [85]. A smaller survey of four strains of fast growers and three strains of slow growers failed to confirm these findings. After 10 days of drying in a silt loam, survival rates were 0.39–0.92 per cent and 0.39–0.47 per cent for fast and slow growers respectively [84]. The discrepancy may have resulted from even slightly different clay contents of the two soils, since the addition of montmorillonite to a sandy soil improved the survival of fast-growing rhizobia, but had an adverse effect on slow growers [85]. This critical role of clay content in rhizobial survival in soils during hot, dry weather was the key factor in explaining "second year mortality" which occurred with clovers and medics but not lupins, in some soils of Western Australia [86, 87].

Moisture level was the dominant factor influencing short and long term survival of *R. japonicum* strains inoculated into a loamy sand. Different serotypes responded differently but in general, survival over a nine-week period was best at soil water potentials of -0.3 to -1.0×10^5 Pa. Cataclysmic declines in population sizes of 2-5 logarithms occurred over the first week at -5 to -15×10^5 Pa, water potential levels which were found to be very common in cultivated soils in North Carolina [88]. These data clearly illustrated the important role of soil moisture in influencing rhizobial inoculant establishment and performance. Soil type was confirmed as an important interacting factor. Even at low water potentials, rhizobial survival was better in sandy loams, silt loams and sandy clay loams than in sands or clay loams [89]. Differences in desiccation sensitivity among four strains of *R. japonicum* were much more pronounced in sand than in soil [62]. Further to rhizobial inoculant performance, as evidenced by subsequent soybean nodulation, survival was better after seven days in the field at moisture levels below wilting point when applied to the soil as a granular peat inoculant rather than to the seed as a powdered peat inoculant or as a liquid inoculant to the seed or soil [90].

The greater susceptibility to desiccation of the fast-growing rhizobia was explained not in terms of differences in internal solute concentrations or in water permeability, but rather that the fast growers had a greater affinity for water because of higher surface energy [91]. The lower internal water content at low relative vapour pressures of the slow growers was believed to impart greater survival capacity by reducing the activity of those enzymes capable of functioning at low moisture. However it has been suggested that if desiccation is relatively mild, the lower internal water-retaining ability of the slow-growing rhizobia may be disadvantageous because of insufficient moisture for the functioning of vital enzymes [92]. Within the slow-growing rhizobial grouping, those which produced copious extracellular polysaccharide were found to be more desiccation-sensitive than strains which produced little or none [93].

There are very few reports on the effects of low moisture on the nodulation process. Reduction of soil moisture in an Australian sandy soil of only 2 per cent from 5.5 to 3.5 per cent represented a large drop in water potential from -0.36 to -3.6×10^5 Pa, and had very severe effects on root hair infection of *Trifolium subterraneum*. Nodulation failure at the low moisture level was not a result of the death of rhizobia; in fact rhizosphere populations of *R. trifolii* were not adversely affected. The root hairs assumed a short, stubby form which appeared to be immune to rhizobial penetration, but which became amenable to infection after re-watering [94]. The water potential levels within which nodulation would be expected to occur have not been defined for other legumes. Soybean was relatively unaffected by withholding water at the "nodule initiation stage," but it appears likely that infection had occurred before moisture reached a critically low level. Nodule number and dry weight of cowpea and hyacinth bean were directly proportional to soil moisture level, but caution is required in interpreting these data since nodulation and plant growth were very poor in both species in all moisture treatments [96].

Pot grown siratro and *Desmodium intortum* subjected to weekly cycles of water stress suffered more in terms of shoot and root growth than in nodule mass or acetylene reduction activity, although the latter was measured under non-stressed conditions. However, nodulation and nitrogen fixing activities were very low even in the non-stressed treatments indicating that plants were utilizing soil nitrogen [97].

The subject of moisture stress effects on the whole legume has been neglected. No reports of comparison of moisture stress effects on nodulated and nitrogen-fed legumes were found. However it appears that root nodules are particularly sensitive to changes in soil water potential. A drop from -0.35 to -0.7×10^5 Pa resulted in significant reduction in nitrogen fixation in soybean [98]. Moisture loss down to 80 per cent of the maximum resulted in proportionate inhibition of acetylene reduction by detached soybean nodules. Below 80 per cent there occurred irreversible damage which correlated with loss of structural integrity in vacuolated non-infected nodule cells [99]. The adverse effects of moderate water stress on nodule activity indicated that considerable day-to-day variation in nitrogen fixation rates occur in field grown legumes simply from changes in whole plant water content with important implications for effects on yields [100]. The decreases observed in acetylene reduction by field-grown cowpeas in Nigeria during daylight hours were believed to have been caused by low relative humidity [101].

Nodules on water-stressed clover plants resumed meristematic activity and quickly increased in fresh weight and nitrogen fixing activity after rewatering [102]. Clovers may therefore recover more rapidly from drought than, for example, soybeans whose nodules do not have a localized meristem. Soybean nodules were shed when soil moisture approached the permanent wilting point [103]. The superior drought tolerance of *Lotus corniculatus* in comparison with *Medicago lupulina* and *Trifolium repens* was correlated with its ability to retain rather than shed its root nodules under dry conditions [104].

MOISTURE EXCESS

Soil waterlogging commonly occurs in the humid and subhumid tropics and even in the more arid tropics during and after prolonged and intense rainfall during the growing season. Rotations of legumes with lowland rice are used in many farming systems in the East. However there have been few studies on the effects of soil inundation on rhizobial survival, and available information is apparently conflicting. Excess moisture had a more deleterious effect than did desiccation on numbers of *R. leguminosarum* in soil [105]. Likewise, the populations of *cicer* rhizobia and *R. japonicum* were depleted after paddy rice in India and Thailand [106, 107], and prior flooding adversely affected nodulation of *Centrosema* and soybean in Brazil [108]. A comparison of a strain of *R. trifolii* and of *R. japonicum* indicated that the former was the more flooding-sensitive, but although populations of both were reduced in soils at high moisture levels, the rhizobial numbers were adequate for maximum nodulation after six weeks waterlogging [109]. A study of four strains of cowpea rhizobia gave a similar result; following an initial drop in numbers after inoculation, populations were stable at 10^3–10^4/g soil over a 30-day period of flooding [110].

Since rhizobia are aerobic organisms, the anoxic conditions of waterlogged soils would be expected to have a detrimental effect on their survival. However, many rhizobia, particularly among the slow growers, can utilize nitrate as an electron acceptor under anaerobiosis [111, 112]. Some of these rhizobia not only survive anaerobic conditions but can even increase in numbers [112]. A *Rhizobium's* potential to survive inundation therefore probably depends on whether it possesses a dissimilatory nitrate reductase. The possible importance of other limiting factors such as protozoan predation and organic acid toxicity should, however, be borne in mind [109].

Oxygen deficiency, which accompanies waterlogging of soil, inhibits root nodulation and nodule development [113–115]. Even in the aquatic legume *Aeschynomene scabra*, root nodulation was almost totally eliminated by flooding [116]. In contrast, nodulation of submerged stems of *Aeschynomene* and *Sesbania* was not adversely affected by waterlogging [117, 118] probably because of their air-conducting lacunae [118]. Severely restricted root growth when submerged doubtless contributes to decreased legume root nodulation, although in microaerobic conditions the former may be more adversely affected than the latter [113]. Ethylene, produced in waterlogged organic soils [119], is a powerful inhibitor of nodulation [120].

The waterlogging of nodulated roots of pea plants inhibited nodule development and acetylene reduction activity [115]. Within a day of flooding, acetylene reduction by soybean root nodules fell to zero [121]. Again in contrast, *Aeschynomene* root nodules retained the potential to reduce nitrogen even after long-term flooding, as shown by acetylene reduction activity when assayed in air [122]. The inhibitory effects of waterlogging on nodule activity in non-aquatic legumes can be attributed to anoxia. Acetylene reduction by soybean roots was directly dependent on adequate oxygen supply [123] and was inhibited at oxygen levels in air below 5 per cent [124]. Even a thin film of water on nodules can be a significant barrier to oxygen diffusion [125, 126]. Depletion of ATP concomitant with limiting oxygen is probably the main reason for decreased acetylene reduction, at least during short-term waterlogging with rhizobia not possessing dissimilatory nitrate reductase [127]. Accumulation of ethanol to toxic levels [126] probably contributed to the cowpea nodule loss observed when waterlogging was of several days duration [128]. However, when waterlogging of cowpea was prolonged, adaptive responses were observed. Although 60 per cent of the nodule tissue was lost after eight days of flooding, acetylene reduction efficiency (μ moles/unit dry nodule/h, assayed in air) of persistent nodules was reduced by only 18 per cent. Enlarged nodule lenticels and increased nodule cortication facilitated gaseous diffusion and increased nitrogen fixation and vegetative growth such that plants harvested at 46 days which had been waterlogged for 32 days were as well developed as those which had been waterlogged for only 16 days [128]. After raising the water table to within three centimetres of the soil surface, soybeans quickly became chlorotic, but recovered and although they had poor vegetative growth they formed as much nodule tissue as unstressed plants [129]. The adverse effect of the acclimation period was reduced by a small application of fertilizer nitrogen [130].

Cowpea seed yields were more adversely affected when short-term waterlogging occurred before flowering rather than after, and this pertained whether plants were dependent on nodule-fixed nitrogen or fertilizer nitrogen [131]. In contrast, final yields of soybean were more adversely affected when inundation for five days occurred at flowering rather than before or after [132]. Recovery of nitrogen-fixing activity was rapid in cowpea root nodules after draining. At four days after an eight-day waterlogging period, acetylene reduction efficiency was 42 per cent higher than in non-waterlogged plants [132]. After being flooded for a day-and-a-half, acetylene reduction activity in soybean nodules rose from zero to full recovery in less than 12 hours [121].

An examination of flooding tolerance in nine temperate legume species revealed a broad range of sensitivities as judged by leaf yellowing. Although growth of each species ceased shortly after waterlogging, the period of time for appearance of chlorosis varied from five days (sainfoin) to 20 days (strawberry clover, white clover and birdsfoot trefoil [134]). *Lupinus luteus* was found to be consistently more flooding tolerant than *L. albus, L. angustifolius* or *L. mutabilis*, and differences in tolerance were observed between cultivars within species [135]. Unfortunately, in these studies no correlation was drawn between flooding tolerance and nodule activity or persistence. Flooding tolerance in plants has been linked to the ability to utilize nitrate as a terminal electron acceptor [136] and to the ability to accumulate malate rather than ethanol in their roots when anoxic [137]. Manganese toxicity can be a further constraint to plant growth when some soils are inundated [138]. The suggestion that tropical legumes should be screened for flooding tolerance [133] ought to be less emphasized.

On the other hand,combination of superior lines with rhizobia possessing dissimilatory nitrate reductase may lead to significantly increased yields in hydromorphic soils.

HIGH TEMPERATURE

Mean seasonal temperatures are higher in the drier than in the more humid zones of the tropics. In the latter, air temperatures exceeding 38°C are rare [56], but when the bare surface of a cultivated soil is exposed to direct sunlight its temperature can greatly exceed the air temperature. In seasonal rainfall areas of the tropics, crops are sown at the end of the hot dry period at the start of the rains. After planting there may be dry hot days when germinating seeds and the roots of developing seedlings are exposed to elevated temperatures. In soils low in availabe nitrogen, nodules of cowpea and soybean are normally initiated within six days of planting and are visible by nine days within a few centimetres of the soil surface [139, 140]. Temperatures in excess of 40°C are common at planting depth [80, 87, 141-144]. The upper limits of temperature within which normal patterns of legume nodulation may be expected are still poorly understood for the agriculturally important tropical legumes. The availabe data indicate that rhizobial survival and/or early nodulation are likely to be adversely affected by tropical soil temperatures.

In a screening of 68 rhizobial isolates from tropical legumes for temperature tolerance, only four were able to grow at temperatures in excess of 38°C, and the maximum temperature for growth was 42°C. When incubated in a sandy soil at 40°C most of these tropical rhizobia died within ten hours [141]. The moist conditions in this experiment possibly contributed to the rapid death of the rhizobia, since it has been shown that rhizobia are likely to be less tolerant of high temperatures in moist soil than in dry [145], although it is not always the case [110]. Strains of *R. trifolii*, *R. japonicum* and *R. lupini* survived five to six hours at 70-80°C in dry, sandy soils [87, 146]. The addition of clay to sandy soils improved the high temperature tolerance of fast-growing rhizobia, but not of slow-growing rhizobia. Concomitantly the survival of *R. trifolii* at 70°C in dry soil was better in loamy sands and sandy loams than in red, grey or yellow sands [146]. Rhizobial tolerance of such high temperatures was suggested to be the result of spore formation [147], but this was later disproved [148].

Twenty strains of the cowpea miscellany were examined for the ability to grow in a mannitol amended silt loam at various temperatures in the range 29-40°C. None grew at 40°C but all grew at 29, 31, 33 and 35° C with the fastest growth rate occurring at 33°C with 16 of them. However all of these strains survived seven days at 42°C in the dry silt loam without significant reduction in numbers [93]. The growth patterns of 35 strains of *R. japonicum* were examined in liquid media at constant temperatures up to 54°C. The maximum temperature for growth was 38°C shown by four strains, whereas the maximum survival temperature was 48.7°C shown by two other strains. Although significant positive correlations existed between maximum temperature permissive of growth and optimum growth temperature and between maximum permissive temperature and maximum survival temperature, there were however notable exceptions. For example, one strain had the lowest optimal temperature of 27.4°C and the highest survival temperature, of 48.7°C [149].

Cowpea rhizobia isolated from the hot, dry environment of the sahel-savannah of West Africa were all high temperature tolerant, showing good growth on yeast mannitol agar at 37°C [150]. More than 90 per cent of these rhizobia grew well also at 40°C, whereas rhizobia from more humid zones of West Africa generally grew poorly or not at all at this temperature (Table 2). No similar correlation between location of rhizobial origin and tolerance to high temperature was found in Australia [141] possibly because temperatures were less extreme. The growth of soybean was more adversely affected by elevated temperatures when dependent on symbiotically fixed nitrogen than when using fertilizer nitrogen [144]. The same was true of cowpea [152].

Table 2. Survey of cowpea rhizobial isolates from three West African locations for ability to grow at 40°C on yeast extract mannitol agar

Origin of rhizobia	No. of isolates tested	Fraction (%) with growth scores of		
		None	Trace[1]	Good[1]
Maradi[2]	44	0	5	95
Ibadan[3]	46	98	2	0
Onne[4]	42	81	19	0

[1]Visual scoring.
[2]Maradi is in the sahel-savannah in Niger.
[3]Ibadan is in the savannah/tropical forest transition zone in Nigeria.
[4]Onne is in the tropical forest zone in Nigeria.
Source: Stowers, Goldman and Eaglesham [151].

Legume species show varying tolerance to high temperature in their nodulating abilities. Three cultivars of chickpea (*Cicer arietinum*) inoculated with any one of five rhizobial strains failed to nodulate at a constant temperature of 33°C [153]. By comparison, cowpea cultivar VITA-3 at a constant 35°C nodulated well with all of ten rhizobial strains, the overall mean nodule number being reduced by only 9.5 per cent in comparison to plants at a constant 25°C [154]. Rhizobial strain to strain variation exists in nodulation potential under temperature stress conditions; this has been demonstrated with several host species including chickpea [155], soybean [144, 155], cowpea [139, 150] and lotus [156].

The most convenient approach to the investigation of temperature effects on nodulation is by the constant temperature technique [144, 155, 156]. One important criticism of this technique is that it does not allow ready extrapolation to field conditions. All of thirteen strains of *R. japonicum* were severely inhibited in their soybean nodulation potential at constant temperatures in excess of 33°C [144]. However, using a diurnal cycling technique, raising the air temperature from 25°C to 41°C for nine hours per day, which produced a maximum root temperature of 40°C, resulted in partially inhibited, but nevertheless profuse and effective, nodulation with four diverse strains (Table 3). Clearly the diurnal cycling

Table 3. Temperature effects on the nodulation and growth of soybean cv. Wilkin inoculated with four strains of *Rhizobium japonicum* at 39 days after planting

Air temperature regime (°C, 9h/day)	Strain	Nodule No. (No./plant)	Nodule dry wt. (mg/plant)	Shoot dry wt. (g/plant)
30	SM 31	214±82[1]	881±150	7.10±1.24
	USDA 83	264±106	838±146	8.47±1.41
	RCR 3407	229± 45	723±100	7.27±0.30
	IRc 20	316± 60	1031± 65	7.79±0.90
	(uninoc.)	0	0	4.03±0.53
36	SM 31	133 ± 59	133 ± 59	10.13±1.99
	USDA 83	87± 26	626±109	12.20±1.81
	RCR 3407	156± 36	649±104	10.53±2.21
	IRc 20	106± 19	741±114	10.29±1.15
	(uninoc.)	0	0	5.93±0.74
41	SM 31	96± 37	567±138	6.33±1.19
	USDA 83	47± 11	465±133	6.37±2.16
	RCR 3407	112± 37	356±182	5.71±1.55
	IRc 20	88± 31	526±110	5.95±0.63
	(uninoc.)	0	0	4.78±0.33

[1]Mean of four replicates ± standard deviation.
Source: Koermendy and Eaglesham [157].

technique presents a more realistic impression of the environmental temperature tolerances of *Rhizobium*-legume associations. Another important criticism of the constant temperature technique is that at elevated temperatures, legume growth *per se* may be significantly reduced, calling into question any effectiveness comparison of strains. Cowpeas grown at a constant 35°C with fertilizer-nitrogen supplied after five weeks growth achieved only 35 per cent of the dry weight of plants at a constant 25°C. On the other hand a diurnal cycling of air temperature from 25°C to 41.5°C (maximum root temperature 40°C) produced significantly better cowpea growth (Table 4).

The symbiotic performance of different rhizobial strains under temperature stress has been correlated with their ability to grow in pure culture at elevated temperatures. Strains from Maradi in the sahel-savannah of the

Table 4. Effects of constant temperature regimes compared to diurnal cycling regimes on cowpea cv. VITA-5 supplied with fertilizer nitrogen

Air temperature regime	Shoot dry weight (g/plant)
Constant 25°C	10.9±1.49[1]
Constant 35°C	3.86±0.56
Cycling 25°-30°C	9.69±1.30
Cycling 25°-41°C	8.37±0.80

[1]Mean of four replicates ±standard deviation.
Source: Eaglesham and Hassouna [154].

Niger Republic grew well on yeast extract mannitol agar at 40°C whereas strains from Onne and Ibadan in the cooler more humid areas of Nigeria were partially or completely inhibited in their growth at 40°C (Table 2). Only the Maradi strains were able to retain or increase effectiveness when the air and root temperatures cycled to 41.5 and 40°C respectively, daily (Table 5). It is interesting to note that the better performance of the Maradi

Table 5. Comparison of the nodulation, growth and N accumulation of cowpea cv. VITA-5 grown in normal and high temperature regimes when inoculated with rhizobia from three climatic zones

Component	Origin of rhizobia	Data at 30°/25°C	Data at 41.5°/25°C	% change
Shoot dry weight (g/plant)	Onne[1]	9.02	7.68	- 14.8
	Ibadan[1]	9.42	7.14	- 24.2
	Maradi[1]	7.12	8.22	+ 15.4
N accumulated (mg N/plant)	Onne	355	309	- 7.8
	Ibadan	392	330	- 15.8
	Maradi	269	324	+ 20.4
Nodule number (No./plant)	Onne	185	86	- 53.5
	Ibadan	192	112	- 41.7
	Maradi	91	48	- 47.0
Nodule dry weight (mg/plant)	Onne	579	588	+ 1.6
	Ibadan	612	611	- 0.2
	Maradi	447	595	+ 33.1
Individual nodule weight (mg/nodule)	Onne	3.1	6.8	+119.0
	Ibadan	3.1	5.6	+ 80.6
	Maradi	5.1	14.9	+192.1

[1]Descriptions of locations of origin given in Table 2.
Source: Eaglesham and Hassouna [154].

strains at high temperature was not the result of relatively higher nodule number but of larger nodules (Table 5). The link between the ability to grow and ability to retain effectiveness, also shown by *R. japonicum* [144], indicates that *Rhizobium* survival at high temperature may be more critical than the nodulation potential of the host legume. In other words, some legumes may have the potential to nodulate at temperatures above the limit of rhizobial growth or survival. Reduction to almost total nodulation failure in cowpeas resulted from exposure of roots to temperatures of 40–44°C for five hours per day and was directly linked to inoculum death, the first few days after planting being the most critical phase [139].

In the early weeks of growth and before canopy closure, exposure of nodulated roots to high temperatures is likely to have relatively less effect on the functioning of nodules than on their establishment [155, 158]. When the soil surface is shaded, after canopy closure, high soil temperatures are unlikely to be a problem. Therefore the time during which high temperature is a critical factor in nodulation is short, between planting and the initial rhizobial infection of roots at the crown. At greater depths, rhizobia and secondary nodulation are less likely to be exposed to temperature extremes.

Although high temperatures in field conditions are often associated with soil desiccation, we found no report on the combined effects of these stresses on legume nodulation in controlled conditions.

THE FUTURE

An essential aspect of the strategy to improve the yields of tropical legumes in stressed environments must involve combining stress-tolerant cultivars with stress-tolerant rhizobia. The latter may exist already in the soil or be introduced as seed- or soil-applied inoculants [159]. Legume screening techniques should be inexpensive and simple in design so that large numbers of plant types can be checked quickly. Of course, preliminary work is required to determine the limits of environmental stress necessary to distinguish types of different tolerances. For most of the stresses under consideration here these limits are only poorly understood. For example, unexpectedly high temperatures have been found necessary to examine tolerance in soybean and *Rhizobium japonicum* [157]; cultivar Wilkin can nodulate and fix nitrogen with some strains when subjected daily to an air temperature of 45°C for eight hours, a root temperature maximum of 45°C.

Combinations of stresses have received almost no attention. There is a need to identify legume/*Rhizobium* combinations tolerant of high temperature with low moisture, high temperature with high salt, high temperature with low moisture and high salt, excess moisture with low pH, excess moisture with excess manganese etc. We also require a better understanding of the degree to which the legume rhizosphere may ameliorate various soil stress conditions. The screening of rhizobia for stress tolerances in isolation from its host symbiont might produce misleading information.

Possibly the best source of superior rhizobia for use in stressed environments is the soil in that or a similar environment. Having isolated rhizobia from nodules of a plant growing in, or inoculated with, the soil, and after verifying stress resistance by the appropriate screening technique the most effective and competitive isolates may be cultured and added back to the soil in an inoculant to greatly increase their numbers and therefore also their nodulation potential. The rhizobial characteristics of three tropical soils have been examined with the aim of identifying superior strains and of making comparisons of rhizobial diversity in the three soils [160]. The cowpea-nodulating rhizobia were found to be diverse, both within and between locations in terms of colony morphology, salt tolerance, high temperature tolerance, host promiscuity, serological and biochemical characteristics [160, 162]. More studies of this kind are required for the better understanding of the role of soil stresses on rhizobial populations.

The factors which confer nodulation competitiveness are not presently understood and this important aspect should be the subject of more intense research effort in the future. Meanwhile rapid methods are required for typing nodules in competitiveness screenings. For example, the unusually dark-coloured nodules produced by some rhizobia on cowpeas are a particularly convenient marker for typing nodules in both pot and field experiments [161].

Possibilities exist for the beneficial use of vesicular-arbuscular mycorrhizal fungi as inoculants, particularly if *in vitro* culturing becomes possible. Positive results in the field, although not as predictable as in pot experiments, have been demonstrated [163–165]. The degree of dependence on mycorrhiza for uptake of phosphate varies with species and with phosphate availability [10] and different mycorrhizas vary in their functional efficiency [166]. Whether mycorrhizas of superior efficiency have the potential to improve legume phosphorus nutrition in phosphate-fixing acid soils remains to be determined. It should be borne in mind that unlike atmospheric nitrogen for rhizobia, phosphate is not a renewable resource for mycorrhiza. The introduction of more efficient endophytes for maximum exploitation of soil phosphate will inevitably lead to soil impoverishment unless fertilizer is used. Rock phosphates and other inexpensive forms of phosphatic fertilizers can be used in conjunction with mycorrhizas [167] even in alkaline soils in which the phosphate would be unavailable directly to the plant [41].

As far as salinity tolerance is concerned, the available information indicates that legume growth is generally more sensitive to saline conditions than rhizobial growth. In breeding salt tolerance into legumes, allowance should be made for the possibility that the nodulation process may be particularly saline sensitive. Breeding material should be screened with low levels of mineral nitrogen available in the presence of appropriate rhizobia.

The ability of the aquatic legume species of *Aeschynomene* and *Sesbania* to form stem nodules on submerged stems and branches can be viewed as an adaptive response to waterlogging. The stem nodules originate at

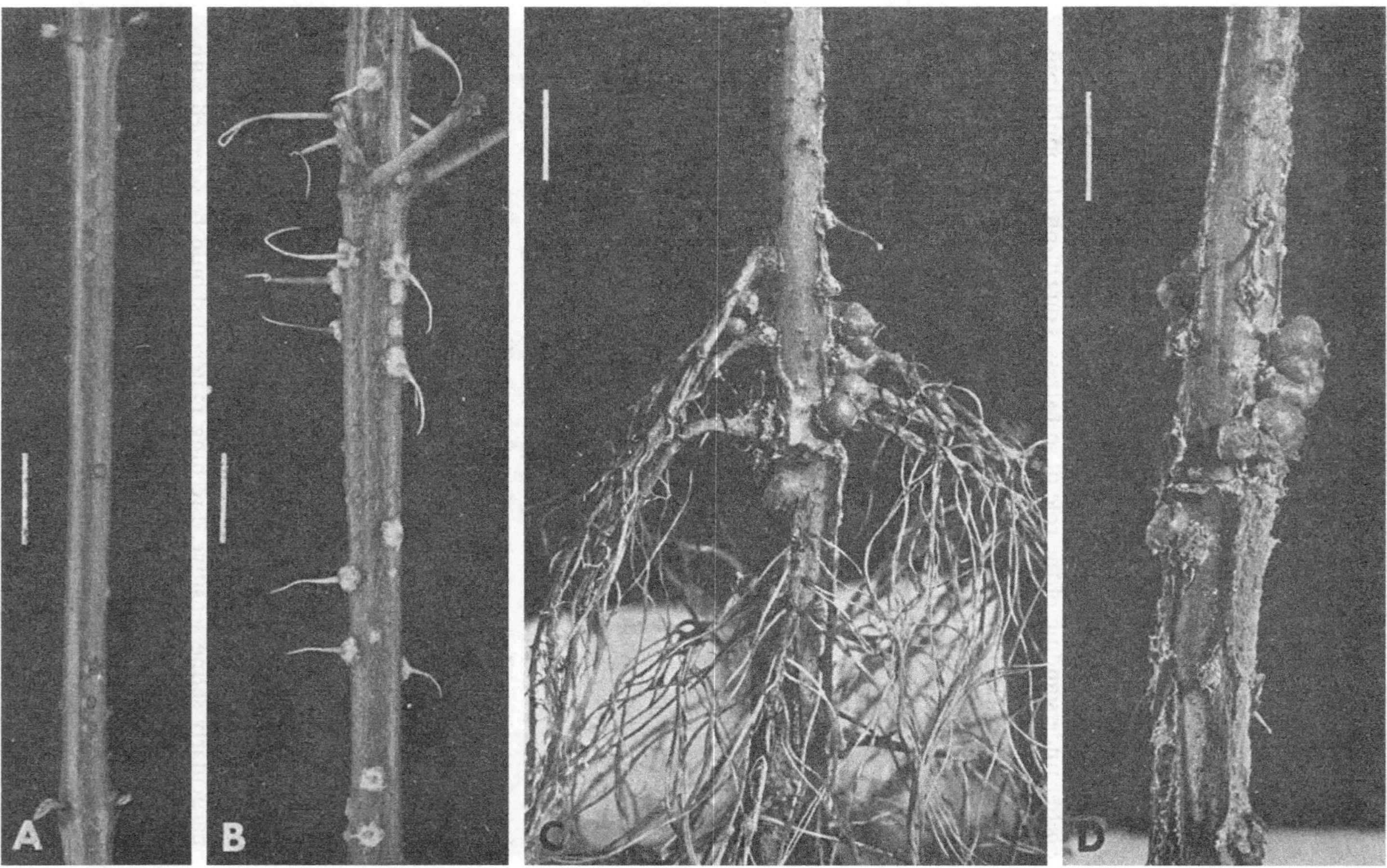
A
B
C
D

previously dormant adventitious root initials [168] and have the advantage over the root nodules in being closer to the water surface where oxygen and nitrogen availability is highest. Some peanut and field-bean cultivars can form nodules on below-soil stems [169, 170] but whether this ability lends tolerance to waterlogging has not been examined. Breeding for the production of adventitious root initials on legume stems may be a viable strategy for conferring long-term waterlogging tolerance. Some soybean types of Asian origin under normal growing conditions produce a large number of dormant adventitious root initials on the lower stem (Fig. 1 A). Pot experiments demonstrated that within a day of inundation the initials on submerged stems grew out and within a few days produced a new root system capable of utilizing the higher levels of dissolved oxygen close to the water surface. These new roots were susceptible to rhizobial invasion and new nodules were produced. After drainge of the floodwater new roots which had penetrated the soil surface appeared to remain functional while the others died back. Nodules which formed at the junction of the stem and adventitious roots retained structural integrity and were functional even after root dieback and removal. These may be regarded as stem nodules (Fig. 1 B, C) and would be a valuable resource to the plant should inundation re-occur.

Rapid progress in recent years in the fields of *Rhizobium* genetics and recombinant DNA technology may make it possible in the foreseeable future to actually construct novel rhizobia possessing a number of desirable characteristics including stress tolerance.

In conclusion, as an aid to the improvement of agricultural productivity in the tropics a large range of types of legumes and rhizobia are available. With the selection of the appropriate legume and, if necessary, combining it with the appropriate rhizobial inoculant and by paying due heed to mineral nutrition and agronomy, legume nitrogen fixation can be increased and concomitantly food production can be improved even under environmentally stressed conditions.

ACKNOWLEDGEMENTS

The authors thank Barbara J. Goldman for her willing assistance in the literature search, Suzette Payne for typing the manuscript and Wm. G. Smith, Jr. for photography. The sponsorship of the United Nations Development Programme, grant No. GLO/77/013, is gratefully acknowledged.

Fig. 1. The lower stems of Asian soybean (*Glycine max* cv. TGm 120). (A) Dormant adventitious root initials visible. (B) After two days of flooding, adventitious roots developing, subtended by outgrowths of large epiderminal cells. (C) Four weeks after inundation and inoculation of flood water with *Rhizobium japonicum*; several days after drainage of flood water. (D) Three weeks after drainage of flood water and ten days after root excision.

(Scale bars = 1 cm)

REFERENCES

1. Eaglesham, A.R.J., Ayanaba, A., Ranga Rao, V. and Eskew, D.L. Mineral N effects on cowpea and soybean crops in a Nigerian soil II. Amounts of N fixed and accrual to the soil, *Plant and Soil 68*, 183-192 (1982).
2. App, A. and Eaglesham, A.R.J. Biological nitrogen fixation—problems and potential, pp. 1-7, in *Biological Nitrogen Fixation Technology for Tropical Agriculture* (Editors, P.H. Graham and S.C. Harris), CIAT, Colombia (1982).
3. Parker, C.A., Trinick, M.J. and Chatel, D.L. Rhizobia as soil and rhizosphere inhabitants, pp. 311-352, in *A Treatise on Dinitrogen Fixation,* Section IV: *Agronomy and Ecology* (Editors, R.W.F. Hardy and A.H.Gibson), John Wiley and Sons, New York (1977).
4. Dart, P.J. The infection process, pp. 381-429, in *The Biology of Nitrogen Fixation* (Editor, A. Quispel), North Holland Publishing Co., Amsterdam (1974).
5. Dart, P.J. Infection and development of leguminous nodules, pp. 519-556, in *A Treatise on Dinitrogen Fixation,* Section III: *Biology* (Editors, R.W.F. Hardy and W.S. Silver), John Wiley and Sons, New York (1977).
6. Pate, J.S. Functional biology of dinitrogen fixation by legumes, pp. 473-518, in *A Treatise on Dinitrogen Fixation,* Section III: *Biology* (Editors, R.W.F. Hardy and W.S. Silver), John Wiley and Sons, New York (1977).
7. Gibson, A.H. The influence of the environment and managerial practices on the legume-*Rhizobium* symbiosis, pp. 393-450, in *A Treatise on Dinitrogen Fixation,* Section IV: *Agronomy and Ecology* (Editors, R.W.F. Hardy and A.H. Gibson), John Wiley and Sons, New York (1977).
8. Andrew, C.S. Legumes and acid soils, pp. 135-160, in *Limitations and Potentials for Biological Nitrogen Fixation in the Tropics* (Editors, J. Dobereiner, R.H. Burris and A. Hollaender), Plenum Press, New York (1978).
9. Munns, D.N. Soil acidity and related problems, pp. 211-236, in *Exploiting the Legume-Rhizobium Symbiosis in Tropical Agriculture* (Editors, J.M. Vincent, A.S. Whitney and J. Bose), NifTAL, Hawaii (1977).
10. Mosse, B. Role of mycorrhiza in legume nutrition, pp. 275-292, in *Exploiting the Legume-Rhizobium Symbiosis in Tropical Agriculture* (Editors, J.M. Vincent, A.S. Whitney and J. Bose), NifTAL, Hawaii (1977).
11. Munns, D.N. and Mosse, B. Mineral nutrition of legume crops, pp. 115-126, in *Advances in Legume Science* (Editors, R.J. Summerfield and A.H. Bunting), Royal Botanic Gardens, Kew (1980).

12. Franco, A.A. Nutritional restraints for tropical grain legume symbiosis, pp. 237–252, in *Exploiting the Legume-Rhizobium Symbiosis in Tropical Agriculture* (Editors, J.M. Vincent, A.S. Whitney and J. Bose), NifTAL, Hawaii (1977).
13. Franco, A.A. Micronutrient requirements of legume-*Rhizobium* symbosis in the tropics, pp. 161–172, in *Limitations and Potentials for Biological Nitrogen Fixation in the Tropics* (Editors, J. Dobereiner, R.H. Burris and A. Hollaender), Plenum Press, New York (1978).
14. Dobereiner, J. and Campelo, A.B. Importance of legumes and their contribution to tropical agriculture, pp. 191–220, in *A Treatise on Dinitrogen Fixation,* Section IV: *Agronomy and Ecology* (Editors, R.W.F. Hardy and A.H. Gibson), John Wiley and Sons, New York (1977).
15. Mulongoy, K., Ayanaba, A. and Pulver, E. Exploiting the diversity in the cowpea-rhizobia symbosis for increased cowpea production, pp. 119–125, in *Global Impacts of Applied Microbiology Sixth International Conference* (Editors, S.O. Emejuaiwe, O. Ogunbi and S.O. Sanni), Academic Press, London (1981).
16. Munns, D.N., Keyser, H.H., Fogle, V.W., Hohenberg, J.S., Righetti, T.L., Lauter, D.L., Zaroug, M.G., Clarkin, K.L. and Whitacre, K.W. Tolerance of soil acidity in symbioses of mung bean with rhizobia, *Agronomy Journal 71,* 256–260 (1979).
17. Vincent, J.M. *A Manual for the Practical Study of Root-Nodule Bacteria,* Blackwell, Oxford (1970).
18. Date, R.A. and Halliday, J. Selecting *Rhizobium* for acid infertile soils of the tropics, *Nature 277,* 62–64 (1979).
19. Keyser, H.H. and Munns, D.N. Tolerance of rhizobia to acidity, aluminum, and phosphate, *Soil Science Society of America Journal 43,* 519–523 (1979).
20. Graham, P.H. and Parker, C.A. Diagnostic features in the characterization of the root-nodule bacteria of legumes, *Plant and Soil 20,* 383–396 (1964).
21. Rice, W.A., Penney, D.C. and Nyborg, M. Effects of soil acidity on rhizobia numbers, nodulation and nitrogen fixation by alfalfa and red clover, *Canadian Journal of Soil Science 57,* 197–203 (1977).
22. Keyser, H.H. and Munns, D.N. Tolerance of rhizobia to acidity, aluminum and phosphate, *Soil Science Society of America Journal 43,* 519–523 (1979).
23. Keyser, H.H. and Munns, D.N. Effects of calcium, manganese and aluminum on growth of rhizobia in acid media, *Soil Science Society of America Journal 43,* 500–503 (1979).
24. Munns, D.N. and Keyser, H.H. Response of *Rhizobium* strains to acid and aluminum stress, *Soil Biology and Biochemistry 13,* 115–118 (1981).

25. Ayanaba, A., Asanuma, S. and Munns, D.N. An agar plate method for rapid screening of *Rhizobium* for tolerance to acid-aluminum stress, *Soil Science Society of America Journal 47,* (in press) (1983).
26. Keyser, H.H., Munns, D.N. and Hohenberg, J.S. Acid tolerance of rhizobia in culture and in symbiosis with cowpea, *Soil Science Society of America Journal 43,* 719-722 (1979).
27. Andrew, C.S. Effect of calcium, pH and nitrogen on the growth and chemical composition of some tropical and temperate pasture legumes. I. Nodulation and growth, *Australian Journal of Agricultural Research 27,* 611-623 (1976).
28. Munns, D.N., Fox, R.L. and Koch, B.L. Influence of lime on nitrogen fixation by tropical and temperate legumes, *Plant and Soil 46,* 591-601 (1977).
29. Mengel, D.B. and Kamprath, E.J. Effect of soil pH and liming on growth and nodulation of soybeans in histosols, *Agronomy Journal 70,* 959-963 (1978).
30. Lie, T.A. The effect of low pH on different phases of nodule formation in pea plants, *Plant and Soil 31,* 391-406 (1969).
31. Franco, A.A. and Munns, D.N. Acidity and aluminum restraints on nodulation, nitrogen fixation, and growth of *Phaseolus vulgaris* in solution culture, *Soil Science Society of America Journal 46,* 296-301 (1982).
32. Sartrain, J.B. and Kamprath, E.J. Effect of liming a highly Al-saturated soil on the top and root growth and soybean nodulation, *Agronomy Journal 67,* 507-510 (1975).
33. Munns, D.N., Hohenberg, J.S., Righetti, T.L. and Lauter, D.J. Soil acidity tolerance of symbiotic and nitrogen-fertilized soybeans, *Agronomy Journal 73,* 407-410 (1981).
34. Bromfield, E.S.P. and Ayanaba, A. The efficacy of soybean inoculation on acid soil in tropical Africa, *Plant and Soil 54,* 95-106 (1980).
35. Pearson, R.W. *Soil Acidity and Liming in the Humid Tropics,* Cornell University, Ithaca (1975).
36. Van Schreven, D.A. Some factors affecting the uptake of nitrogen by legumes, pp. 137-163, in *Nutrition of the Legumes* (Editor, E.G. Hallsworth), Butterworths, London (1958).
37. Cassman, K.G., Whitney, A.S. and Fox, R.L. Phosphorus requirements of soybean and cowpea as affected by mode of N nutrition, *Agronomy Journal 73,* 17-22 (1981).
38. Zaroug, M.G. and Munns, D.N. Nodulation and growth of *Lablab purpureus (Dolichos lablab)* in relation to *Rhizobium* strain, liming and phosphorus, *Plant and Soil 53,* 329-339 (1979).
39. Stalder, L. Über Dispositionsverschiebungen bei der Bildung von Wurzelknollchen, *Phytopathologische Zeitschrift 18,* 376-403 (1952).
40. Diener, T. Über die Bedingungen der Wurzelknollchenbildung bei *Pisum sativum, Phytopathologische Zeitschrift 16,* 129-170 (1950).

41. Mosse, B., Powell, C. Ll. and Hayman, D.S. Plant growth responses to vesicular-arbuscular mycorrhiza IX. Interactions between VA mycorrhiza, rock phosphate and symbiotic nitrogen fixation, *New Phytologist 76,* 331-342 (1976).
42. Ross, J.P. and Harper, J.A. Effect of endogone mycorrhiza on soybean yields, *Phytopathology 60,* 1552-1556 (1970).
43. Schenk, N.C. and Hinson, K. Response of nodulating and non-nodulating soybeans to a species of *Endogone* mycorrhiza, *Agronomy Journal 65,* 849-850 (1973).
44. Crush, J.R. Plant growth responses to vesicular-arbuscular mycorrhiza VII. Growth and nodulation of some herbage legumes, *New Phytologist 73,* 743-749 (1974).
45. Mosse, B. Plant growth responses to vesicular-arbuscular mycorrhiza X. Responses of *Stylosanthes* and maize to inoculation in unsterile soils, *New Phytologist 78,* 277-288 (1977).
46. Waidyanatha, U.P. de S., Yogaratnam, N. and Ariyaratne, W.A. Mycorrhizal infection of growth and nitrogen fixation of *Pueraria* and *Stylosanthes* and uptake of phosphorus from two rock phosphates, *New Phytologist 82,* 147-152 (1979).
47. Skipper, H.D. and Smith, G.W. Influence of soil pH on the soybean-endomycorrhiza symbiosis, *Plant and Soil 53,* 559-563 (1979).
48. Munns, D.N. Mineral nutrition and the legume symbiosis, pp. 353-392, in *A Treatise on Dinitrogen Fixation,* Section IV: *Agronomy and Ecology* (Editors, R.W.F. Hardy and A.H. Gibson), John Wiley and Sons, New York (1977).
49. Dobereiner, J. Manganese toxicity effects on nodulation and nitrogen fixation of beans *(Phaseolus vulgaris* L.) in acid soils, *Plant and Soil 24,* 153-166 (1966).
50. Franco, A.A. and Day, J.M. Effects of lime and molybdenum on nodulation and nitrogen fixation of *Phaseolus vulgaris* L. in acid soils of Brazil, *Turrialba 30,* 99-105 (1980).
51. Franco, A.A. and Munns, D.N. Response of *Phaseolus vulgaris* L. to molybdenum under acid conditions, *Soil Science Society of America Journal 45,* 1144-1148 (1981).
52. Munns, D.N. and Fox, R.L. Comparative lime requirements of tropical and temperate legumes, *Plant and Soil 46,* 533-548 (1977).
53. Munns, D.N. and Fox, R.L. Depression of legume growth from liming, *Plant and Soil 45,* 701-705 (1970).
54. Edwards, D.G., Kang, B.T. and Danso, S.K.A. Differential response of six cowpea (*Vigna unguiculata* (L.) Walp.) cultivars to liming in an ultisol, *Plant and Soil 59,* 61-73 (1981).
55. Hayward, H.E. and Wadleigh, C.H. Plant growth on saline and alkali soils, *Advances in Agronomy 1,* 1-38 (1949).
56. Webster, C.C. and Wilson, P.N. *Agriculture in the Tropics,* Longman, London (1980).

57. Singleton, P.W., El Swaify, S.A. and Bohlool, B.B. Effect of salinity on *Rhizobium* growth and survival, *Applied and Environmental Microbiology 44,* 884-890 (1982).
58. Pillai, R.N. and Sen, A. Salt tolerance of *Rhizobium trifolii, Indian Journal of Agricultural Science 36,* 80-84 (1966).
59. Ethiraj, S., Sharma, H.R. and Vyas, S.R. studies on salt tolerance of rhizobia, *Indian Journal of Microbiology 12,* 87-91 (1972).
60. Yadav, N.K. and Vyas, S.R. Salt and pH tolerance of rhizobia, *Folia Microbiologica 18,* 242-247 (1973).
61. Pillai, R.N. and Sen, A. Salt tolerance of *Rhizobium* from *Dolichos lablab, Zentralblatt für Bakteriologie II, 128,* 538-542 (1973).
62. Al-Rashidi, R.K., Loynachan, T.E. and Frederick, L.R. Desiccation tolerance of four strains of *Rhizobium japonicum, Soil Biology and Biochemistry 14,* 489-493 (1982).
63. Hua, S.-S.T., Tsai, V.Y., Lichens, G.M. and Noma, A.T. Accumulation of amino acids in *Rhizobium* sp. strain WR1001 in response to sodium chloride salinity, *Applied and Environmental Microbiology 44,* 135-140 (1982).
64. Upchurch, R.G. and Elkan, G.H. Comparison of colony morphology, salt tolerance, and effectiveness in *Rhizobium japonicum, Canadian Journal of Microbiology 23,* 1118-1122 (1977).
65. Eaglesham, A.R.J. Personal observation.
66. Stowers, M.D., Koermendy, A. and Eaglesham, A.R.J. Unpublished data.
67. Subba Rao, N.S., Lakshmi-Kumari, M., Singh, C.S. and Magu, S.P. Nodulation of lucerne (*Medicago sativa* L.) under the influence of sodium chloride, *Indian Journal of Agricultural Science 42,* 384-386 (1972).
68. Lakshmi-Kumari, M., Singh, C.S. and Subba Rao, N.S. Root hair infection and nodulation in lucerne (*Medicago sativa* L.) as influenced by salinity and alkalinity, *Plant and Soil 40,* 261-268 (1974).
69. Andrew, C.S. Nutritional restraints on (forage) legume species, pp. 253-274, in *Exploiting the Legume-Rhizobium Symbiosis in Tropical Agriculture* (Editors, J.M. Vincent, A.S. Whitney and J. Bose), NifTAL, Hawaii (1977).
70. Russell, J.S. Comparative salt tolerance of some tropical and temperate legumes and tropical grasses, *Australian Journal of Experimental Agriculture and Animal Husbandry 16,* 103-109 (1976).
71. Lauter, D.J., Munns, D.N. and Clarkin, K.L. Salt response of chickpea as influenced by N supply, *Agronomy Journal 73,* 961-966 (1981).
72. Wilson, J.R. Response to salinity in *Glycine* VI. Some effects of a range of short-term salt stresses on the growth, nodulation, and nitrogen fixation of *Glycine wightii* (formerly *javanica*), *Australian Journal of Agricultural Research 21,* 571-582 (1970).

73. Bernstein, L. and Ogata, G. Effects of salinity on nodulation, nitrogen fixation, and growth of soybeans and alfalfa, *Agronomy Journal 58,* 201-203 (1966).
74. Singh, C.S., Lakshmi-Kumari, M., Biswas, A. and Subba Rao, N.S. Nodulation of lucerne (*Medicago sativa* L.) under the influence of chlorides of magnesium and potassium, *Proceedings of the Indian National Academy of Sciences B 76,* 90-96 (1972).
75. Balasubramanian, V. and Sinha, S.K. Effects of salt stress on growth, nodulation and nitrogen fixation in cowpea and mung beans, *Physiologia Plantarum 36,* 197-220 (1976).
76. Tu, J.C. Effect of salinity on *Rhizobium*-root-hair interaction, nodulation and growth of soybean, *Canadian Journal of Plant Science 61,* 231-239 (1981).
77. Sprent, J.I. The effects of water stress on nitrogen-fixing root nodules III. Effects of osmotically applied stress, *New Phytologist 71,* 451-460 (1972).
78. Abel, G.H. and MacKenzie, A.J. Salt tolerance of soybean varieties (*Glycine max* L. Merrill) during germination and later growth, *Crop Science 4,* 157-161 (1964).
79. Bhardwaj, K.K.R. Growth and symbiotic effectiveness of indigenous *Rhizobium* species of a saline-alkali soil, *Proceedings of the Indian National Academy of Sciences B 40,* 540-543 (1974).
80. Diatloff, A. Relationship of soil moisture, temperature and alkalinity to a soybean nodulation failure, *Queensland Journal of Agricultural and Animal Sciences 27,* 279-293 (1970).
81. Yadav, N.K. and Vyas, S.R. Response of root-nodule rhizobia to saline, alkaline and acid conditions, *Indian Journal of Agricultural Sciences 41,* 875-881 (1971).
82. Subba Rao, N.S., Lakshmi-Kumari, M., Singh, C.S. and Biswas, A. Salinity and alkalinity in relation to legume-*Rhizobium* symbiosis, *Proceedings of the Indian National Academy of Sciences B 40,* 544-547 (1974).
83. Chen, M. and Alexander, M. Survival of soil bacteria during prolonged desiccation, *Soil Biology and Biochemistry 5,* 213-221 (1973).
84. Pena-Cabriales, J.J. and Alexander, M. Survival of *Rhizobium* in soils undergoing drying, *Soil Science Society of America Journal 43,* 962-966 (1979).
85. Bushby, H.V.A. and Marshall, K.C. Some factors affecting the survival of root nodule bacteria on desiccation, *Soil Biology and Biochemistry 9,* 143-147 (1977).
86. Marshall, K.C., Mulcahy, M.J. and Chowdhury, M.S. Second-year clover mortality in Western Australia—a microbiological problem, *Journal of the Australian Institute of Agricultural Science 29,* 160-164 (1963).

87. Chatel, D.L. and Parker, C.A. Survival of field-grown rhizobia over the dry summer period in Western Australia, *Soil Biology and Biochemistry 5,* 415-423 (1973).
88. Mahler, R.L. and Wollum, A.G. Influence of water potential on the survival of rhizobia in a Goldsboro loamy sand, *Soil Science Society of America Journal 44,* 988-992 (1980).
89. Mahler, R.L. and Wollum, A.G. The influence of soil water potential and soil texture on the survival of *Rhizobium japonicum* and *Rhizobium leguminosarum* isolates in the soil, *Soil Science Society of America 45,* 761-766 (1981).
90. Smith, R.S. and del Rio Escurro, G.A. Soybean inoculant types and rates evaluated under dry and irrigated field conditions, *Journal of Agriculture of the University of Puerto Rico 66,* 241-249 (1982).
91. Bushby, H.V.A. and Marshall, K.C. Water status of rhizobia in relation to their susceptibility to desiccation and to their protection by montmorillonite, *Journal of General Microbiology 99,* 19-27 (1977).
92. Jansen van Rensburg, H. and Strijdom, B.W. Survival of fast- and slow-growing *Rhizobium* spp. under conditions of relatively mild desiccation, *Soil Biology and Biochemistry 12,* 353-356 (1980).
93. Osa-Afiana, L.O. and Alexander, M. Differences among cowpea rhizobia in tolerance to high temperature and desiccation in soil, *Applied and Environmental Microbiology 43,* 435-439 (1982).
94. Worrall, V.S. and Roughley, R.J. The effect of moisture stress on infection of *Trifolium subterraneum* L. by *Rhizobium trifolii* Dang, *Journal of Experimental Botany 27,* 1233-1241 (1976).
95. Rathore, T.R., Chhonkar, P.K., Sachan, R.S. and Ghildyal, B.P. Effect of soil moisture stress on legume-*Rhizobium* symbiosis in soybeans, *Plant and Soil 60,* 445-450 (1981).
96. Habish, H.A. and Mahdi, A.A. Effect of soil moisture on nodulation of cowpea and hyacinth bean, *Journal of Agricultural Science,* Cambridge *86,* 553-560 (1976).
97. Ahmed, B. and Quilt, P. Effect of soil moisture stress on yield, nodulation and nitrogenase activity of *Macroptilium atropurpureum* cv. Siratro and *Desmodium intortum* cv. Greenleaf, *Plant and Soil 57,* 187-194 (1980).
98. Kuo, T. and Boersma, L. Soil water suction and root temperature effects on nitrogen fixation in soybeans, *Agronomy Journal 63,* 901-904 (1971).
99. Sprent, J.I. The effects of water stress on nitrogen-fixing root nodules, *New Phytologist 70,* 9-17 (1971).
100. Sprent, J.I. The effects of water stress on nitrogen-fixing root nodules IV. Effects on whole plants of *Vicia faba* and *Glycine max, New Phytologist 71,* 603-611 (1972).
101. Ayanaba, A. and Lawson, T.L. Diurnal changes in acetylene reduc-

tion in field-grown cowpeas and soybeans, *Soil Biology and Biochemistry 9,* 125-129 (1977).

102. Engin, M. and Sprent, J.I. Effects of water stress on growth and nitrogen-fixing activity of *Trifolium repens, New Phytologist 72,* 117-126 (1973).
103. Diatloff, A. Effects of soil moisture on the success of soil applied inocula for soybeans, *Queensland Journal of Agricultural and Animal Sciences 36,* 163-165 (1979).
104. Foulds, W. Response to soil moisture supply in three leguminous species II. Rate of N_2 (C_2H_2) fixation, *New Phytologist 80,* 547-555 (1978).
105. Vandecaveye, S.C. Effect of moisture, temperature, and other climatic conditions on *R. leguminosarum* in the soil, *Soil Science 23,* 355-362 (1927).
106. ICRISAT. *Annual Report 1979/80,* ICRISAT, Patancheru (1981).
107. Rerkasem, B. and Tongkumdee, D. Legume-*Rhizobium* symbiotic development in rice-based multiple cropping systems, p. 435, in *Current Perspectives in Nitrogen Fixation* (Editors, A.H. Gibson and W.E. Newton), Australian Academy of Science, Canberra (1981).
108. De-Polli, H., Franco, A.A. and Dobereiner, J. Survival of *Rhizobium* in flooded soils, *Pesquisa Agropecuaria Brasileira, Serie Agronomia 8,* 133-138 (1973).
109. Osa-Afiana, L.O. and Alexander, M. Effect of moisture on the survival of *Rhizobium* in soil, *Soil Science Society of America Journal 43,* 925-930 (1979).
110. Boonkerd, N. and Weaver, R.W. Survival of cowpea rhizobia in soil as affected by soil temperature and mositure, *Applied and Environmental Microbiology 43,* 585-589 (1982).
111. Zablotowicz, R.M., Eskew, D.L. and Focht, D.D. Denitrification in *Rhizobium, Canadian Journal of Microbiology, 24,* 757-760 (1978).
112. Daniel, R.M., Limmer, A.W., Steele, K.W. and Smith, I.M. Anaerobic growth, nitrate reduction and denitrification in 46 *Rhizobium* strains, *Journal of General Microbiology 128,* 1811-1815 (1982).
113. Ferguson, T.P. and Bond, G. Symbiosis of leguminous plants and nodule, *Annals of Botany 18,* 385-397 (1954).
114. Loveday, J. Influence of oxygen diffusion rate on nodulation of subterranean clover, *Australian Journal of Science 26,* 90-91 (1963).
115. Minchin, F.R. and Pate, J.S. Effects of water, aeration, and salt regime on nitrogen fixation in a nodulated legume—definition of an optimum root environment, *Journal of Experimental Botany 26,* 60-69 (1975).
116. Eaglesham, A.R.J. and Szalay, A.A. Aerial stem nodules on *Aeschynomene* spp., *Plant Science Letters* (in press) (1983).
117. Arora, N. Morphological development of the root and stem nodules of *Aeschynomene indica* (L.), *Phytomorphology 4,* 211-216 (1954).

118. Eaglesham, A.R.J. Personal observation.
119. Jackson, M.B. and Campbell, D.J. Movement of ethylene from roots to shoots, a factor in the responses of tomato plants to waterlogged soil conditions, *New Phytologist 74,* 397-406 (1975).
120. Grobbelaar, N., Clarke, B. and Hough, M.C. The nodulation and nitrogen fixation of isolated roots of *Phaseolus vulgaris* III. The effect of carbon dioxide and ethylene, *Plant and Soil, Special Volume,* 215-224 (1971).
121. Huang, C.-Y., Boyer, J.S. and Vanderhoef, L.N. Acetylene reduction (nitrogen fixation) and metabolic activities of soybean having various leaf and nodule water potentials, *Plant Physiology 56,* 222-227 (1975).
122. Albrecht, S.L., Bennet, J.M. and Quesenberry, K.H. Growth and nitrogen fixation of *Aeschynomene* under water stressed conditions, *Plant and Soil 60,* 309-315 (1981).
123. Sprent, J.I. Prolonged reduction of acetylene by detached soybean nodules, *Planta 88,* 372-375 (1969).
124. Dart, P.J. and Day, J.M. Effects of incubation temperature and Oxygen tension on nitrogenase activity of legume root nodules, *Plant and Soil Special Volume,* 167-184 (1971).
125. Schwinghamer, E.A., Evans, H.J. and Dawson, M.D. Evaluation of effectiveness in mutant strains of *Rhizobium* by acetylene reduction relative to other criteria of N_2 fixation, *Plant and Soil 33,* 192-212 (1970).
126. Sprent, J.I. and Gallacher, A. Anaerobiosis in soybean root nodules under water stress, *Soil Biology and Biochemistry 8,* 317-320 (1976).
127. Zablotowicz, R.M. and Focht, D.D. Denitrification and anaerobic, nitrate-dependent acetylene reduction in cowpea *Rhizobium, Journal of General Microbiology III,* 445-448 (1979).
128. Minchin, F.R. and Summerfield, R.J. Symbiotic nitrogen fixation and vegetative growth of cowpea (*Vigna unguiculata* (L.) Walp.) in waterlogged conditions, *Plant and Soil 45,* 113-127 (1976).
129. Hunter, M.N., de Jabrun, P.L.M. and Byth, D.E. Response of nine soybean lines to soil moisture conditions close to saturation, *Australian Journal of Experimental Agriculture and Animal Husbandry 20,* 339-345 (1980).
130. Troedson, R.J., Lawn, R.J. and Byth, D.E. Growth and nodulation of soybeans in high water table (HWT) culture, p. 464, in *Current Perspectives in Nitrogen Fixation* (Editors, A.H. Gibson and W.E. Newton). Australian Academy of Science, Canberra (1981).
131. Minchin, F.R., Summerfield, R.J., Eaglesham, A.R.J. and Stewart, K.A. Effects of short-term waterlogging on growth and yield of cowpea (*Vigna unguiculata*), *Journal of Agricultural Science,* Cambridge *90,* 355-366 (1978).

132. Fakui, I. and Ito, R. Fertility of the soybean as affected in short period by the excessive soil moisture content at different growing periods, *Proceedings of the Crop Science Society of Japan 20*, 271-273 (1952), (Japanese with English summary).
133. Hong, T.D., Minchin, F.R. and Summerfield, R.J. Recovery of nodulated cowpea plants (*Vigna unguiculata* (L.) Walp.) from waterlogging during vegetative growth, *Plant and Soil 48,* 661-672 (1977).
134. Heinrichs, D.H. Flooding tolerance of legumes, *Canadian Journal of Plant Science 50,* 435-438 (1970).
135. Broue, P., Marshall, D.R. and Munday, J. Response of lupins to waterlogging, *Australian Journal of Experimental Agriculture and Animal Husbandry 16,* 549-554 (1975).
136. Garcia-Novo, F. and Crawford, R.M.M. Soil aeration, nitrate reduction and flooding tolerance in higher plants, *New Phytologist 72,* 1031-1039 (1973).
137. McManmon, M. and Crawford, R.M.M. A metabolic theory of flooding tolerance: the significance of enzyme distribution and behaviour, *New Phytologist 70,* 299-306 (1971).
138. Siman, A., Cradock, F.W. and Hudson, A.W. The development of manganese toxicity in pasture legumes under extreme climatic conditions, *Plant and Soil 41,* 129-140 (1974).
139. Day, J.M., Roughley, R.J., Eaglesham, A.R.J., Dye, M. and White, S.P. Effect of high soil temperatures on nodulation of cowpea, *Vigna unguiculata, Annals of Applied Biology 88,* 445-487 (1978).
140. Eaglesham, A.R.J., Ayanaba, A., Ranga Rao, V. and Eskew, D.L. Mineral N effects on cowpea and soybean crops in a Nigerian soil I. Development, nodulation, acetylene reduction and yield, *Plant and Soil 68,* 183-192 (1982).
141. Bowen, G.D. and Kennedy, M.M. Effect of high soil temperatures on *Rhizobium* spp., *Queensland Journal of Agricultural Science 16,* 177-197 (1959).
142. Philpotts, H. The effect of soil temperature on nodulation of cowpeas (*Vigna sinensis*), *Australian Journal of Experimental Agriculture and Animal Husbandry 7,* 372-376 (1967).
143. IITA, *Annual Report for 1975,* Ibadan, Nigeria (1976).
144. Munevar, F. and Wollum, A.G. Effect of high root temperature and *Rhizobium* strain on nodulation, nitrogen fixation and growth of soybeans, *Soil Science Society of America Journal 45,* 1113-1120 (1981).
145. Wilkins, J. The effects of high temperatures on certain root-nodule bacteria, *Australian Journal of Agricultural Research 18,* 299-304 (1967).
146. Marshall, K.C. Survival of root-nodule bacteria in dry soils exposed to high temperatures, *Australian Journal of Agricultural Research 15,* 273-281 (1964).

147. Bissett, K.A. Complete and reduced life cycle in *Rhizobium*, *Journal of General Microbiology 7,* 233-242 (1952).
148. Graham, P.H., Parker, C.A. Oakley, A.E., Lange, R.T. and Sanderson I. J.V. Spore formation and heat resistance in *Rhizobium*, *Journal of Bacteriology 86,* 1353-1354 (1963).
149. Munevar, F. and Wollum, A.G. Growth of *Rhizobium japonicum* strains at temperatures above 27°C, *Applied and Environmental Microbiology 42,* 272-276 (1981).
150. Eaglesham, A., Seaman, B., Ahmad, H., Hassouna, S., Ayanaba, A. and Mulongoy, K. High temperature tolerant "cowpea" rhizobia, p. 436, in *Current Perspectives in Nitrogen Fixation* (Editors, A.H. Gibson and W.E. Newton), Australian Academy of Sciences, Canberra (1981).
151. Stowers, M.D., Goldman, B.J. and Eaglesham, A.R.J. Unpublished data.
152. Eaglesham, A.R.J. Unpublished data.
153. Dart, P.J. Islam, R. and Eaglesham, A.R.J. The root nodule symbiosis of chickpea and pigeon pea, pp. 63-83, in *Proceedings of the International Workshop on Grain Legumes,* ICRISAT, Hyderabad (1976).
154. Eaglesham, A.R.J. and Hassouna, S. Unpublished data.
155. Dart, P., Day, J., Islam, R. and Dobereiner, J. Symbiosis in tropical grain legumes: some effects of temperature and the composition of the rooting medium, pp. 361-384, in *Symbiotic Nitrogen Fixation in Plants* (Editor, P.S. Nutman), Cambridge University Press, Cambridge (1976).
156. Ranga Rao, V. Effect of temperature on the nitrogenase activity of intact and detached nodules in *Lotus* and *Stylosanthes*, *Journal of Experimental Botany 28,* 261-267 (1977).
157. Koermendy, A. and Eaglesham, A.R.J. Unpublished data.
158. Eaglesham, A., Day, J. and Dart, P. Effects of inoculation, root temperature and nitrogenous fertilizer on grain legume symbiosis, pp. 84-85, in *Rothamsted Report for 1973,* Part 1, Bartholomew Press, Dorking (1974).
159. Brockwell, J. Inoculation methods for field experimenters and farmers, pp. 211-227, in *Nitrogen Fixation in Legumes* (Editor, J.M. Vincent), Academic Press, Sydney (1982).
160. Ahmad, M.H., Eaglesham, A.R.J., Hassouna, S., Ayanaba, A., Mulongoy K., Pulver, E.L. Examining the potential for inoculant use with cowpeas in West African soils, *Tropical Agriculture* (Trinidad) 58, 325-335 (1981).
161. Eaglesham, A.R.J., Ahmad, M.H., Hassouna, S. and Goldman, B.J. Cowpea rhizobia producing dark nodules: use in competition studies, *Applied and Environmental Microbiology 44,* 611-618 (1982).

162. Ahmad, M.H., Eaglesham, A.R.J. and Hassouna, S. Examining serological diversity of "cowpea" rhizobia by the ELISA technique, *Archives of Microbiology 130,* 281-287 (1981).

163. Khan, A.C. The effect of vesicular arbuscular mycorrhizal associations on growth of cereals II. Effects on wheat growth, *Annals of Applied Biology 80,* 27-36 (1975).

164. Islam, R., Ayanaba, A. and Sanders, F.E. Response of cowpea (*Vigna unguiculata*) to inoculation with VA-mycorrhizal fungi and to rock phosphate fertilization in some unsterilized Nigerian soils, *Plant and Soil 54,* 107-117 (1980).

165. Islam, R. and Ayanaba, A. Effect of seed inoculation and pre-infecting cowpea (*Vigna unguiculata*) with *Glomus mosseae* on growth and seed yield of the plants under field conditions, *Plant and Soil 61,* 341-350 (1981).

166. Powell, C. Mycorrhizal fungi stimulate clover growth in New Zealand hill country soils, *Nature 264,* 436-438 (1976).

167. Mosse, B. Advances in the study of vesicular-arbuscular mycorrhiza, *Annual Review of Phytopathology 11,* 171-196 (1973).

168. Duhoux, E. and Dreyfus, B. Nature of symbiotic infection sites on the stem of the legume *Sesbania rostrata, Comptes Rondus Academie des Sciences 294,* 407-411 (1982).

169. Nambiar, P.T.C., Dart, P.J., Srinivasa Rao, B. and Ramanatha Rao, V. Nodulation in the hypocotyl region of groundnut (*Arachis hypogaea*), *Experimental Agriculture 18,* 203-207 (1982).

170. Fyson, A. and Sprent, J.I. A light and scanning electron microscope study of stem nodules in *Vicia faba* L., *Journal of Experimental Botany 31,* 1101-1106 (1980).

2. Interaction of Nitrogen-Fixing Microorganisms with Other Soil Microorganisms

N.S. Subba Rao

Like other members of the microbial community in soil, nitrogen-fixing microorganisms are also subject to the antagonistic and associative effects of other species of bacteria, fungi, actinomycetes, protozoa and bacteriophages. Besides the specific effects of these microorganisms, nematodes and insect predators of nodules also diminish the symbiotic effects in legumes. Nodulation failures or the inability to obtain the desired responses to seed inoculation of legumes with rhizobia have been frequently observed by agronomists and one of the possible reasons for this lack of response in a particular soil environment may be due to a shift in the microbiological equilibrium towards the creation of a rhizosphere microflora predominated by microorganisms antagonistic to rhizobia. It is however difficult to conclusively prove this point *in situ* in the rhizosphere for various reasons. The main reason is the non-availability of quick and reliable methods to accurately count the various groups of microorganisms in soil.

Apart from the detrimental effects of nitrogen fixation attributable to the inhibitory action of microorganisms towards rhizobia, it would be worthwhile to exploit some of the observed beneficial influence of some microorganisms or their products towards improving plant growth and increasing the number of viable nitrogen-fixing bacteria in the rhizosphere. In this context, the interaction between nitrogen-fixing bacteria and vesicular arbuscular mycorrhizae (VAM) merits consideration. This review highlights some ecological aspects of microbial interactions in soil with reference to nitrogen-fixing bacteria and points out the importance of these studies in inoculant preparation and application.

INTERACTION BETWEEN *RHIZOBIUM* AND OTHER MICROORGANISMS

There have been several reports on the inhibitory effects of bacteria, actinomycetes and fungi on the growth of *Rhizobium* spp. in culture media

Table 1. Inhibitory effects of interaction of different bacteria, actinomycetes and fungi on *Rhizobium* and root nodulation

Rhizobium sp./host plant	Other microorganisms involved/nature of interaction	Results of interaction	Reference
1	2	3	4
Several species of *Rhizobium*	*Actinomyces flavus, Bacillus mesentericus, B. subtilis, Escherichia coli* var. *Communicor* and certain fungi	Inhibited the growth of rhizobia	Konishi, K. [1]
Several species of *Rhizobium*	Gram negative, non-spore forming soil bacteria	Two bacteria inhibited the multiplication of *Rhizobium* whereas thirteen of them stimulated the growth of *Rhizobium*	Krasilnikov, N.A. and Korenyako, A.I. [2]
Several species of *Rhizobium*/leguminous plants	Six antagonists of rhizobia comprising bacteria, fungi and actinomycetes	In sterilized soil five *Rhizobium* inoculated legumes were raised in the presence of antagonists when the latter interfered with nodulation	Robinson, R.S. [3]
R. trifolii/*Trifolium repens* (red clover)	*Aspergillus wentii*	Reduced the number of nodules of clover	Robinson, R.S. [4]
Several species of *Rhizobium*	*Streptomyces* and *Penicillium*	Both the genera inhibited the growth of rhizobia in culture media	Thornton G.D. et al. [5]
R. lupini, R. japonicum, cowpea rhizobia/*Lupinus lupini*	*Streptomyce* spp., a spore forming bacteria	*R. lupini, R. japonicum* growth was inhibited by *Streptomyces* sp. whereas cowpea rhizobia were least susceptible. Nodulation in plants was in no way affected. A spore forming bacterium showed marked inhibition upon nodulation and nitrogen fixation of lupine	Abdel Ghaffar, A.S. and Allen, O.N. [6]
R. trifolii/*T. subterraneum* (subterraneum clover)	Several microorganisms	Poor nodulation was due to microbial antagonism	Hely, F.N. et al. [7]
R. trifolii/clover	Bacteria and fungi	14 organisms were harmful and 8 beneficial for plant-*Rhizobium* association. *In vitro* inhibition of rhizobia had no relation with plant effects	Anderson, K.J. [8]

R. trifolii/T. subterraneum (red clover)	*Penicillium* spp. and *Pythium* spp.	Inhibited nodulation	Holland, A.A. [9]
R. trifolii and *R. meliloti*	*Streptomyces* spp.	All rhizobia sensitive to *Streptomyces.* Exposure of rhizobia to antagonistic actinomycetes produced variants whose nodulation abilities differed from the parent	van Schreven, D.A. [10]
T. alexandrium, Trifolium sp., *Melilotus parviflora, Trigonella foenum-graecum, Cicer arietinum, Pisum sativum*	*Cephalosporium* predominant on nodules	Inhibited the growth of *Rhizobium*	Subba Rao, N.S. and Vasantha, P. [11]
Several common legumes/rhizobia	*Cephalosporium, Alternaria, Aspergillus, Penicillium, Rhizopus, Acrothecium, Fusarium, Rhizoctonia, Curvularia, Pythium, Trichoderma*	*R. trifolii, R. phaseoli, R. leguminosarum, R. japonicum,* were more susceptible to fungal antibiotics than *R. meliloti* or *R.* sp. (cowpea group). *Cephalosporium* reduced nitrogen content and weight of *T. alexandrinum*	Chhonkar, P.K. and Subba Rao, N.S. [12]
R. japonicum	Soil actinomycetes	20 isolates of soil actinomycetes produced no inhibition of rhizobia. Two isolates inhibited the growth of two strains of *R. japonicum* and one isolate reduced the number of nodules of Kent variety	Damirgi, S.M. and Johnson, H.W. [13]
R. trifolii	246 bacterial and actinomycete isolates and 107 fungal isolates from clover rhizosphere	24% of bacteria and actinomycetes and 6 fungi showed antagonistic influences towards one or more rhizobia. One of the powerful antagonist was identified as *Pseudomonas*	Hattingh, M.J. and Louw, H.A. [14]
R. trifolii/T. subterraneum	Soil fungi, bacteria and actinomycetes	31 fungi produced detectable antibiotics towards *Rhizobium* and inhibited the growth of seedlings	Holland, A.A. and Parker, C.A. [15]

(Contd.)

1	2	3	4
Rhizobium spp.	Several soil fungi	*Aspergillus, Cephalosporium, Chaetomium, Cladosporium, Fusarium, Mortierella, Penicillium, Rhizoctonia, Scolecobasidium, Sordaria* and *Thielavia*, were inhibitory and *Paecilomyces, Phoma* and *Rhizopus* were stimulatory to *Rhizobium*	Sethi, R.P. and Subba Rao, N.S. [16]
R. trifolii	Soil bacteria	*Pseudomonas, Xanthomonas, Achromobacter, Flavobacterium, Alcaligenes, Erwinia, Aerobacter, Bacillus, Streptomyces, Nocardia, Corynebacterium, Arthrobacter* and *Brevibacterium* (in all 83 isolates) inhibited *R. trifolii*	Hattingh, M.J. and Louw, H.A. [17]
Rhizobium (sp.) for chickpea (*Cicer arietinum*)		Certain rhizosphere isolates of *Achromobacter, Aerobacter, Agrobacterium, Bacillus, Micrococcus, Arthrobacter, Sarcina* and *Pseudomonas* inhibited rhizobial growth while other isolates of the same species of bacteria stimulated growth	Bhalla, H. and Sen, A.N. [18]
R. phaseoli	Soil bacteria, fungi, actinomycetes	*Streptomyces, Aspergillus,* and *Bacillus subtilis* inhibited the growth of *Rhizobium*	Ibrahim, A.M. et al. [19]
R. sp. for groundnut (*Arachis hypogea*)	Soil microorganisms	Inhibition of *Rhizobium* is a factor to be reckoned with	Rao, J.V.D.K. and Shende, S.T. [20]
Cassia fistula, C. occidentalis, Leuceaena leucocephala	Root extracts and rhizosphere microorganisms	Inhibited the growth of rhizobia	Ranga Rao, V. and Subba Rao, N.S. [21]
Several rhizobia	Actinomycetes	23–70% of the actinomycetes isolates showed inhibitory activity on	Patel, J.J. [22]

		Rhizobium; non-acid producers of *Rhizobium* were less susceptible than the acid producers	
R. japonicum/Glycine max (soybean)	Rhizosphere bacteria	8 to 9 rhizosphere bacteria inhibited the growth of *R. japonicum* but nodulation was not affected in spite of severe tap root injury by one isolate	Smith, R.S. and Miller, R.H. [23]
Rhizobia/*Vigna mungo* and *Glycine max*	Bacteria, fungi and actinomycetes on seed	The seed invariably had isolates which inhibited *Rhizobium* growth	Jain, M.K. and Rewari, R.B. [24]
Rhizobium spp.	Rhizosphere bacteria from *Arstida oligantha*	All the 6 *Rhizobium* spp. were inhibited by *Xanthomonas azonopodis, Arthrobacter citreus, A. simplex, Enterobacter aerogenes, Micrococcus luteus* inhibited one or more *Rhizobium* sp. *Bacillus cereus* and *B. megaterium* stimulated growth	Leuck, E.E. and Rice, E.L. [25]
R. japonicum/Glycine max (soybean)	*Rhizoctonia solani*	*R. solani* significantly reduced the top and nodule weights of Lee and Kent varieties of soybean	Orellana, R.G. et al. [26]
R. japonicum/Glycine max (soybean)	*Azotobacter*	*Azotobacter* inoculation did not help nodulation or efficiency of *Rhizobium*	Rao, J.V.D.K. and Patil, R.B. [27]
Rhizobium japonicum/Glycine max (soybean)	17 common soil fungi	*Trichoderma viride, Rhizopus nigricans* and *Mucor vesiculosis* were antagonistic to *R. japonicum* in culture. In perlite sand and also in unsterilized soil *T. viride* reduced nodulation and acetylene reduction value and number of *R. japonicum*. The effects of other two fungi were not as drastic as that of *T. viride*	Angle et al. [28]
Rhizobium sp./*Vigna radiata* (mung)	*Rhizoctonia bataticola*	The infected roots showed distortion of outer layers of root	Kush, A.K. [29]
R. japonicum/Glycine max (soybean)	Actinomycetes, fungi and bacteria in field plots	Establishment of inoculated soybean depends on antagonistic actinomycetes in soil	Pugashetti, B.K. et al. [30]

(Table 1). The inhibitory microorganisms were isolated from soil, rhizosphere, seed coat and root nodule surface. These microorganisms were cultured on agar medium along with rhizobia or their culture filtrates were used to test the inhibitory effects [1-30]. Some of the bacterial species which were inhibitory towards one or more species of *Rhizobium* included species of *Bacillus, Escherichia, Pseudomonas, Xanthomonas, Achromobacter, Flavobacterium, Alcaligenes, Erwinia, Aerobacter, Corynybacterium, Arthrobacter, Brevibacterium, Agrobacterium, Micrococcus, Sarcina* and *Enterobacter*. Some of the actinomycetes which proved inhibitory to rhizobia belonged to the genera *Actinomyces, Streptomyces* and *Nocardia*. Among the inhibitory fungi, mention may be made of *Cephalosporium, Alternaria, Aspergillus, Penicillium, Rhizopus, Acrothecium, Fusarium, Rhizoctonia, Curvularia, Pythium, Trichoderma, Chaetomium, Cladosporium, Mortierella, Scolecobasidium, Sordaria, Thielavia, Aspergillus* and *Mucor*. It must be pointed out however, that many investigators merely counted the gross number of different groups of microorganisms which proved inhibitory towards rhizobia without identifying the microorganisms even up to the generic level. Furthermore, these studies are less meaningful because the observed antibiotic effects in culture media appear to have no relevance to soil conditions. In some experiments, *Rhizobium*-inoculated legumes were grown in the presence of antagonistic microorganisms in soil under potted conditions and the results were invariably detrimental to nodulation and plant growth. For instance, mycocolonization in soil with *Cephalosporium* sp. markedly reduced the nitrogen content and weight of *Rhizobium trifolii* inoculated Egyptian clover (*Trifolium alexandrinum*) [12]. In another instance, *Rhizoctonia solani* application to soil significantly reduced the top and nodule weights of Lee and Kent varieties of soybean [26]. Similarly, *Trichoderma viride* in soil reduced the acetylene reduction abilities of nodules as well as the numbers of *R. japonicum* in soil when planted with soybean [28]. Actinomycetes (notably *Streptomyces* isolates) had invariably deleterious effect on nodulation when co-inoculated with soil [30]. In many instances, the non-establishment of clovers or soybean in field conditions has been attributed to the antagonistic effects of soil microorganisms towards rhizobia. There is need, however, for caution in extrapolating the results of pot culture studies to field effects unless the *in situ* occurrence of a large number of antagonistic microorganisms in soil can be established quantitatively by reliable methods. Moreover, a secondary build-up of *Rhizobium* populations after the primary antibiotic effects have weakened in soil, cannot be ruled out. Nevertheless, the use of strains of *Rhizobium* resistant to a wide spectrum of antibiotics could prove useful in endemic areas where nodulation difficulties are encountered.

While inhibitory effects of microorganisms on rhizobia have been frequently reported, occasional specific observations on the stimulatory

Table 2. Stimulatory effects of different bacteria, actinomycetes and fungi on *Rhizobium* and root nodulation

Rhizobium sp./host plant	Other microorganisms involved/ nature of interaction	Results of interaction	Reference
1	2	3	4
R. Meliloti/*medicago sativa* (lucerne) *R. Trifolii*/*Trifolium repens* (white clover)	*Azotobacter*	Did not significantly stimulate nitrogen fixation	Jensen, H.L. [31]
Several species of *Rhizobium*	Gram negative, non-spore forming soil bacteria	Two bacteria inhibited the multiplication of *Rhizobium* whereas thirteen of them stimulated the growth of *Rhizobium*	Krasilnikov, N.A. and Korenyakò, A.I. [2]
R. trifolii/clover	Certain bacteria and fungi	Certain bacteria and fungi markedly increased nodulation of clover by a poorly virulent strain but had little effect upon the behaviour of a highly virulent strain	Harris, J.R. [32]
R. trifolii/clover	Bacteria and fungi	14 organisms were harmful and 8 beneficial for plant-*Rhizobium* association. *In vitro* inhibition of rhizobia had no relation with plant effects	Anderson, K.J. [8]
Rhizobium sp.	Several soil fungi	*Aspergillus, Cephalosporium, Chaetomium, Cladosporium, Fusarium, Mortierella, Penicillium, Rhizoctonia, Scolecobasidium, Sordaria* and *Thielavia* were inhibitory and *Paecilomyces, Phoma* and *Rhizopus* were stimulatory to *Rhizobium*	Sethi, R.P. and Subba Rao, N.S. [16]
Rhizobium sp. for chickpea (*Cicer arietinum*)		Certain rhizosphere isolates of *Achromobacter, Aerobacter, Agrobacterium, Bacillus, Micrococcus, Arthrobacter, Sarcina* and	Bhalla, H. and Sen, A.N. [18]

(*Contd.*)

1	2	3	4
		Pseudomonas inhibited rhizobial growth while other isolates of the same species of bacteria stimulated growth	
R. japonicum/Glycine max (soybean)	*Beijerinckia, Azotobacter*	Peat based *R. japonicum* + *Beijerinckia* and *Azotobacter* increased yield and nodulation in potted plants	Apte, R. and Iswaran, V. [33]
Rhizobium spp.	*Azotobacter chroococcum, Bacillus megaterium* var. *phosphaticum*	In nitrogen-free medium *A. chroococcum* fixed appreciable amount of nitrogen and stimulated the growth of *B. megaterium* and *Rhizobium* sp.	Shende et. al. [34]
Phaseolus vulgaris (French bean)	Phosphate solubilizing bacteria	*Agrobacterium*, a phosphate solubilizer enhanced *Rhizobium* inoculation effect	Barea, J.M. et al. [35]
Rhizobium spp.	Rhizosphere bacteria from *Arstida oligantha*	All the 6 *Rhizobium* sp. were inhibited by *Xanthomonas axonopodis, Arthrobacter citreus, A. simplex, Enterobacter aerogenes. Micrococcus luteus* inhibited one or more *Rhizobium* spp. *Bacillus cereus* and *B. megaterium* stimulated growth	Leuck, E.E. and Rice, E.L. [25]
Rhizobium sp./*Cicer arietinum* (chickpea)	*Azotobacter, Beijerinckia*	*Azotobacter* + *Rhizobium* inoculation increased the mass of nodules and grain yield but combination of *Beijerinckia* + *Rhizobium* did not have the same effect	Ravat, A.K. and Sanoria, C.L. [36]
R. japonicum/Glycine max (soybean)	*Azospirillum brasilense*	*Azospirillum* + *Rhizobium* combination increased the number of nodules and yield of soybean in pots, in 3 successive experiments	Singh, C.S. and Subba Rao, N.S. [37]
Rhizobium spp./*Vigna radiata* (mung) *Glycine max* (soybean) and *Pisum sativum* (pea)	*Azotobacter*	Combination of *Rhizobium* + suitable strain of *Azotobacter* gave higher yields of legumes	Jauhri, K.S. et al. [38]

Rhizobium spp./*Glycine max* (soybean) *Vigna unguiculata* (cowpea) and *Trifolium repens*	*Azotobacter vinelandii*	*Azotobacter* increased root nodules of all the 3 plants in green house and in field grown soybean. Non-nitrogen fixing mutants of *Azotobacter* increased the number of nodules. Culture filtrates of *Azotobacter* were of no avail in increasing nodules. Hence, it is believed that a cell-bound protein is responsible for stimulation of nodulation	Burns et al. [39]
Rhizobium spp./*Trifolium*, lucerne and chickpea (*Cicer arietinum*)	*Azospirillum brasilense*	Field trials showed *Rhizobium* + *Azospirillum* generally increased nodulation and yield	Tilak, K.V.B.R. et al. [40]
Rhizobium spp./*Trifolium alexandrinum, Medicago sativa, Vigna mungo, V. unguiculata, Glycine max, Leuceana leucocephala*	*Saccharomyces cerevisiae*	Improved root nodulation and yield of legumes cited, in pot trials.	Tuladhar and Subba Rao, unpublished

effects of other microorganisms on rhizobia and root nodulation (Table 2) have also been reported [2, 8, 16, 18, 25, 31-40]. In cultural studies on agar and in liquid media, it has been reported that *Paecilomyces, Phoma* and *Rhizopus* stimulated the growth of *Rhizobium* [16]. Of special significance is the beneficial effect of *Saccharomyces cerevisiae* on nodulation and yield of *Trifolium alexandrinum, Medicago sativa, Vigna mungo, V. unguiculata, Glycine max* and *Leuceana leucocephala* (Tuladhar and Subba Rao, unpublished). In one instance, the behaviour of isolates of the same bacterium were different in their inhibitory or stimulatory effects—some isolates of *Achromobacter, Aerobacter, Agrobacterium, Bacillus, Micrococcus, Arthrobacter, Sarcina* and *Pseudomonas* were inhibitory towards *Rhizobium* while others were stimulatory [18]. On the contrary, certain strains of bacteria and fungi markedly increased the nodulation of clover by a poorly virulent *R. trifolii* but a similar effect could not be seen in clover nodulated by a highly virulent *R. trifolii* [32].

Whether non-symbiotic and symbiotic nitrogen fixers like *Azotobacter* and *Rhizobium* and phosphate solubilizing microorganisms such as *Bacillus megaterium* and *Agrobacterium* contribute towards a synergistic effect on symbiotic nitrogen fixation in legumes has been the subject of investigation by several workers but the results have not been very consistent [31, 33-40]. In one of the earliest investigations, *Azotobacter* was found to have no effect on nodulating lucerne [31] and white clover whereas in one of the later experiments, intact cells but not the culture filtrates of *Azotobacter* increased root nodules of soybean, cowpea and clover [39], thereby suggesting that a cell-bound proteinaceous material may have been responsible for the observed improved nodulation. In field experiments with chickpea (*Cicer arietinum*), the *Azotobacter* + *Rhizobium* combination increased the mass of nodules and grain yield but similar results were not observed with a *Beijerinckia* + *Rhizobium* combination [36]. On the contrary, in potted experiments peat based *R. japonicum* + *Beijerinckia* as well as *Azotobacter* increased yield and nodulation of soybean [33]. Similarly, French bean plants (*Phaseolus aureus*) established well and produced more nodule when inoculated with *Agrobacterium*, a phosphate solubilizer [35]. These conflicting reports can be reconciled only on the basis of variations in soil types, inoculum load and environmental factors.

INTERACTION OF PROTOZOA WITH *RHIZOBIUM*

The decline in rhizobial numbers in soil is accompanied by a fall in the density of protozoa which feed on *Rhizobium* (Table 3) and *Bdellovibrio* is a widespread genus of the protozoan predators in soil. Rhizobia however, are not entirely eliminated by protozoa and the numbers of rhizobia attain normalcy when they survive the attack of the predators. These observations have been made by studies with *R. meliloti* and *Rhizobium* sp. (cowpea type). In other studies, is was apparent that antibiotic resistant *R. phaseoli*

Table 3. Inhibitory effects of protozoa on *Rhizobium* and root nodulation

Rhizobium sp./host plant	Other microorganisms involved/ nature of interaction	Results of interaction	Reference
Rhizobia	Soil protozoa	The numbers of *R. meliloti*, *Rhizobium* sp. (cowpea) decreased with increase in the numbers of protozoa	Danso, S.K.A. et al. [41]
Rhizobium	*Bdellovibrio*	*Rhizobium* cells survived the attack of the predator in liquid media and did not multiply in sterile soil. Large numbers of rhizobia in soil encouraged the predator	Keya, S.O. and Alexander, M. [42]
Rhizobium spp.	Microorganisms in soil in general, particularly protozoa	Protozoa in soil, particularly *Bdellovibrio* influenced the numbers of rhizobia in soil	Alexander, M. [43]
Rhizobium spp.	"	"	Danso, S.K.A. [44]
Rhizobium phaseoli	Interaction with protozoa in soil	Antibiotic resistant *R. phaseoli* numbers were reduced in soil accompanied by rise in protozoal numbers and similar effects were not seen in *R. meliloti* inoculated sterilized soil without protozoa	Chao, W.L. and Alexander, M. [45]

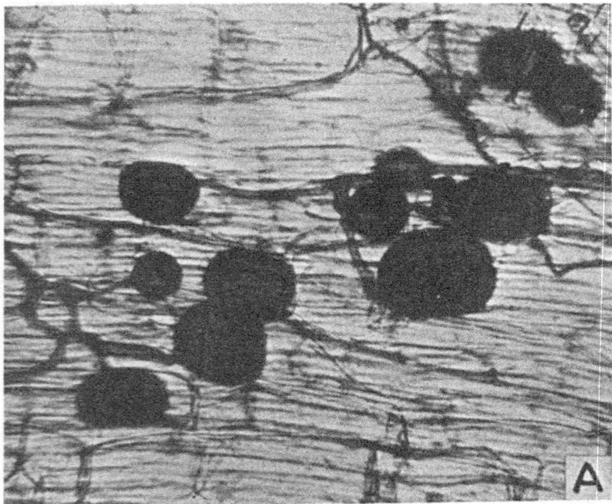
A

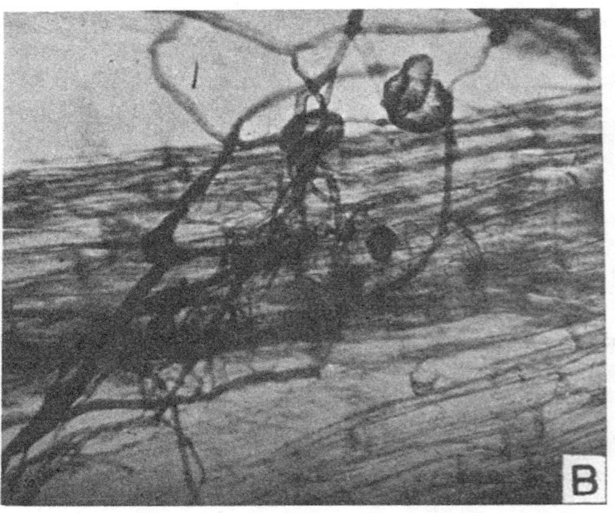
B

O kg P_2O_5/ha
(a)
(b)
(c)
(d)
C

diminished in numbers in natural soil accompanied by a concomitant increase in protozoal numbers although similar effects were not seen in *R. meliloti* inoculated sterilized soil [41-45].

RHIZOBIOPHAGES IN RELATION TO LEGUME SYMBIOSIS

There has been no clear-cut evidence to demonstrate the inhibitory effects of phages on root nodules even though a number of workers, from time to time, have isolated and demonstrated the non-specific lytic action of phages on rhizobia. Some of these reports relate to methods of isolation of phages from soil and plant roots including nodules and morphological features, distribution in soils, properties and cross-reaction of phages between different strains of rhizobia [46-65]. The plants investigated for phage reaction were invariably cultivated grain and fodder legumes such as clovers, *Pisum sativum, Vicia faba, Vicia sativa, Lens esculenta*, soybean and chickpea.

Reports attributing positive detrimental effects of phage on symbiotic nitrogen fixation have come by studies on alfalfa and peas [52, 54, 55]. On the other hand, doubts have been cast on the role of bacteriophages in minimizing the benefits of legume symbiosis based on the following observations: (1) a strain of rhizobiophage could attack only 10-15 per cent of the number of pea and clover rhizobia collected over a large area; (2) the absence of specificity between a phage and a strain or species of *Rhizobium*; (3) the presence of strains of *Rhizobium* both susceptible as well as resistant to the action of phages in the same soil sample; and (4) the rapid development of strains of rhizobia resistant to phage action, thereby affording a chance for the natural build up of strains capable of nodulating the host in spite of the presence of lytic phages [61].

In another study, it was observed that mutations in *Rhizobium* to phage resistance may coincide with its morphology and effectiveness in nitrogen fixation. The frequency of occurrence of mutation and stability of the acquired features, however, depended on the variations between strains of *Rhizobium*. It was further observed that ineffective phage resistant mutants developed rather more readily from effective strains of *Rhizobium* than vice versa, indicating the possibility that the proportion of ineffective strains of *Rhizobium* may increase in soil at the expense of effective strains [58].

LEGUME-*RHIZOBIUM* SYMBIOSIS IN VIRUS-INFECTED PLANTS

Nodulation and nitrogen fixation in clover (*Trifolium repens*) were

---

Fig. 1. Interaction of *Rhizobium* and VAM fungus on groundnut (*Arachis hypogea*). A and B—Photomicrographs showing vescicles and hyphae of *Glomus* sp. on roots; C—Synergestic effects of *Rhizobium* and *Glomus* sp. on groundnut grown in unsterilized soil in pots: (a) control without *Rhizobium* or *Glomus* sp.; (b) with *Glomus* sp.; (c) with *Rhizobium* and (d) with both *Rhizobium* and *Glomus* sp. (Photographs of potted plants, courtesy of Dr. J. Raj, University of Agricultural Sciences, Bangalore).

reduced on infection by clover phyllody virus [66–68]. Similarly, soybean mosaic virus and bean pod mottle virus influenced nodulation in soybean [69]. In field bean (*Dolichos lablab*) infected with *Dolichos* enation mosaic virus (DEMV), a reduction in nodule numbers has been reported in natural soil whereas increase in nodule numbers could be observed in infected plants grown in nitrogen-free sand cultures [70]. On the other hand, in a detailed study on *Dolichos* enation mosaic virus (DEMV) and symbiotic nitrogen fixation in field bean (*Dolichos lablab*), it was observed that infection decreased the levels of total carbohydrates and reducing sugars in roots and nodules of these plants after 24 days of growth. However, the total nitrogen content in these nodulated virus infected plants was more than that of the healthy plants. There was a higher rate of transfer of fixed nitrogen from nodules to the tops of plants as well as roots consequent upon virus infection of the host. Soluble proteins, total soluble nitrogen constituents and the total insoluble nitrogen increased in nodules following virus infection of the legume. The infected plants showed increased nitrogen fixation during 24 to 31 days which was higher than healthy plants and this high level was maintained throughout the growth period [71]. These results point out that virus infection promotes nitrogen fixation. Therefore, there is need for renewed investigation on nodulation in plants infected by plant pathogens, especially in view of the contradicting results obtained so far.

DUAL EFFECT OF *RHIZOBIUM* AND VAM FUNGI ON LEGUME SYMBIOSIS

Several legumes have been found to benefit by the combined inoculation of *Rhizobium* and VAM fungi (Fig. 1). The legumes studied extensively in this regard are soybean, lucerne and clovers but other plants such as French bean, groundnut, *Stylosanthes* and cowpea have also been used as test plants to understand this interaction. The VAM fungi used were species of *Endogone* and *Glomus* and in certain cases vaguely described as VAM fungi. *Glomus* spp. appears to be a predominant colonizer of roots and is non-specific with regard to host plants which it infects. The characteristic features taken into account in understanding the VAM effects are plant size, yield and *p*-content in plants grown in fumigated or non-fumigated soil and sterilized or unsterilized soil in pots. In some instances, field-plot experiments have been carried out to verify the results. There appears to be a consensus of opinion that VAM inoculation to nodulating legumes generally increases the plant size and yield, although contradictory results have come forth with regard to increased *p*-uptake by plant tops (Table 4). The availability of *p* in soil diminishes with time of incubation of added phosphates in soil and the beneficial effects of VAM inoculation seem to be better when phosphates are freshly added to soil. Apart from the observed VAM effects on *p*-uptake, the nitrogen content of plants appears to be enhanced due to better nodulation and symbiotic effects [72–88].

Table 4. Dual effects of *Rhizobium* and VAM fungi on legume symbiosis

Rhizobium/host plant	VAM fungi	Results of interaction	Reference
1	2	3	4
R. japonicum/*Glycine max* (soybean)	*Endogone*	In fumigated plots, infestation with VAM fungus resulted in 34–40% increase in growth and yield of soybean accompanied by higher content of P, N, Ca, Cu and Mn than VAM free plots, although similar effects were not noticeable in non-fumigated plots	Ross, J.P. and Harper, J.A. [72]
R. japonicum/*Glycine max* (soybean)	*Endogone*	Mycorrhizal plants do not dissolve insoluble native soil p more efficiently than non-mycorrhizal plants	Ross, J.P. and William, J.W. [73]
R. japonicum/*Glycine max* (soybean)	*Endogone*	VAM fungus increased the yield of nodulating soybean in fumigated plots while no such increase was noticeable in non-fumigated plots or with non-nodulating soybean	Schenk, N.C. and Hinson, K. [74]
R. phaseoli/French bean (*Phaseolus vulgaris*)	*Endogone*	The effects of VAM on legume growth, nodulation and N_2-fixation are similar to those of adding phosphate	Daft, M.J. and El-Giahmi, A.A. [75]
Rhizobium sp./*Arachis hypogea* (groundnut)	*Glomus mossae*	Plant size, yield of nuts and p content increased by dual inoculation with *Rhizobium* + VAM	Daft, M.J. and El-Giahmi, A.A. [76]
R. trifolii/*Trifolium subterraneum* (clover)	VAM fungi	When phosphates are added to soil, availability of p diminishes with time of incubation and temperature up to 80°C. However, VAM fungi inoculation more than doubled in freshly phosphate added soil than in soil incubated for long with added phosphate	Barrow, N.J. et al. [77]

(*Contd.*)

1	2	3	4
R. trifolii/Trifolium subterraneum	*Glomus mossae* and *G. fasciculatus*	In superphosphate added steamed soil, VAM fungi increased nodulation, growth and p-content of clover markedly with *G. mossae* but not in untreated soil	Abbot, L.K. and Robson, A.D. [78]
Rhizobium sp./*Stylosanthes guyanensis*	VAM endophytes	In 12 unsterilized soils, nodulation and p uptake were often stimulated by VAM but no such effects were seen in irradiated soil	Mosse, B. [79]
R. meliloti/Medicago sativa (lucerne)	VAM fungi	VAM brought about more extensive nodulation, nodule weight, nitrogenase activity and p-content than non-mycorrhizal plants	Smith, S.E. and Daft, M.J. [80]
R. meliloti/Medicago sativa (lucerne)	VAM fungi	Satisfactory nodulation was dependent on VAM fungi	Azcon-G de Anguilar, C. and Barea, J.M. [81]
R. meliloti/Medicago sativa (lucerne)	*Glomus mossae*	VAM inoculation improved the total N, P, and K uptake and the effect of *Rhizobium* was only effective when applied together with *Glomus*	Azcon-G de Anguilar, C., Azcon, R. and Barea, J.M. [82]
R. japonicum/Glycine max (soybean)	*Glomus fasciculatus*	The number, dry weight and N content of root nodules in plants inoculated with *Glomus*+*Rhizobium* were significantly more than *Rhizobium* inoculated ones	Bagyraj, D.J., Manjunath, A. and Patil, P.B. [83]
R. trifolii/Trifolium pratense	VAM and phospho-bacteria	Inoculation with *Rhizobium* +VAM+ Phosphobacteria was better than individual inoculation and *Rhizobium* + phosphobacteria treatments in P and N uptake	Delorengiri, C., Barea, J.M. and Olivares, J. [84]
R. trifolii/Trifolium repens (white clover)	*Glomus tenuis* and *Gigaspora margarita*	Increased shoot yield up to 91% in pots and 37% in the field by inoculation with VAM fungi	Powell, C.Ll. [85]

Rhizobium sp./*Vigna unguiculata* (cowpea)	*Glomus fasciculatum*	Growth of crop was encouraged by VAM fungi	Islam, R. et al. [86]
Rhizobium japonicum/*Glycine max* (soybean)	*Glomus fasciculatum*	In nutrient solutions, soybean plants inoculated with *Rhizobium* + VAM had significantly higher rates of N_2 fixation, plant and nodule mass and p content	Belhlenfalvay, G.J. and Yoder, J.F. [87]
Rhizobium sp./*Hedysarum boreale* (sweetvetch)	*Glomus fasciculatum*	Under the influence of VAM, *Rhizobium* inoculated plants grew better and registered higher p uptake than control plants.	Redente, E.F. and Reeves, F.B. [88]

Table 5. Some interactions between *Azotobacter* and fungi

Nature of interactions	Reference
Azotobacterin decreased the incidence of viral and bacterial disease of potato	Dorosinskii, L.M. [89]
Azotobacter suppressed the growth of fungi on germinating seeds of wheat. *Alternaria* infection of maize decreased depending upon the strain of *Azotobacter*	Mishustin, E.N. and Naumova, A.N. [90]
Azotobacter decreased the infection of wheat and *Medicago luperina* seeds by *Fusarium* and *Ascochyta imperfecta*	Brown, M.E. and Burlingham, S.K. [91]
Azotobacter decreased the incidence of *Sclerotinia libertina* in grass, of smut in millets and brown rust in wheat	Menkina, R.A. [92]
Azotobacter spp. produced a thermostable, ether soluble fungistatic substance which inhibited the growth of *Fusarium moniliforme*	Lakshmi Kumari et al. [93]
Azotobacter inhibited the growth of several fungi	Shende, S.T. et al. [94]

INTERACTION BETWEEN *AZOTOBACTER* AND OTHER MICROORGANISMS

The antifungal properties of *Azotobacter* towards *Fusarium, Alternaria, Ascochyta, Sclerotinia* and other fungi, and the amelioratory influence of *Azotobacter* on several seedling diseases of plants (Table 5) have been reported from time to time [89-94]. The antifungal antibiotic principle from *Azotobacter* has been determined as a thermostable, ether-soluble substance which inhibited *Fusarium moniliforme* in culture media [93].

Several investigations deal with the associative beneficial effects of *Azotobacter* with other bacteria, especially with phosphate-dissolving bacteria in improving nitrogen fixation [95-100]. In some experiments, the fixation of nitrogen by *Azotobacter* was related to the action of cellulolytic and pectinolytic bacteria [101-103] growing in association with *Azotobacter* or to the influence of mixed cultures with other bacteria such as *Rhodopseudomonas* and *Azospirillum* (Table 6) [104-107].

POSSIBLE BENEFITS OF INTERACTION

The associative effects of other microorganisms in soil and on roots with nitrogen-fixing microorganisms could be harnessed in the preparation of carrier-based inoculants. A mixture of beneficial microorganisms and rhizobia could be used in the carrier to obtain better and longer survival of rhizobia both in the carrier and on the inoculated seed. Bacteria which produce high amounts of extracellular polysaccharide (slime) and increase the growth of rhizobia can prove extremely beneficial to augment the moisture-retaining capacity of the carrier material during harsh storage conditions.

Phosphorus is essential for successful nodulation and nitrogen fixation in legumes. The dual effect of phosphorus mobilizing VAM fungi and specific nitrogen-fixing rhizobia is yet another instance of a naturally occurring interrelationship among two beneficial microorganisms on roots which could be exploited for better nitrogen as well as phosphorus nutrition of legumes.

REFERENCES

1. Konishi, K. Effects of certain soil bacteria on the growth of root nodule bacteria, *Memoirs of College of Agriculture, Kyoto Imperial University Chemical Series 16,* 17 (1931).
2. Krasilnikov, N.A. and Koreyanko, A.I. Influence of soil bacteria on the virulence and activity of nodule bacteria, *Mikrobiologiya (USSR) 19,* 39-44 (1944).
3. Robinson, R.S. The antagonistic action of the by-products of several soil microorganisms on the activity of legume bacteria, *Proceedings of Soil Society of America 10,* 206-210 (1945).

Table 6. Some interactions between *Azotobacter* and other bacteria

Microorganisms involved	Nature of interaction	Reference
1	2	3
Combinations of *Azotobacter* with *Cytophaga, Cellulomonas, Corynebacterium*, unidentified spore-forming and nonspore-forming bacteria, fungi and actinomycetes with cellulose as the sole carbon source	*Azotobacter* was able to fix nitrogen only in association with *Corynebacterium* and *Cellulomonas*	Jensen, H.L. [101]
Combinations of *Azotobacter vinelandii* with *Bacillus circulans, B. mesentericus*	*A. vinelandii* fixed more nitrogen when contaminated with *B. circulans* but *B. mesentericus* inhibited the growth of *Azotobacter* in mixed culture	Lind, L.J. and Wilson, P.W. [95]
Combinations of *Azotobacter* and actinomycetes	*Azotobacter* fixed more nitrogen in mixed cultures	Lal, A. and Achari, T.K.T. [102]
Combinations of *Azotobacter chroococcum, A. agilis* with actinomycetes and *Rhizobium* spp.	Some actinomycetes stimulated nitrogen fixation while others depressed the process. *Rhizobium* greatly stimulated nitrogen fixation	Gadgil, P.D. and Bhide, V.P. [96]
Combination of *Azotobacter vinelandii* with *Rhodospeudomonas capsulatus*	Mixed cultures increased nitrogen fixation, probably due to slime formation	Okuda, Y. and Kobayashi, M. [104]
Combination of *A. chroococcum* with pectinolytic *Bacillus*	Stimulated the population of *Azotobacter* in the rhizosphere of barley	Remacle, J. [103]
Combination of *A. chroococcum* with *Pseudomonas* sp.	Growth of *Azotobacter* was inhibited	Chan, E.C.S. et al. [97]
Combination of *A. chroococcum* with *Pseudomonas* sp.	Inhibited the growth of *Azotobacter*	Ostwal, K.P. and Bhide, V.P. [98]
Combinations of *A. chroococcum* with *Bacillus megaterium* var. *phosphaticum* and *Rhizobium* sp.	*Azotobacter* did not fix more nitrogen in the presence of *B. megaterium*	Shende, S.T. et al. [34]
Combinations of *Azotobacter* and phosphobacteria	The number of each organism in the rhizosphere of *Lavandula spica* was markedly stimulated when inoculated as mixed culture	Ocampo, J.A. et al. [100]

(Contd.)

1	2	3	4
Combinations of *Azotobacter* and *Azospirilium*		Mixed culture inoculation increased growth of *Cynodon dactylon* by 17% and helped in increased total N in the top	Baltensperger, A.A. et al. [105]
Combinations of *Azospirillum brasilense* with *Azotobacter chroococcum*		Inoculation in unsterilized soil increased the root biomass of rice seedlings	Dewan, G.I. and Subba Rao, N.S. [106]
"		Inoculation significantly increased the dry matter production of maize (*Zea mays*) and sorghum (*Sorghum bicolor*)	Tilak, K.V.B.R. et al. [107]

4. Robinson, R.S. The antagonistic action of the by-products of a culture of *Aspergillus wentii* on legume bacteria, *Journal of Bacteriology 51,* 129 (1946).
5. Thornton, G.D., Alenar, J.D. and Smith, F.B. Some effects of *Streptomyces albus* and *Penicillium* spp. on *Rhizobium meliloti, Proceedings of Soil Society of America 14,* 188-191 (1949).
6. Abdel-Ghaffar, A.S. and Allen, O.N. The effect of certain microorganisms on the growth of rhizobia, *Fourth International Soil Science Congress Amsterdam Transactions 3,* 93-96 (1950).
7. Hely, F.W., Bergersen, F.J. and Brockwell, J. Microbial antagonism in the rhizosphere as a factor in the failure of inoculation of subterranean clover, *Australian Journal of Agricultural Research 8,* 24-44 (1957).
8. Anderson, K.J. The effect of soil microorganisms on the plant-rhizobia association, *Phyton 8,* 59-73 (1957).
9. Holland, A.A. The effect of indigenous saprophytic fungi upon nodulation and establishment of subterranean clover pp. 147-164, in *Antibiotics in Agriculture*, (Editor, M. Woodbine), Butterworths, London, (1962).
10. van Schreven, D.A. The effect of some actinomycetes on rhizobia and *Agrobacterium radiobacter, Plant and Soil 21,* 283-302 (1964).
11. Subba Rao, N.S. and Vasantha, P. Fungi on nodular surface of some legumes, *Naturwissenshaften 52,* 44-45 (1965).
12. Chhonkar, P.K. and Subba Rao, N.S. Fungi associated with legume root nodules and their effect on rhizobia, *Canadian Journal of Microbiology 12,* 1253-1261 (1966).
13. Demirgi, S.M. and Johnson, H.W. Effect of soil actinomycetes on strains of *Rhizobium japonicum, Agronomy Journal 58,* 223-224 (1966).
14. Hattingh, M.J. and Louw, H.A. The antagonistic effects of soil microorganisms isolated from the root region clovers on *Rhizobium trifolii, South African Journal of Agricultural Science 9,* 239-251 (1966).
15. Holland, A.A. and Parker, C.A. Studies on microbial antagonism in the establishment of clover pasture. II. The effect of saprophytic soil fungi upon *Rhizobium trifolii* and growth of subterranean clover, *Plant and Soil 25,* 329-340 (1966).
16. Sethi, R.P. and Subba Rao, N.S. Inhibitory and stimulatory effects of soil fungi on rhizobia, *Journal of General Applied Microbiology 14,* 325-327 (1968).
17. Hattingh, M.J. and Louw, H.A. Clover rhizosphere bacteria antagonistic to *Rhizobium trifolii, Canadian Journal of Microbiology 15,* 361-364 (1969).
18. Bhalla, H. and Sen, A.N. Note on the stimulatory and inhibitory

effects of some rhizosphere bacteria of Bengal gram (*Cicer arietinum*) on its specific *Rhizobium, Indian Journal of Agricultural Science 41,* 1126–1127 (1971).

19. Ibrahim, A.M., Kamal, M. and Shata, A.M. Occurrence of antagonistic microorganisms in the rhizosphere of *Phaseolus vulgaris, Agrokem Talajtan 20,* 36–44 (1971).
20. Rao, J.V.D.K., Sen, A. and Shende, S.T. Inhibition of groundnut *Rhizobium* in Indian soils. *Proceedings of Indian National Science Academy 40*(B) *5,* 535–539 (1974).
21. Ranga Rao, V. and Subba Rao, N.S. Factors responsible for non-nodulating nature of some legumes, *Proceedings of Indian National Science Academy 40*(B) *6,* 613–617 (1974).
22. Patel, J.J. Antagonism of actinomycetes against rhizobia. *Plant and Soil 41,* 395–402 (1974).
23. Smith, R.S. and Miller, R.H. Interactions between *Rhizobium japonicum* and soybean rhizosphere bacteria, *Agronomy Journal 66* (4), 564–567 (1974).
24. Jain, M.K. and Rewari, R.B. Isolation of seed-borne microflora from leguminous crops and their antagonistic effect of *Rhizobium, Current Science 43,* 151 (1974).
25. Leuck, E.E and Rice, E.L. Effect of rhizosphere bacteria of *Arstida oligantha* on *Rhizobium* and *Azotobacter, Botanical Gazette 137*(2), 160–164 (1976).
26. Orellana, R.G., Sloger, C. and Miller, V.L. *Rhizoctonia-Rhizobium* interaction in relation to the yield parameters of soybeans, *Phytopathology 66* (4), 464–467 (1976).
27. Rao, J.V.D.K., Patil, R.B. Effect of inoculation with *Rhizobium* and *Azotobacter* on nodulation, growth and yield of soybean, *Current Science 45* (14), 523–524 (1976).
28. Angle, J.S., Pugashetti, B.K. and Wagner, G.H. Fungal effects on *Rhizobium japonicum*-soybean symbiosis, *Agronomy Journal 73,* 301–306 (1981).
29. Kush, A.K. Interaction between symbiosis and root pathogenesis in green gram (*Vigna radiata*), *Plant and Soil 65,* 133–135 (1982).
30. Pugashetti, B.K., Angle, J.S. and Wagner, G.H. Soil microorganisms antagonistic towards *Rhizobium japonicum, Soil Biology and Biochemistry 14,* 45–49 (1982).
31. Jensen, H.L. Nitrogen fixation in leguminous plants. II. Is symbiotic nitrogen fixation influenced by *Azotobacter*? *Proceedings of the Linnaen Society of New South Wales 67,* 205–212 (1942).
32. Harris, J.R. Influence of rhizosphere microorganisms on the virulence of *R. trifolii, Nature 172,* 507–508 (1953).
33. Apte, R. and Iswaran, V. Cultures of *Rhizobium* inoculants with those of *Beijerinckia* and *Azotobacter, Proceedings of Indian National Science Academy 40,* 482–485 (1971).

34. Shende, S.T., Arora, C.K. and Sen, A. Interactions between *A. chroococcum, B. megaterium* var. *phosphaticum* and *Rhizobium* species, *Zentrallblatt Bakteriology, Parasitenk Infektionskr Hygiene* II *128* (7-8), 668-677 (1973).
35. Barea, J.M., Azcon, R., Gomez, M. and Callao, V. Inoculation of microorganisms solubilizing phosphate and *Rhizobium* in sand-organic matter cultures of *Phaseolus vulgaris:* Effects on flowering, fruit development and yield, *Microbiology Espano 26*(4), 135-147 (1973).
36. Ravat, A.K. and Sanoria, C.L. Effect of *Rhizobium, Azotobacter* and *Beijerinckia* inoculation on *Cicer arietinum* var. type 1. *Current Science 45*(8), 665-666 (1976).
37. Singh, C.S. and Subba Rao, N.S. Associative effect of *Azospirillum brasilense* with *Rhizobium japonicum* on nodulation and yield of soybean (*Glycine max*), *Plant and Soil 53,* 387-392 (1979).
38. Jauhri, K.S., Bhatnagar, R.S. and Iswaran, V. Associative effect of inoculation of different strains of *Azotobacter* and homologous *Rhizobium* on the yield of moong (*Vigna radiata*), soybean (*Glycine max*) and pea (*Pisum sativum*), *Plant and Soil 53,* 105-108 (1979).
39. Burns, T.A., Bishop, P.E. and Israel, D.W. Enhanced nodulation of leguminous plant roots by mixed cultures of *Azotobacter vinelandii* and *Rhizobium, Plant and Soil 62,* 399-412 (1981).
40. Tilak, K.V.B.R., Singh, C.S. and Rana, J.P.S. Effects of combined inoculation of *Azospirillum brasilense* with *Rhizobium trifolii, Rhizobium meliloti* and *Rhizobium* sp. (cowpea miscellany) on nodulation and yield of clover (*Trifolium repens*), lucerne (*Medicago sativa*) and chickpea (*Cicer arietinum, Zentrallblatt Bakteriologie* II Abt. *136,* 117-120 (1981).
41. Denso, S.K.A., Keya, S.O. and Alexander, M. Protozoa and decline of *Rhizobium* populations added to soil, *Canadian Journal of Microbiology 21,* 884-895 (1975).
42. Keya, S.O. and Alexander, M. Regulation of parasitism by host density: The *Bdellovibrio-Rhizobium* interrelationship, *Soil Biology and Biochemistry 7,* 231-237 (1975).
43. Alexander, M. Ecology of N_2-fixing organisms, pp. 94-114, in *Biological Nitrogen Fixation in Farming Systems in the Tropics,* (Editors, A. Ayanaba and P.J. Dart), John Wiley and Sons, U.K. (1977).
44. Danso, S.K.A. Recent advances in the study of the ecology of *Rhizobium,* pp. 115-125, in *Biological Nitrogen Fixation in Farming Systems* (Editors, A. Ayanaba and P.J. Dart), John Wiley and Sons, U.K. (1977).
45. Chao, W.L. and Alexander, M. Interaction between protozoa and *Rhizobium* in chemically amended soil, *Soil Science Society of America Journal 45,* 48-50 (1981).

46. Almon, L. and Wilson, P.W. Bacteriophage in relation to nitrogen fixation by red clover, *Archives of Microbiology 4*, 209-221 (1933).
47. Hitchner, E.R. The isolation of bacteriolytic principles from the root nodules of red clover, *Journal of Bacteriology 19*, 191-201 (1930).
48. Desai, S.V. Studies on bacteriophages of the root nodule organisms, *Indian Journal of Agricultural Science 2*, 138-156 (1932).
49. Laird, D.G. Bacteriophages and root nodule bacteria, *Archives of Microbiology 3*, 189-193 (1932).
50. Laird, D.G. A study of strains of the rhizobia with particular reference to the bacteriophage, *Proceedings of the World Grain Exhibition and Conference, Regina, Canada 2*, 362-369 (1933).
51. Vandecaveye, S.C. and Katznelson, H. Bacteriophage as related to the root nodule bacteria of alfalfa, *Journal of Bacteriology 31*, 465-477 (1936).
52. Vandecaveye, S.C., Fuller, W.H. and Katznelson, H. Bacteriophage of rhizobia in relation to symbiotic nitrogen fixation by alfalfa, *Soil Science 50*, 15-28 (1940).
53. Fuller, W.H. and Vandecaveye, S.C. Isolation and identification of rhizobia bacteriophage, *Proceedings of Soil Science Society of America 6*, 197-199 (1941).
54. Hoffer, A.W. Bacteriophage as a possible factor in reducing yields of canning peas, *Journal of Bacteriology 45*, 39 (1943).
55. Vandecaveye, S.C. and Moodie, C.D. Effect of *Rhizobium meliloti* bacteriophage on alfalfa, *Proceedings of Soil Science Society of America 8*, 241-247 (1944).
56. Datta, S.C. On the bacteriophage of root nodule organisms, *Indian Journal of Agricultural Science 14*, 272-276 (1944).
57. Kleczkowska, J. The production of plaques by *Rhizobium* bacteriophage in poured plates and its value as a counting method, *Journal of Bacteriology 50*, 71-79 (1945).
58. Kleczkowska, J. A study of phaze-resistant mutants of *Rhizobium trifolii, Journal General Microbiology 4*, 298-310 (1950).
59. Parker, D.T. and Allen, O.N. Characteristics of four rhizobiophages, active against *Rhizobium meliloti, Canadian Journal of Microbiology 3*, 651-668 (1957).
60. Bruch, C.W. and Allen, O.N. Host specificities of four lotus rhizobiophages, *Canadian Journal of Microbiology 3*, 181-189 (1957).
61. Kleczkowska, J. A study of the distribution and the effects of bacteriophage of root nodule bacteria in the soil, *Canadian Journal of Microbiology 3*, 171-180 (1957).
62. Takahashi, I. and Quadling, C. Lysogeny in *Rhizobium trifolii, Canadian Journal of Microbiology 7*, 455-465 (1961).
63. Schwinghamer, E.A. and Reinhardt, D.J. Lysogeny in *Rhizobium leguminosarum* and *R. trifolii, Australian Journal of Biological Sciences 16*, 597-605 (1963).

64. Kowalski, M., Ham, G.E., Frederick, L.R. and Anderson, I.C. Relationship between strains of *Rhizobium japonicum* and their bacteriophages from soil and nodules of field-grown soybeans, *Soil Science 118*, 221-228 (1974).
65. Singh, R.B., Dhar, B., Singh, B.D., Singh, R.M. and Srivastava, J.S. Rhizobiophages in Indian soils: Distribution, morphology and general characteristics, *Current Perspectives in Nitrogen Fixation* (Editors, A.H. Gibson and W.E. Newton), Australian Academy of Science, Canberra (1981).
66. Vanderveken, J. Influence d'um virus sur la nodulation chez *Trifolium repens, Annals Institute Pasteur,* (Paris) *107*, 143-148 (1964).
67. Joshi, H.U. and Carr, A.J.H. and Jones, D.G. Effect of clover phyllody virus on nodulation of white clover (*Trifolium repens*) by *Rhizobium trifolii, Journal of General Microbiology 47,* 139-151 (1967).
68. Joshi, H.U. and Carr, A.J.H. Effect of clover phyllody virus on nodulation of white clover (*Trifolium repens*) by *Rhizobium trifolii, Journal of General Microbiology 49,* 385-392 (1967).
69. Tu, J.C., Ford, R.E. and Quiniones, S.S. Effects of soybean mosaic virus and/or bean pod mottle virus infection on soybean nodulation, *Phytopathology 60,* 518-523 (1970).
70. Rajagopalan, N. and Raju, P.N. The influence of infection by *Dolichos* enation mosaic virus on nodulation and nitrogen fixation by field bean (*Dolichos lablab*), *Phytopathology Zeitshrift 73,* 285-309 (1972).
71. Raju, P.N. *Dolichos* enation mosaic virus (DEMV) and symbiotic nitrogen fixation in field bean (*Dolichos lablab*), *Proceedings of the Indian National Science Academy 40*(B), 629-635 (1974).
72. Ross, J.P. and Harper, J.A. Effect of *Endogone* on soybean yields, *Phytopathology 60,* 1552-1556 (1970).
73. Ross, J.P. and William, J.W. Effect of *Endogone* mycorrhiza on phosphate uptake by soybean from inorganic phosphates, *Soil Science Society of America Proceedings 37,* 237-239 (1973).
74. Schenk, N.C. and Hinson, K. Response of nodulating and non-nodulating soybeans to a species of *Endogone* mycorrhiza, *Agronomy Journal 65*, 849-850 (1973).
75. Daft, M.J. and El-Giahmi, A.A. Effect of *Endogone* mycorrhiza on plant growth. VII. Influence of infection on the growth and nodulation in French bean (*Phaseolus vulgaris*), *New Phytologist 73,* 1139-1147 (1974).
76. Daft, M.J. and El-Giahmi, A.A. Studies on nodulated and mycorrhizal peanuts, *Annals of Applied Biology 83,* 273-276 (1976).
77. Barrow, N.J., Malajczuk, N. and Shah, T.C. A direct test of the ability of vesicular-arbuscular mycorrhiza to help plants take up fixed soil phosphate, *New Phytologist 78,* 269-276 (1977).

78. Abbot, L.K. and Robson, A.D. Growth stimulation of subterranean clover with vesicular-arbuscular mycorrhizas, *Australian Journal of Agricultural Research 28,* 639-649 (1977).
79. Mosse, B. Plant growth responses to vesicular-arbuscular mycorrhizae: 10. Responses of *Stylosanthes* and maize to inoculation in unsterile soils, *New Phytologist 78,* 277-288 (1977).
80. Smith, S.E. and Daft, M.J. Interactions between growth, phosphate content and N_2 fixation in mycorrhizal *Medicago sativa, Australian Journal of Plant Physiology 4*(3), 403-413 (1977).
81. Azcon-G de Anguilar, C. and Barea, J.M. Effects of interactions between different culture fractions of 'phosphobacteria' and *Rhizobium* on mycorrhizal infection, growth and nodulation of *Medicago sativa, Canadian Journal of Microbiology 24,* 520-524 (1978).
82. Azcon-G de Anguilar, C., Azcon, R. and Barea, J.M. Endomycorrhizal fungi and *Rhizobium* as biological fertilizers for *Medicago sativa* in normal cultivation, *Nature 279,* 325-327 (1979).
83. Bhagyaraj, D.J., Manjunath, A. and Patil, P.B. Interaction between vesicular-arbuscular mycorrhizae and *Rhizobium* and their effect on soybean in the field, *New Phytologist 82,* 141-145 (1979).
84. Delorengini, C., Barea, J.M. and Olivares, J. Biological fertilization (mycorrhizac+*Rhizobium*+phosphobacteria) of *Trifolium pratense* in different cultural conditions, *Microbiology 21*(3), 129-134 (1979).
85. Powell, C.Ll. Inoculation of white clover and rye grass seed with mycorrhizal fungi, *New Phytologist 83,* 81-85 (1979).
86. Islam, R., Ayanaba, A. and Sanders, F.E. Response of cowpea (*Vigna unguiculata*) to inoculation with VA mycorrhizal fungi and to rock phosphate fertilization in some unsterilized Nigerian soils, *Plant and Soil 54,* 107-117 (1980).
87. Bethlenfalvay, G.J. and Yoder, J.F. The Glycine-*Glomus-Rhizobium* symbiosis: I. Phosphorus effect on nitrogen fixation and mycorrhizal infection, *Physiologia Plantarum 52,* 141-145 (1981).
88. Redente, E.F. and Reeves, F.B. Interaction between vesicular-arbuscular mycorrhiza and *Rhizobium* and their effect on sweet vetch growth, *Soil Science 132,* 410-415 (1981).
89. Dorosinskii, L.M. Some questions on the use of bacterial fertilizers, *Mikrobiologiya 31,* 738-744 (1962).
90. Mishustin, E.N. and Naumova, A.N. Bacterial fertilizers, their effectiveness and mode of action, *Mikrobiologiya 31,* 543-555 (1962).
91. Brown, M.E. and Burlingham, S.K. *Azotobacter* and plant diseases, *Annual Report of Rothamsted Experimental Station* (1963).
92. Menkina, R.A. Bacterial fertilizers and their importance for agricultural plants, *Mikrobiologiya 19,* 308-316 (1963).
93. Lakshmi Kumari, M., Vijayalakshmi, M. and Subba Rao, N.S. Interaction between *Azotobacter* species and fungi. I. *In vitro* studies with

Fusarium moniliforme sheld, *Phytopathology Zeitshrift 75*, 27-30 (1972).

94. Shende, S.T., Apte, R.G. and Singh, T. Multiple action of *Azotobacter, Indian Journal of Genetics and Plant Breeding 35*, 314-315 (1975).
95. Lind, L.J. and Wilson, P.W. Nitrogen fixation by *Azotobacter* in association with other bacteria, *Soil Science 54*, 105-112 (1949).
96. Gadgil, P.D. and Bhide, V.P. Nitrogen fixation by *Azotobacter* in association with some associated soil microorganisms, *Proceedings of the National Institute of Sciences of India 26*(B), 60-63 (1960).
97. Chan, E.C.S., Basavanand, P. and Liivak, T. The growth inhibition of *Azotobacter chroococcum* by *Pseudomonas* sp., *Canadian Journal of Microbiology 16*, 9-16 (1970).
98. Ostwal, K.P. and Bhide, V.P. *In vitro* effect of soil *Pseudomonas* on the growth of *Azotobacter chroococcum* and *Rhizobium* spp. *Science and Culture 38*, 288-290 (1972).
99. Shende, S.T., Arora, C.K. and Sen, A. Interaction between *Azotobacter chroococcum, Bacillus megaterium* var. *phosphaticum* and *Rhizobium* sp. *Zentrallblatt Bacteriology* Abt. II. Bd. *128*, 668-677 (1973).
100. Ocanpo, J.A., Barea, J.M. and Montoya, E. Interaction between *Azotobacter* and phosphobacteria and their establishment in the rhizosphere as affected by soil fertility, *Canadian Journal of Microbiology 21*, 1160-1165 (1975).
101. Jensen, H.L. Nitrogen fixation and cellulose decomposition by soil microorganisms. I. Aerobic cellulose decomposers in association with *Azotobacter, Proceedings of the Linnaen Society of New South Wales 65*, 543-55 (1940).
102. Lal, A. and Achari, T.K.T. Microbiological studies. VII. Studies on *Azotobacter* and actinomycetes in relation to nitrogen fixation and cellulose utilization, *Proceedings of the National Academy of Sciences, India 23*, 137-149 (1953).
103. Remacle, J. The development of *Azotobacter chroococcum* population in the barley rhizosphere in the presence of a pectinolytic *Bacillus, Supplement Annals of Institute of Pasteur* (Paris) *III*(3), 149-154 (1966).
104. Okuda, Y. and Kobayashi, M. Production of slime substances in mixed cultures of *Rhodopseudomonas capsulatus* and *Azotobacter vinelandii, Nature 192*, 1207-1208 (1961).
105. Baltensperger, A.A., Schank, S.C., Smith, R.L., Littell, R.C., Button, J.H. and Dudeck, A.E. Effect of inoculation with *Azospirillum* and *Azotobacter* on turf type genotypes, *Crop Science 18*, 103-104 (1978).
106. Dewan, G.I. and Subba Rao, N.S. Seed inoculation with *Azospirillum brasilense* and *Azotobacter chroococcum* and the root biomass of rice (*Oryza sativa* L.) *Plant and Soil 53*, 295-302 (1979).

107. Tilak, K.V.B.R., Singh, C.S., Roy, N.K. and Subba Rao, N.S. *Azospirillum brasilense* and *Azotobacter chroococcum* inoculum: Effect on yield of maize (*Zea mays*) and sorghum (*Sorghum bicolor*), *Soil Biology and Biochemistry 14*, 417-418 (1982).

3. Attachment of Nitrogen-Fixing Bacteria to Roots of Host Plants

F.B. Dazzo and G.L. Truchet

INTRODUCTION

The process of cellular recognition between microorganisms and higher plants is receiving considerable attention in the light of its effect on plant morphogenesis, nutrition, and protection against infectious disease. These positive cellular recognitions are believed to arise from a specific union, reversible or irreversible, between chemical receptors on the surface of interacting cells [1]. This hypothesis implies that communication occurs when cells that recognize one another come into contact, and therefore the complementary components of the cell surfaces have naturally been the focus for most biochemical studies. Such is the case for studies on the infection of legume roots by the nitrogen-fixing symbiont, *Rhizobium*. According to the lectin-recognition hypothesis [2–4], specific, complementary lectin-polysaccharide interactions serve as the basis of host specificity in this nitrogen-fixing symbiosis.

There are many cellular recognition phenomena which occur during the infection of legume root hairs by *Rhizobium*, for example, bacterial attachment, root-hair deformation, infection thread formation and infection thread growth directed by the root hair nucleus. Lectin-mediated specific attachment of rhizobia to legume host root hairs has been the step of cellular recognition studied in greatest detail. During the early stages of the infection process, the bacteria attach via hapten-reversible interactions, and then later become irreversibly anchored to the host cell. The ability of the bacteria to attach to root hairs is controlled by Roa (root hair attachment) genes [5], which occur on large, transmissible plasmids [6]. Roa phenotype is illustrated by non-infective mutant strains which are defective in hapten-specific attachment steps [6–9]. Recent studies, however, have clearly shown that successful infection of root hairs by rhizobia requires additional events of cellular recognition. An understanding of the recognition code to host specificity in the *Rhizobium*-legume symbiosis could provide ways to broaden the range of agricultural crops which can enter into efficient nitrogen-fixing associations. Studies on lectin-*Rhizobium*

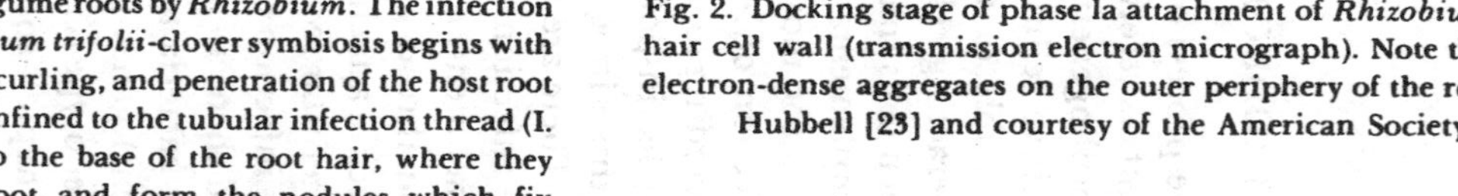

Fig. 1. Infection of legume roots by *Rhizobium*. The infection process in the *Rhizobium trifolii*-clover symbiosis begins with bacterial attachment, curling, and penetration of the host root hair. The bacteria, confined to the tubular infection thread (I. T.) are transported to the base of the root hair, where they penetrate into the root and form the nodules which fix nitrogen into ammonia (phase contrast micrograph). 630×.

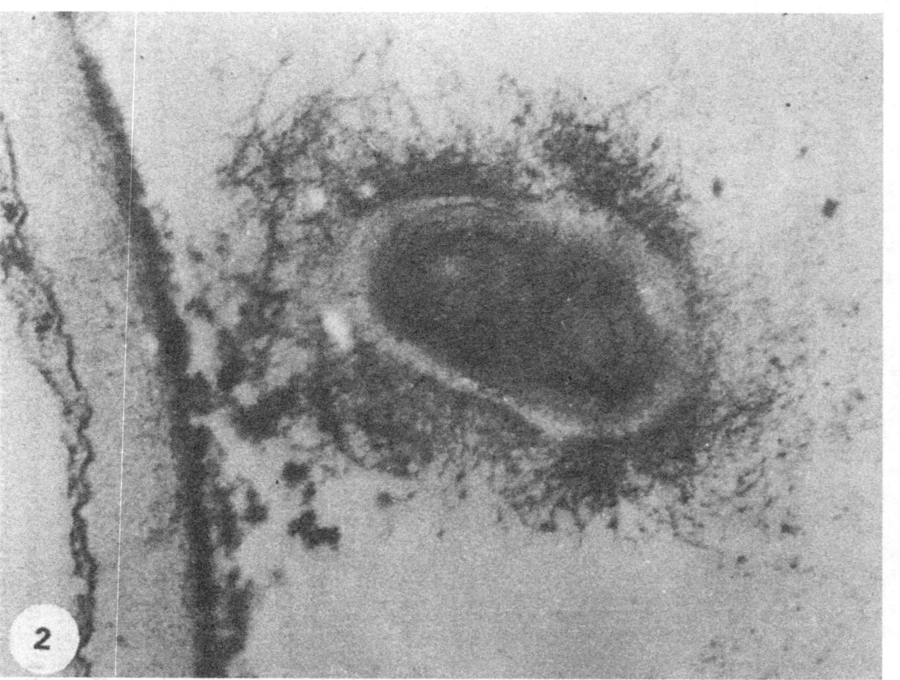

Fig. 2. Docking stage of phase Ia attachment of *Rhizobium trifolii* NA-30 to the clover root hair cell wall (transmission electron micrograph). Note the fibrillar capsule which contacts electron-dense aggregates on the outer periphery of the root hair cell wall. From Dazzo and Hubbell [23] and courtesy of the American Society for Microbiology. 41,000×.

interactions have been exhaustively reviewed [10-24]. This article highlights some of the earlier work (1970's), and expands on more recent studies which focus on lectin-saccharide interactions in specific rhizobial attachment, with particular emphasis on how this process of cellular recognition is regulated.

INFECTION OF LEGUME ROOT HAIRS BY *RHIZOBIUM*

Rhizobium is a genus of Gram negative bacteria that selectively infects legume roots and then forms root nodules that "fix" atmospheric nitrogen into ammonia, which then becomes immediately available to the plant for growth. The infection process is very selective for certain combinations of rhizobia and legume, and this high degree of host-range specificity is used to define the various species of *Rhizobium*. For example, *R. trifolii* infects clover root hairs (Fig. 1) and *R. meliloti* infects alfalfa root hairs. Neither species infects root hairs of the heterologous plant host. Successful infection of legume roots by these nitrogen-fixing bacteria in soil is of immense importance in the nitrogen cycle on earth.

Bacterial attachment is an early recognition step of root-hair infection

Rhizobium attaches to the root-hairs that are later infected. Quantitative light microsopic assays [7, 13], and transmission electron microscopic studies [15, 25-27] of the *Rhizobium*-clover symbiosis have revealed multiple mechanisms of bacterial attachment to the root hairs. A non-specific mechanism allows all species of rhizobia to attach in low numbers (two to four cells per 200 μm root hair length per 12 hours using low inoculum per seedling). In addition, a specific mechanism allows selective attachment in significantly (P=0.005) larger numbers (22-27 cells per 200 μm root hair length per 12 hours) under identical conditions [7]. Host-specific attachment has also been demonstrated in *R. japonicum*-soybean [28] and *R. leguminosarum*-pea root systems [29] *R. japonicum* also selectively attaches to soybean root cells in suspension culture [30]. However, specificity was not found in quantitative root attachment studies [31] employing very high densities of radiolabelled rhizobia (10^9-10^{10} cells per seedling) but many unattached bacteria could have contributed to this result. In a recent study employing marble chips to dislodge bacteria attached to the root system and quantitative plating assays, "firm" attachment was found to be host-specific in *R. trifolii*-clover and *R. meliloti*-alfalfa systems, and "loose" attachment was non-specific [32]. It is therefore apparent that bacterial attachment to host roots is an early expression of cellular recognition in the *Rhizobium*-legume symbiosis.

PHASE I ATTACHMENT

Transmission electron microscopy [25] disclosed that the initial bacterial attachment step consisted of contact between the fibrillar capsule of *R. trifolii* and electron-dense globular aggregates lying on the outer periphery of the clover root hair cell wall (Fig. 2). This "docking" stage is the first point of physical contact between the microbe and the host (Phase I Attachment), and occurs within minutes after inoculation of encapsulated cells of *R. trifolii* on the host clover.

To identify the cell surface molecules involved in Phase I attachment the strategy adopted in our laboratory has been to examine the surface components of the bacterium and the host that interact with the same order of specificity as is observed with the adhesion of the bacterial cells. Immunochemical and genetic studies have demonstrated that the surfaces of *R. trifolii* and clover epidermal cells contain a unique carbohydrate antigen that is immunochemically cross-reactive [25, 33], suggesting its structural relatedness on both symbionts. This antigen contains receptors that bind hapten-reversibly to a multivalent clover lectin called trifoliin A (originally trifoliin) which has been isolated from seeds and seedling roots [33, 34]. A specific hapten inhibitor of trifoliin A binding to these receptors is 2-deoxy-D-glucose [25, 35]. The first clue that trifoliin A on the root may be involved in rhizobial attachment came from the observation that 2-deoxy-D-glucose specifically inhibited the attachment of *R. trifolii* to clover root hairs [7], thereby reducing the high level of bacterial adhesion to that characteristic of background. Subsequent studies showed that 2-deoxy-D-glucose specifically facilitated the elution of trifoliin A from the intact clover root [34, 35], and inhibited the binding of *R. trifolii* capsular polysaccharide to clover root hairs [35]. In contrast, 2-deoxy-D-glucose did not inhibit adsorption of *R. meliloti* or its capsular polysaccharide to alfalfa root hairs [35]. Consistent with the above results in the *Rhizobium trifolii*-clover system, lectins on pea, alfalfa, and soybean roots accessible for binding to the appropriate rhizobia have been demonstrated [8, 9, 28, 29, 36–40]. In addition, specific hapten-facilitated elution of lectin from pea, alfalfa, and soybean roots has also been found [36; 39–40], (Kamberger, personal communication). In such hapten-facilitated elution techniques, it is believed that the sugar acts by combining specifically with the site on the lectin which is normally occupied by the natural saccharide receptor. This implies a close but not necessarily identical structure of the hapten and the native determinant. However, some lectins undergo conformational changes when associated with saccharide binding [41], and so this possibility must also be considered in the interpretation of hapten inhibition studies.

LECTIN CROSS-BRIDGING MODEL

Dazzo and Hubbell [25] proposed a model to explain this early recognition event of Phase I attachment on the clover root hair surface prior to infection. According to this hypothesis, the multivalent trifoliin A (it aggluti-

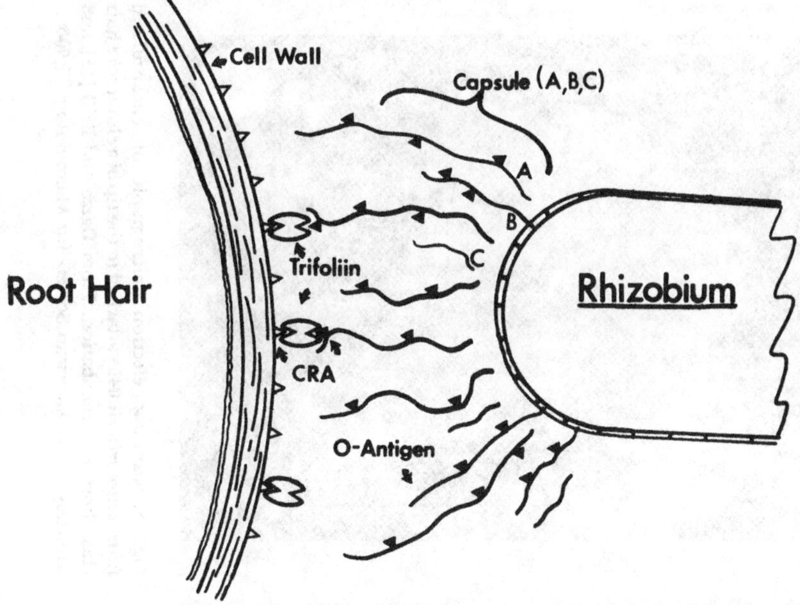

Fig. 3. Proposed cross-bridging of *Rhizobium trifolii* receptors to clover root hairs by the host lectin, trifoliin A. Modified from Dazzo and Hubbell [25].

nates cells) recognizes similar saccharide residues on *R. trifolii* and clover and cross-bridges them. This complementary interaction forms the correct molecular interfacial structure that initiates the preferential and specific adsorption of the bacteria to the root hair surface. Therefore, the lectin may also function as a "cell recognition molecule" since it could feasibly influence which cells associate in sufficient proximity to the root hairs to allow subsequent specific recognition steps to occur. Fig. 3 is a revision of the original cross-bridging model, which takes into account the recent finding that *R. trifolii* has multiple receptors for trifoliin A on its surface (described in more detail below).

The model predicts that there exist host-specific receptor sites on the legume root which interact specifically with surface molecules of the rhizobial symbiont. A key experiment that demonstrated these receptor sites and localized them on the root surface was performed by first labelling the trifoliin A-binding capsular polysaccharide of *R. trifolii* with the fluorescent dye fluorescein isothiocyanate (FITC), incubating the conjugate with sterile seedling roots for a brief period, and then examining the roots by epifluorescence microscopy [35]. The receptor sites on clover roots that immediately bound the FITC-capsular polysaccharide from *R. trifolii* were located at discrete root hair sites that had differentiated on the epidermal root surface (Fig. 4). They accumulated at root hair tips and dimin-

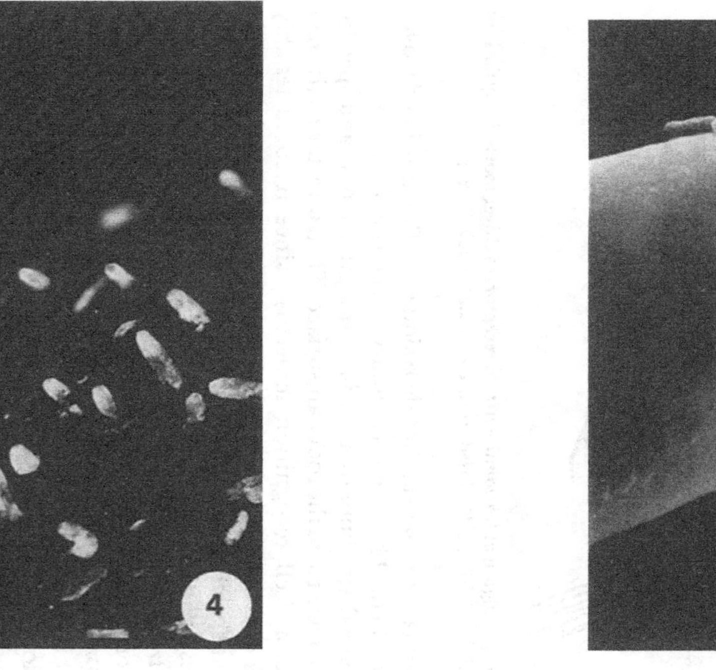

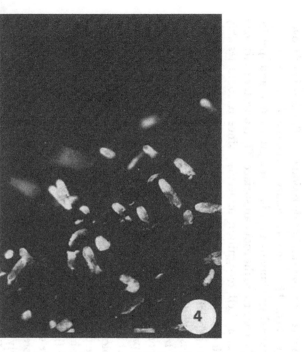

Fig. 4. Specific binding of FITC-labelled capsular polysaccharide from *Rhizobium trifolii* 0403 to clover root hairs (epifluorescence micrograph). From Dazzo and Brill [35], and courtesy of the American Society for Microbiology. 321×.

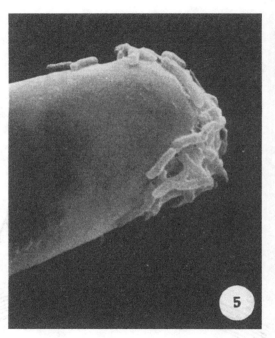

Fig. 5. Scanning electron micrograph of encapsulated *Rhizobium trifolii* 0403 attached to the tip of a clover root hair after short-term incubation. From Dazzo and Brill [33], and courtesy. of the American Society for Microbiology. 7,500×.

ished toward the base of the root hair. This unique location exactly matched both the distribution of trifoliin A on the surface of the clover seedling root [34] and the sites which immediately bound encapsulated *R. trifolii* in Phase I attachment [33] (Fig. 5). The result also highlighted the importance of epidermal cell differentiation in the development of receptor sites that recognize rhizobia. Close inspection of the photomicrographs revealed that undifferentiated epidermal cells in the root hair region did not bind the bacterial polysaccharide, whereas epidermal root hair primordia had this surface property (Fig. 4).

Specificity of these receptor sites was demonstrated by the ability of unlabelled capsular polysaccharide from *R. trifolii,* but not from *R. meliloti,* to block the binding of the labelled polysaccharide to clover root hairs. Similar specific binding of bacterial polysaccharides to legume host root hairs has been demonstrated in the *R. meliloti*-alfalfa [35], *R. leguminosarum*-pea [29], and *R. japonicum*-soybean systems [42, 43].

The results of immunochemical and genetic studies suggest that trifoliin A and cross-reactive anti-clover root antibody bind to the same or similar overlapping saccharide determinants on *R. trifolii.* First, the

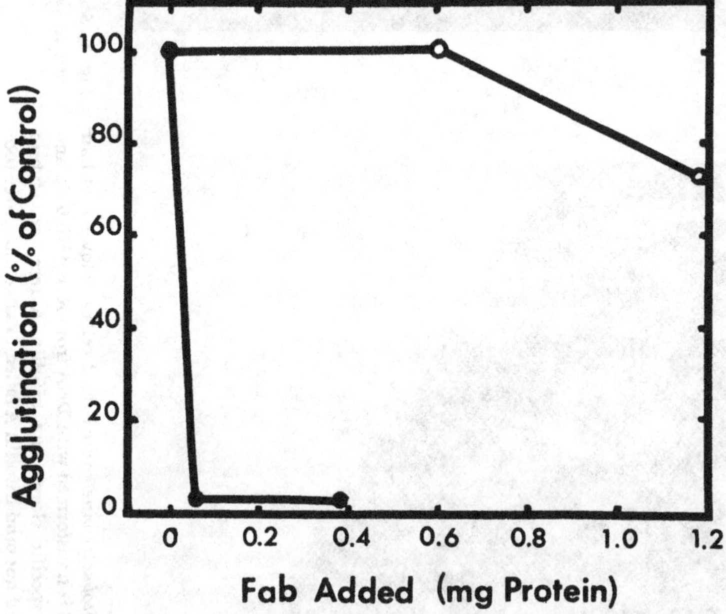

Fig. 6. Effect of Fab of immune anti-clover root antigen (●) and of preimmune serum (○) on trifoliin A-mediated agglutination of *R. trifolii* 0403. Washed cells were pretreated with Fab, washed, and assayed for agglutination with purified trifoliin A. The specific agglutination activity of the uninhibited control was 9,142 units per mg protein. From Dazzo and Brill [33], and courtesy of the American Society for Microbiology.

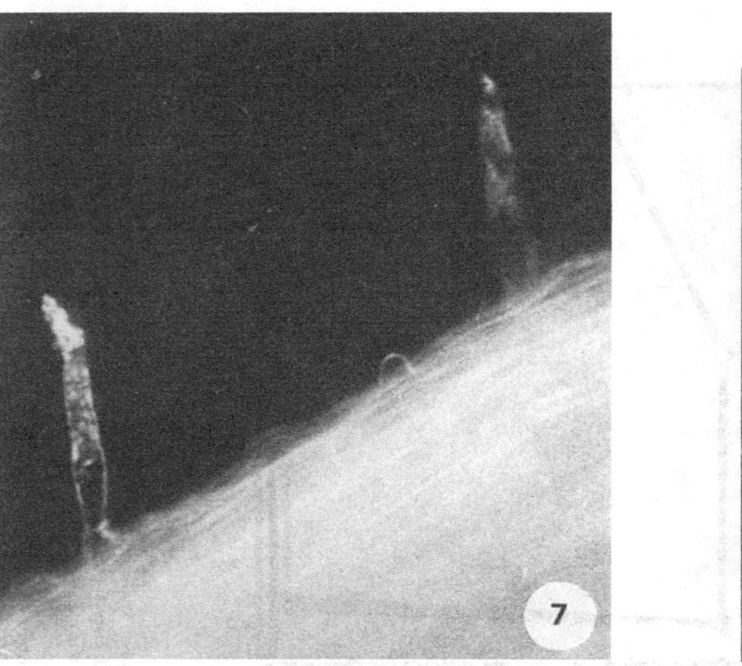

Fig. 7. Binding of *Azotobacter vinelandii* hybrid cells to clover root hair tips. Strain RtAv 10-54 is transformed with DNA from *R. trifolii* 0403 and carries the *R. trifolii*-specific clover root trifoliin A receptor. Hybrid transformants which did not bind trifoliin A failed to attach to clover root hairs. From Dazzo and Brill [33], and courtesy of the American Society for Microbiology. 700×.

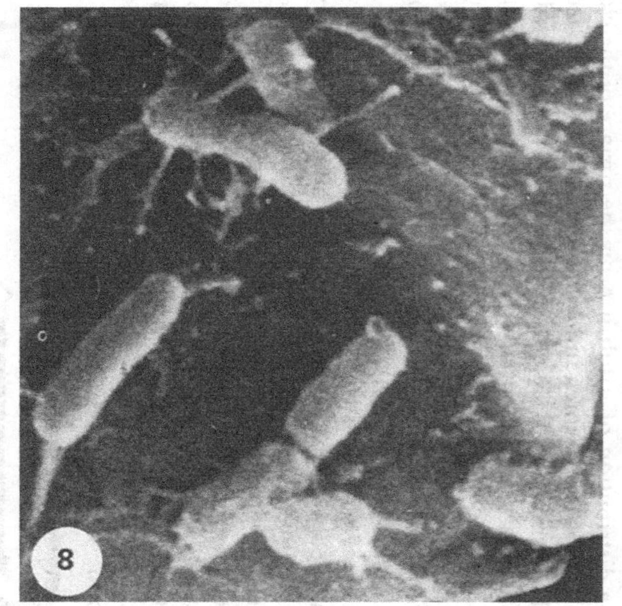

Fig. 8. Scanning electron micrograph of aggregated microfibrils associated with *R. trifolii* 0403 firmly attached to the clover root hair surface after prolonged incubation (Phase 2 adhesion). 20,000×.

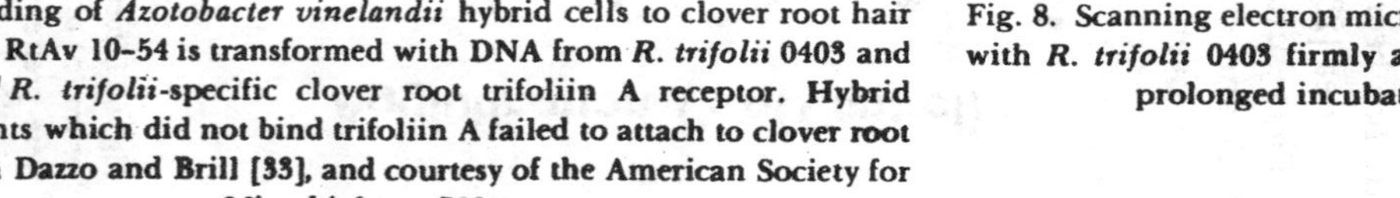

antibody and the lectin bind specifically to the same isolated polysaccharides from *R. trifolii* [33, 34]. Second, this interaction is specifically inhibited by the hapten, 2-deoxy-D-glucose. Third, the genetic markers of *R. trifolii* that bind trifoliin A and the antibody co-transform into *Azotobacter vinelandii* with 100% frequency [45]. Fourth, monovalent Fab fragments of IgG from anti-clover root antiserum strongly block the binding of trifoliin A to *R. trifolii* [33] (Fig. 6). Considered collectively, these studies suggest that *R. trifolii* and clover roots have similar saccharide receptors for trifoliin A. However, the definitive test of their identity as antigenically related structures will require knowledge of the minimal saccharide sequence that binds the clover lectin.

The results of three experiments indicate that trifoliin A and antibody to the cross-reactive antigen bind to the same *R. trifolii* saccharide determinants that bind these bacteria to clover root hairs [33]. Firstly, Fab fragments of anti-clover root IgG blocked Phase I attachment of *R. trifolii* to clover root hairs. Secondly, only the *A. vinelandii* hybrid transformants that carried the trifoliin A receptor bound to clover root hairs in Phase I attachment assays (Fig. 7). Thirdly, competition assays using fluorescence microscopy indicated that the *R. trifolii* polysaccharides that bound trifoliin A had the highest affinity for clover root hairs.

As with the *R. trifolii*-clover system [6, 7, 13], heterologous rhizobia [29] or non-nodulating mutant strains of *R. leguminosarum* which produce less extracellular/capsular polysaccharide [46, 47] adhere in smaller numbers to pea root hairs as compared with the wild-type nodulating strains [8, 29] (C. Napoli, personal communication).

PHASE II ADHERENCE

Phase II adherence is characterized by the firm anchoring of the bacterial cell to the root hair surface [15, 48]. Phase II adherence may be important in maintaining the firm contact between the bacterium and the host root hair necessary for triggering the tight root hair curling (shepherd's crook formation) and successful penetration of the root hair cell wall during infection [49].

During Phase II adherence, fibrillar materials, recognized by scanning electron microscopy, are characteristically found associated with the adherent bacteria (Fig. 8). The nature of these microfibrils is unknown. One possibility is that they are bundles of cellulose microfibrils, known to be produced by many rhizobia [49, 50]. Another possibility is that they are collections of pili, which have been recently demonstrated in *Rhizobium* [51] (see also [37]). Future studies should be directed to isolate and characterize these fibrils associated with the adherent bacteria in order to better understand the Phase II adhesion process. This is particularly important in light of the recent demonstration that the degree of host-specific firm attachment of rhizobial strains to the root shows a significant positive correlation with the degree of their success in interstrain competition for nodule sites on the root [32].

Rhizobial attachment is only a piece of the puzzle

Although attachment of infective rhizobia to target root hairs is a prerequisite for infection, several observations indicate that other undefined events must occur to initiate root hair infection. First of all, very few root hairs to which infective rhizobia attach eventually become infected. This may be due to a transient susceptibility of the root hairs to infection by the rhizobial symbiont [52, 53]. Secondly, genetic hybrids of *Azotobacter vinelandii* which carry the trifoliin A-binding saccharide receptor on their surface as a result of intergeneric transformation with DNA from *R. trifolii* [45], have acquired the ability to adhere specifically to clover root hairs [33] (Fig. 7), but do not infect them. Finally, although mutant strains which fail to bind the host lectin neither attach well to the host root hairs nor infect them [7, 9], another class of non-infective mutant strains has been shown to bind the host lectin and attach to the host root hairs [9, 54]. Each of these cases serves to illustrate the importance of lectin-mediated root hair attachment to the infection process. But it also makes it clear that other post-attachment events of cell recognition must occur to advance the infection process to the stage of root hair penetration. Possible genes or gene products which may have not been expressed in the above situations include those controlling cell-wall hydrolytic enzymes [55, 56], inducers of host polygalacturonase [57, 58], root hair curling factors [59], and periplasmic extrinsic substance ES-6000 which promotes root hair infection [60].

Lectins and their saccharide receptors are regulated by combined nitrogen

Combined nitrogen limits the development of the *Rhizobium*-legume root nodule symbiosis. For instance, white clover becomes resistant to infection by *R. trifolii* when the roots are grown with 15 mM nitrate. In fact, nitrate supplied at critical concentrations inhibits all of the morphogenetic steps of the nodulation process known to require the bacterial symbiont [61]. Microscopic assays indicated that the specific binding of *R. trifolii* 0403 to clover root hairs and the levels of trifoliin A on these epidermal cells declined in parallel as the nitrate concentration was increased from 1 mM to 15 mM in the rooting medium [62] (Fig. 9). The inhibition was due specifically to nitrate ion and 15 mM nitrate did not stunt seedling growth.

How does nitrate modulate levels of trifoliin A on clover roots? The first possibility tested (Dazzo and Hrabak, unpublished) was that nitrate binds to trifoliin A and prevents its interaction with root walls or its detection by homologous antibody. Possible binding of nitrate to trifoliin A was examined by incubating trifoliin A with radioactive 13-N nitrate (produced in a cyclotron), and then testing for the presence of a radioactive 13-N nitrate:

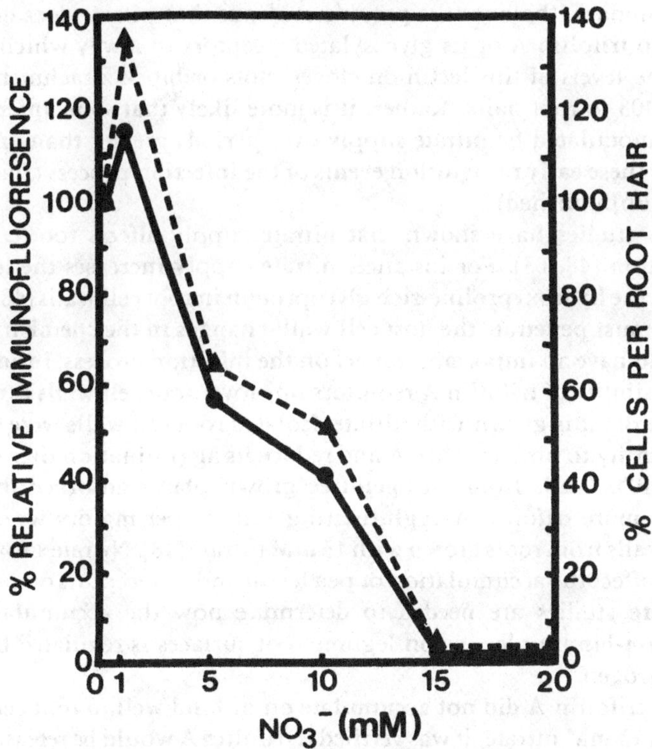

Fig. 9. The effect of NO_3^- on adsorption of *R. trifolii* 0403 to root hairs (solid line) and on immunologically detectable trifoliin A (dotted line) in the root hair region of clover seedlings. Bacterial adsorption was measured by direct microscopic counting and trifoliin A was measured by cytofluorimetry using indirect immunofluorescence. Values from roots grown in nitrogen-free nutrient solution are taken as 100%, and represent 980 photovolts/mm² and 21 cells/root hair 200 μm in length. Points along the curve are means from 10–15 root hairs or seedling roots, standard deviations vary within 10% of the means. Values are corrected for non-specific rhizobial adsorption, root autofluorescence, and non-specific adsorption of conjugated goat antirabbit gamma globulin. From Dazzo and Brill [62], and courtesy of the American Society of Plant Physiologists.

trifoliin A complex by selective molecular ultrafiltration. The results showed that nitrate does not bind to trifoliin A. Also, there was no deviation in the quantitative immunoprecipitin curve using trifoliin A as antigen and homologous anti-trifoliin A IgG as antibody in the presence of 15 mM nitrate. In addition, this concentration of nitrate caused no reduction in rhizobial attachment to clover root hairs in a one-hour assay, in contrast with the significant reduction in rhizobial attachment to clover root hairs if the period of exposure to nitrate was extended to 12 hours [62]. Furthermore, the specific agglutinating activity of trifoliin A was unaffected by 15 mM nitrate.

In summary, these results provide evidence that nitrate does not bind directly to trifoliin A or its glycosylated receptors in a way which would reduce the levels of this lectin on clover roots or block attachment of *R. trifolii* 0403 to root hairs. Rather, it is more likely that some intervening process, modulated by nitrate supply over periods greater than one hour regulates these early recognition events of the infection process (Dazzo and Hrabak, unpublished).

Other studies have shown that nitrate supply affects root cell wall composition [48, 63]. For instance, nitrate supply increases the levels of extensin, the hydroxyproline-rich glycoprotein in root cell walls [48]. Since rhizobia must penetrate the host cell wall, changes in the chemistry of the wall could have an important impact on the infection process. In addition, the accessibility of trifoliin A receptors on clover root cell walls is reduced when the plant is grown with nitrate. Isolated root cell walls were assayed for the ability to bind trifoliin A and reduce its agglutination titer with *R. trifolii* 0403. Walls from nitrogen-free grown plants adsorbed three- to four-fold more trifoliin A agglutinating activity per mg dry wt of walls than of walls from roots grown with 15 mM nitrate [48]. Nitrate supply also seems to affect the accumulation of pea lectin and its receptors on pea roots [63]. More studies are needed to determine how the accumulation of *Rhizobium*-binding lectins on legume root surfaces is regulated by combined nitrogen.

Since trifoliin A did not accumulate on or bind well to root cell walls grown in 15 mM nitrate, it was verified if trifoliin A would be released from the roots and accumulate in root exudate. The presence of trifoliin A in root exudate of two varieties of white clover was detected by an immunofluorescence assay and was confirmed by its purification using immunoaffinity chromatography [64, 65]. Thirty-fold higher levels of trifoliin A (per constant total protein concentration) were detected from root exudate of clovers grown under nitrogen-free conditions than when grown with 15 mM nitrate [64].

The presence of trifoliin A in clover root exudate which can bind to receptors on *R. trifolii* provides supporting evidence for a lectin recognition model proposed by Solheim [66]. According to this model, a glycoprotein lectin excreted from the legume root binds to the rhizobia. This active complex then combines with a receptor site on the root. Thus, both partners in the symbiosis could benefit from the discriminatory reaction of a cross-bridging lectin which could be either bound to a glycosylated receptor on the root hair cell wall or released from the root to bind to the rhizobial cell [66]. This event would help to ensure that only the symbiotic bacterium could establish the proper intimate contact with the host cell required to trigger other recognition events that lead to successful infection. Combined nitrogen (e.g., nitrate) would play a role in regulating the recognition process as proposed by Solheim [66].

Lectin receptors are transient on *Rhizobium*

The selective ability of *R. trifolii* to adhere to clover root hairs is also influenced by the accumulation of the saccharide receptor on the bacterium. Evidence supporting this hypothesis came from data which showed that the transient appearance of trifoliin A receptors on *R. trifolii* may influence the ability of these bacteria to attach to clover root hairs [67]. Cells grown on agar plates of a defined medium were most susceptible to agglutination by trifoliin A when they were harvested at five days of growth. In broth cultures, the antigenic determinants on the bacteria that are cross-reactive with clover roots were "exposed" for only short periods as cultures left their lag phase of growth and as they entered stationary phase. Clover roots adsorbed the bacteria in greatest quantity when the cells were harvested from plate culture incubated for five days and from broth cultures in early stationary phase [67] (Truchet and Dazzo, unpublished).

We recently found that growth-phase dependence for trifoliin A binding to *R. trifolii* in broth culture is related to the appearance of a unique determinant in the lipopolysaccharide (LPS) of the bacteria [44]. As the culture advanced from exponential to early stationary phase, changes in the immunochemistry of the LPS were detected with antisera made specific for lipopolysaccharides of cells in early stationary phase by exhaustive adsorption with exponentially growing cells (Fig. 10). Gas chromatography and combined gas chromatography-mass spectrometry showed culture-phase dependent differences in the quantities of several glycosyl components (e.g., quinovosamine, which is 2-amino-2, 6-dideoxyglucose) in the LPS that bound trifoliin A. D-quinovosamine, N-acetyl-β-D-quinovosamine, and its n-propyl-β-glycoside were found to be effective hapten inhibitors of trifoliin A. In addition, LPS increased in apparent size as the culture aged as shown by gel filtration chromatography. The new immunochemical determinants that occur in LPS as cells enter stationary phase were apparently recognition sites for trifoliin A binding, since immune monovalent Fab fragments of IgG specific for these unique determinants block the agglutination of cells with trifoliin A. The potential importance of this finding to the infection process was suggested by root hair infection studies using standardized inocula. White clover plants had more infected root hairs after incubation with an inoculum of cells in the early stationary phase than with cells in the mid-exponential phase. As previously predicted [33], trifoliin A and the cross-reactive anti-clover root antibody bind to unique determinants in the LPS of *R. trifolii* 0403 which are not immunodominant and which appear for only a transient period on cells in batch culture. In plate culture, the development of the capsule on *R. trifolii* 0403 coincides with the appearance of trifoliin A receptors, and these receptors are transient on the cell since the encapsulated cells loose their ability to bind the lectin uniformly as the culture ages [67] (Truchet and Dazzo, unpublished).

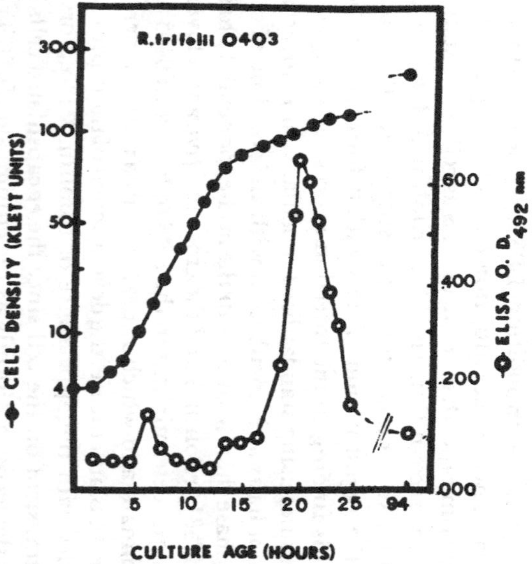

Fig. 10. Effect of culture age on the binding of antibody specific for unique determinants in lipopolysaccharide of *Rhizobium trifolii* 0403 in early stationary phase. Cells were grown in a chemically defined medium, and monitored for cell density with a Klett-Summerson colorimeter (red filter). Samples were adjusted to 10^7 cells, and assayed by ELISA. From Hrabak, Urbano, and Dazzo [44], and courtesy of the American Society for Microbiology.

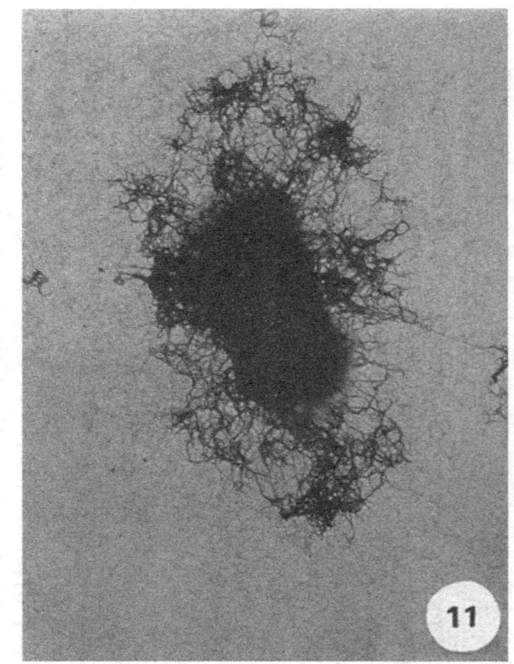

Fig. 11. Transmission electron micrograph of encapsulated cell of *R. trifolii* 0403 from 5 day-old cultures grown on BIII plates (defined medium), and then contrasted by the glutaraldehyde/ruthenium red/uranyl acetate method of Mutaftschiev et al. [81]. 28,000×.

There is a transient appearance and disappearance on *R. japonicum* of the receptor that specifically binds soybean lectin [68] (Truchet et al., unpublished). Most strains of *R. japonicum* have the highest percentage of soybean lectin-binding cells and the greatest number of soybean lectin-binding sites per cell in the early and mid-log phases of growth. The proportion of galactose residues in the capsular polysaccharide is high at a culture age when the cells bind the galactose-reversible soybean lectin [69]. A decline in lectin-binding activity of cells accompanying culture aging is concurrent with a decline in galactose content and a rise in 4-0-methyl galactose residues in the capsular polysaccharides. The latter methylated sugar has low affinity for the galactose-binding soybean lectin. These results suggest that the galactose residues in the capsular polysaccharide become methylated and, as a consequence, the cell loses its ability to combine specifically with the soybean lectin. Shedding of soybean lectin-binding capsular polysaccharides from the cells (Truchet et al., unpublished) explains why the broth culture as a whole continues to bind soybean lectin [70].

The profound influence of the growth phase on the composition of lectin-binding polysaccharides of *Rhizobium* may be a major underlying cause of conflicting data among laboratories testing the lectin-recognition hypothesis. Furthermore, the growth-phase dependent modifications of the lectin-binding polysaccharides of rhizobia [44, 69] may reflect mechanisms which regulate cellular recognition in the *Rhizobium*-legume symbiosis.

There are multiple lectin receptors on *Rhizobium*

Rhizobium produces several different polysaccharides in pure culture, including acidic heteropolysaccharides, lipopolysaccharides, and neutral glucans. Their presence in crude extracts has necessitated the development of complex techniques for purification. Because of this, and the need to know when most cells in culture bind the lectin, most of the earlier work on polysaccharides from *Rhizobium* did not reveal information on the chemical nature of the saccharide receptor which binds the host lectin. A major controversy was whether the lectin receptor on *Rhizobium* was the capsular polysaccharide [25], LPS [71], or glycans [72]. The picture which is now emerging is that *R. japonicum* binds soybean lectin through its capsular and extracellular polysaccharides [69, 73–76], (Truchet et al. unpublished); *R. meliloti* binds alfalfa lectin through its LPS [77] (J. Handelsman, personal communication); and the related species *R. trifolii* and *R. leguminosarum* specifically bind clover and pea lectin, respectively, through both their extracellular/capsular polysaccharide and their LPS at certain culture ages [8, 29, 33, 36, 44, 77, 78]. Similarly, peanut lectin binds LPS and capsular polysaccharide of peanut rhizobia [79]. Since the compositions and immunodominant structures of LPS vary widely among strains of a single *Rhizobium* species [80], the lectins from clover, pea, alfalfa, and

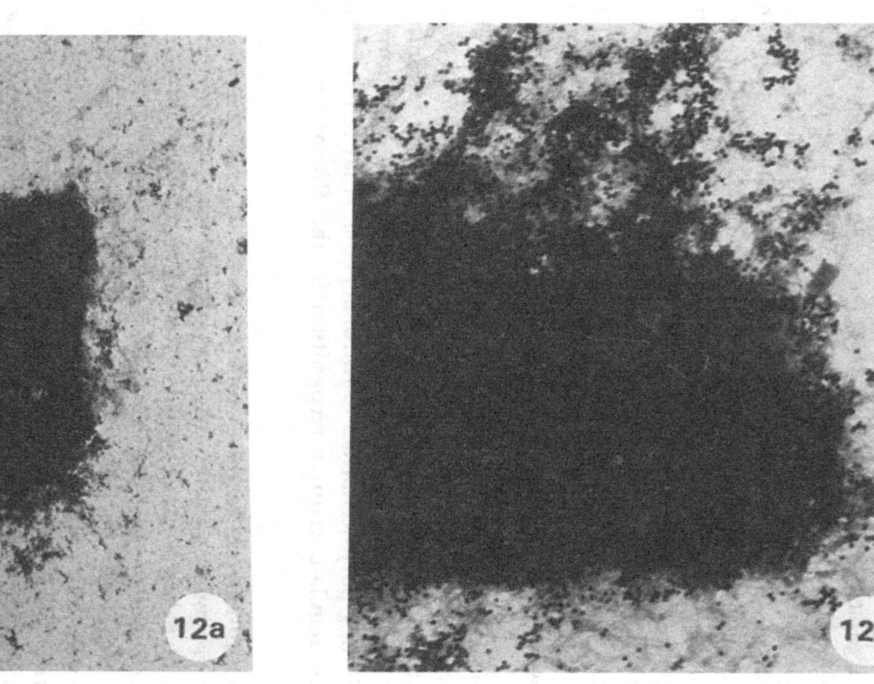

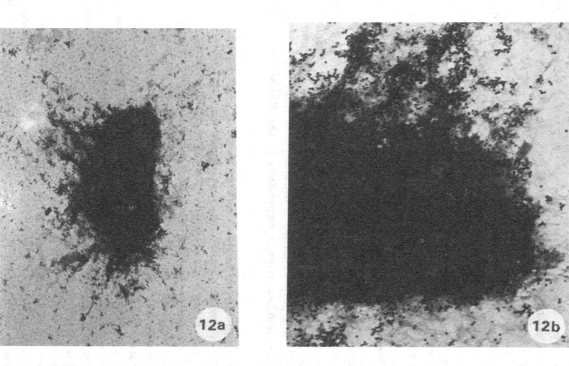

Fig. 12. Binding of trifoliin A-colloidal gold particles to an encapsulated cell of *R. trifolii* 0403 processed as in Fig. 11. Note binding of trifoliin A-gold colloid to the entire cell in *a* (32,000×), and its higher magnification in *b*. 88,000×.

peanut may be interacting specifically with a portion of the symbiont's LPS which is poorly immunogenic and common to different strains of the same *Rhizobium* species. The discovery that the lectin-binding sites on the polysaccharides of *R. trifolii* are not immunodominant is thus of paramount importance [33, 44].

A powerful, new technique of specimen preparation for transmission electron microscopy has recently been developed for *Rhizobium* [81]. This technique reveals the details of acidic polymer exostructures on the cell without introducing the artifacts and loss of capsular material associated with centrifugation (Fig. 11). The technique has been expanded to include lectin-markers in the form of colloidal gold-lectin conjugates (Fig. 12a and 12b, Truchet et al. unpublished) and may be of value in further studies of mutant strains of rhizobia which fail to reveal capsules using the traditional method of negative staining with India ink followed by light microscopy [82, 83].

The presence of multiple lectin receptors on rhizobia raises the question of whether each one has a different role in root hair infection. Infection studies by Kamberger [54] suggest that the lectin cross-bridging hypothesis [25] needs to be modified. For example, the capsular polysaccharides could be responsible for attachment of high numbers of rhizobial cells to the target root hairs via cross-bridging lectins as an early recognition event. This would be followed by secondary recognition events requiring the host-range specific binding of lectin on localized sites of the root hair to LPS, which triggers subsequent invasive steps [54]. One challenge of the coming few years is to test the validity of this hypothesis.

Host-specificity genes are plasmid encoded

Exciting evidence is beginning to emerge that genetic elements important to surface polysaccharides and symbiotic recognition are encoded on very large, transmissible plasmids of *Rhizobium* [6, 84-91]. For instance, the genes responsible for the 2-deoxy-D-glucose inhibitable attachment of *R. trifolii* to clover root hairs are encoded on the large nodulation plasmid designated pWZ2 [6]. Incorporation of the trifoliin A binding sugar, quinovosamine into the LPS of *R. trifolii* is controlled by the clover nodulation plasmid [92]. The ability of *R. meliloti* to induce polygalacturonase production in alfalfa roots also seems to be controlled by a plasmid [93]. Genes required for interstrain competition of rhizobia for root nodulation, and the synthesis of a unique 24,000 dalton protein expressed in the host rhizosphere are located on the pea nodulation plasmid of *R. leguminosarum* [94]. Conjugal transfer of the nodulation plasmid from *R. trifolii* to *Agrobacterium tumefaciens* results in a hybrid which nodulates clover roots [89], and which binds trifoliin A (Dazzo, Truchet, and Hooykaas, unpublished).

Analysis of symbiotically defective mutant strains of rhizobia is complicated by multiple pleiotropic effects when the mutated genes affect production of polysaccharides. For instance, the *R. leguminosarum* mutant strain, EXO-1, does not nodulate peas, and Saunders, Carlson, Napoli, and Albersheim [46] reported that 27 per cent of its total LPS mass is anthrone-reactive carbohydrate, as compared to 63 per cent of the total LPS mass from the wild-type *R. leguminosarum* strain from which it was derived. The glycosyl and antigenic compositions of the O-antigen of the mutant and wild-type strain do not seem to be different, and both strains are lysed by the same bacteriophages. The hypothesis advanced by the authors was that the mutant strain EXO-1 has reduced its production of extracellular polysaccharide (it excretes 5 per cent of the amount of the wild type into the culture medium), but not of LPS [46]. However, an alternative hypothesis is that the EXO-1 phenotype could be due to a defective O-antigen polymerase, which would fail to polymerize in a block fashion the repeating O-antigen on the polyisoprenoid acyl carrier lipid [95]. When the O-antigen oligosaccharide is transferred to the R-core lipid A via the translocase reaction, an LPS with reduced O-antigen polymerization would result. Bacteriophage with receptors for the O-antigens could still recognize those defective LPS structures, and their glycosyl composition would be the same or similar, exactly matching the phenotype of these non-nodulating *R. leguminosarum* mutant strains described by Saunders et al. [46]. Mutations, such as the one described above, which reduce the chain length of the O-antigen, cause pleiotropic negative effects on the biogenesis and assembly of the outer membrane of the Gram negative cell. The reduction in chain length of this carbohydrate moiety of the LPS causes a concomitant decrease and sometimes virtual loss of outer membrane proteins [96]. This pleiotropic negative effect complicates any direct interpretation of the significance of EXO-1 mutant strain to the infection process. Non-nodulating mutant strains of *R. japonicum* have been found which have discrete saccharide changes in the somatic antigens [97], and further analysis of these strains should determine what role LPS plays in root nodulation.

The trifoliin A-binding capsule of *R. trifolii* is altered by enzymes released from clover roots

The ultimate level of regulation of lectin receptors in the root environment is one in which both the bacterium and the host plant play key roles. For instance, Bhuvaneswari and Bauer [98] showed that some strains of *R. japonicum* bind soybean lectin in the root environment but not in pure culture. This suggests that the host plays some role in expression of lectin-binding receptors on the rhizobial cells. Other strains of *R. japonicum* could bind soybean lectin better when grown in soil extract than in standard bacteriological media [99]. These and our finding summarized below illustrate that an understanding of the biochemical basis of *Rhizobium*-legume interactions will require detailed studies of the microorganisms in the rhizosphere of the host root as the normal case.

The first clue that the clover root environment altered the lectin receptors on *R. trifolii* 0403 came from detailed studies on the orientation of attachment of these bacteria to clover root hairs [7, 48, 35, 65] (Dazzo et al. unpublished).

If an inoculum (10^7–10^9) of fully encapsulated cells of *R. trifolii* (which bind trifoliin A uniformly around the cell) were incubated for 15 minutes with clover seedlings, cells attached with no preferred orientation to root hair tips (see Fig. 5). However, after four hours of incubation, additional cells began to attach along the sides of the root hair in a distinct polar orientation (Fig. 13). If lower inoculum densities were used (10^5–10^6 per seedling), most cells attached polarly to root hairs within 12 hours without preference to root hair tips [7].

Why was there a delay in polar attachment of the bacteria when a high inoculum of uniformly encapsulated, lectin-binding cells was used? The answer was obtained by analyzing the effect of concentrated root exudate of clover seedlings on encapsulated cells of *R. trifolii.* We found that the lectin-binding capsule of cells was altered by enzymes released from axenically grown roots into the surrounding environment [65]. A summary of that work is described below.

Fluorescence microscopy showed that trifoliin A in clover root exudate bound uniformly to encapsulated, heat-fixed cells during one hour's incubation on microscope slides. After four to eight hours of incubation with root exudate, trifoliin A was bound to only one pole of the cells. Transmission electron microscopy showed that the capsule itself was altered. The disorganization of the acidic polymers of the capsule began in the equatorial centre of the rod-shaped cell and then progressed towards the poles at unequal rates. Trifoliin A could no longer be detected on heat-fixed cells following 12 hours of incubation with root exudate. However, trifoliin A was detected *in situ* on one pole of cells grown for four days in the clover root environment of Fahraeus slide cultures (Fig. 14). Inhibition studies using the hapten 2-deoxy-D-glucose showed that trifoliin A in root exudate had higher affinity for one of the cell poles. Immunoelectrophoresis was used to monitor the alteration of the extracellular polysaccharides from *R. trifolii* 0403 by concentrated root exudate. These polysaccharides were converted into products which eventually lost their ability to immunoprecipitate with homologous antibody. This progressive loss of antigenic reactivity proceeded more rapidly with root exudate from seedlings grown under nitrogen-free conditions than from plants grown with 15 mM nitrate. The root exudate, depleted of trifoliin A by immunoaffinity chromatography, was still able to alter the capsule of *R. trifolii* 0403. Reconstitution experiments showed that the protein(s) in root exudate which induced this alteration of the capsule were high molecular weight, heat-labile, trypsin-sensitive, and antigenically unrelated to trifoliin A. A variety of glycosidase activities were also detected in the fraction depleted of

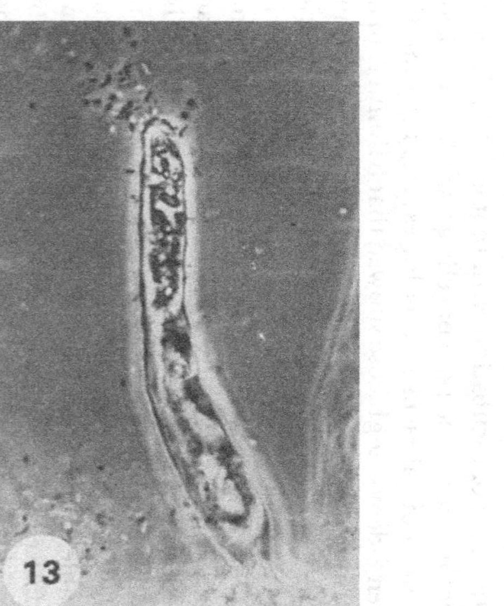

Fig. 13. Attachment of *R. trifolii* 0403 to a clover root hair after 4 hr of incubation with a high inoculum density. Note random attachment of bacteria to the root hair tip, and distinct polar attachment of bacteria along the sides of the root hair (phase contrast micrograph). 2,533×.

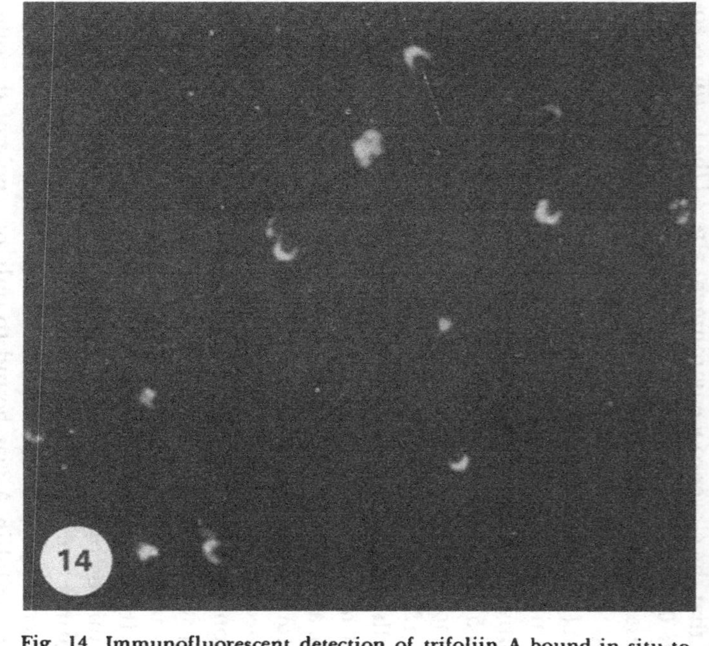

Fig. 14. Immunofluorescent detection of trifoliin A bound in situ to *Rhizobium trifolii* 0403 growing in the clover root environment of Fahraeus slide cultures. From Dazzo, Truchet, Sherwood, Hrabak and Gardiol [65], and courtesy of the American Society for Microbiology. 5,000×.

trifoliin A. These results suggest that enzymes, which are antigenically unrelated to trifoliin A, accumulate in root exudate and alter the trifoliin A-binding capsule in a way which would favour polar attachment of *R. trifolii* to clover root hairs.

Based on these rhizosphere studies, we have subdivided the Phase I attachment process of encapsulated *R. trifolii* to clover root hairs into the following sequential events [48] (Fig. 15): (a) Most encapsulated cells which have trifoliin A receptors around the entire cell surface bind within minutes in a random orientation to clover root hair tips where trifoliin A accumulates. (b) Cells which do not immediately contact the root hairs encounter enzymes in root exudate which modify their surface polysaccharides so that they become progressively less-reactive with trifoliin A. This alteration proceeds less rapidly at one cell pole. Newly synthesized lectin-binding polysaccharide may also be deposited at one pole of non-encapsulated cells at a rate which may keep pace with the exudate enzyme-mediated modifications. (c) Some cells with trifoliin A and/or its saccharide receptors bound to one pole eventually contact the cell wall along the sides of the root hair, where they then attach end-on.

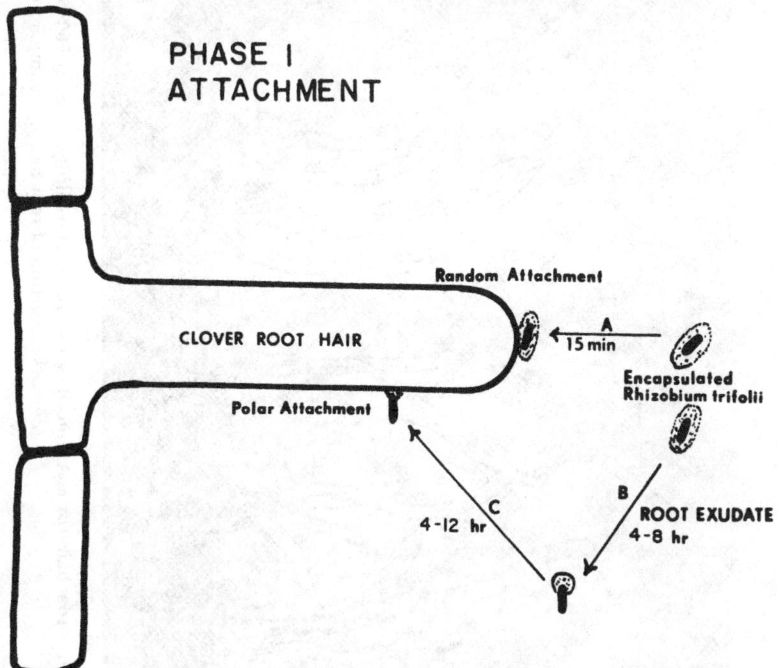

Fig. 15. Schematic diagram of the sequential events of Phase I Attachment of *R. trifolii* to clover root hairs, using an inoculum of fully encapsulated cells.

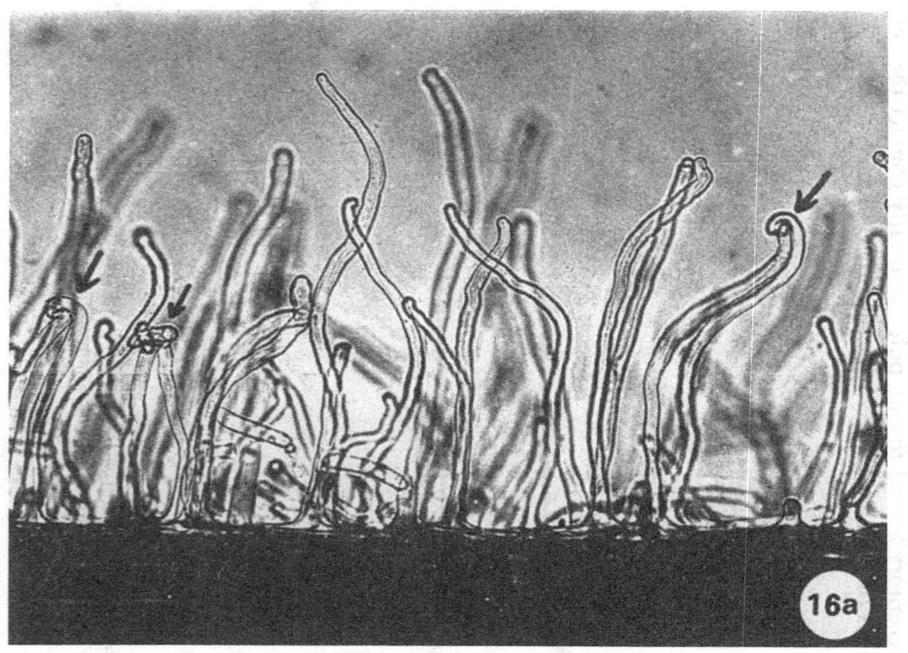

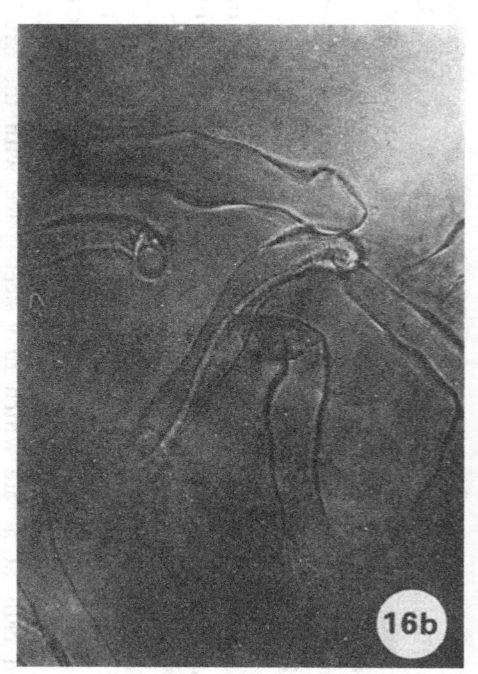

Fig. 16. Adhesion of root hair tips. *a*. Axenically growing clover seedlings (arrows). 500×. *b*. Tip-adhered root hair infected with *R. trifolii* 0403. Note lack of marked curling of root hair tip (compare with Fig. 1). 1,286×.

Lectin is involved in the tip adhesion of root hairs

The phenomenon of adhesion of root hair tips was recognized in early studies on the invasion of legume roots by *Rhizobium* [100, 101]. Such cell-cell adhesions are predicted by the lectin cross-bridging model [25] (Fig. 3), where the multivalent lectin on one root hair tip would bind to complementary receptors accessible on an adjacent root hair. Quantitative microscopic studies [102] have recently shown that root hair tip adhesions on clover seedlings grown under axenic conditions are generally restricted to a zone located 1 mm below the root hair closest to the hypocotyl and 2–3 mm above the meristem (Fig. 16a). Trifoliin A was localized by immunofluorescence at contact points of tip adhesions. Conditions known to reduce the levels of trifoliin A on the root surface (treatment with 2-deoxy-D-glucose or growth in medium containing 15 mM nitrate) significantly reduced the formation of root hair tip adhesions. These results suggest that trifoliin A is involved in the formation and/or stability of tip adhesions.

Root hair tip adhesions may have a physiological significance to the *Rhizobium*-legume symbiosis. Root hairs which sandwich rhizobia between tip adhesions frequently become infected without marked deformations [49, 103] (Fig. 16b). In contrast, infected root hairs which develop separately demonstrate marked deformations such as shepherd's crooks when incubated with homologous rhizobia. In both cases, conditions optimal for rhizobial infection could occur in the microenvironment created by overlapping root hair cell walls.

It was noted in our laboratory that some root hairs grown under axenic conditions developed marked curvatures at tip-to-tip adhesions (arrows, Fig. 16a). These curvatures were less tight than typical shepherd's crooks shown in Fig. 1b, but may be a consequence of the same physiological process. Development of shepherd's crooks requires direct contact with homologous rhizobia [59]. It is possible that shepherd's crooks, as well as the tight curvatures of tip-to-tip root hair adhesions, are a consequence of the growth of root hairs about a fixed surface: the adherent bacterial floc or adjacent root hair tip. This process would be similar to what has been reported as the "contact guidance" system of pollen tube orientation during its passage through the style to the ovule of compatible flowering plants [104].

Root nodulation of soybean lines lacking the 120,000 dalton lectin in their seeds

Pull, Pueppke, Hymowitz and Ord [105] have identified five lines of soybean which lack the N-acetylgalactosamine-specific 120,000 dalton soybean lectin in their seeds (but see [12]). Root hairs of these lines are still infectable by *R. japonicum* and root nodulation is specific for the soybean rhizobia [106]. Thus, the presence of this lectin in soybean seeds is not necessary for root nodulation by *R. japonicum*. It will be important to

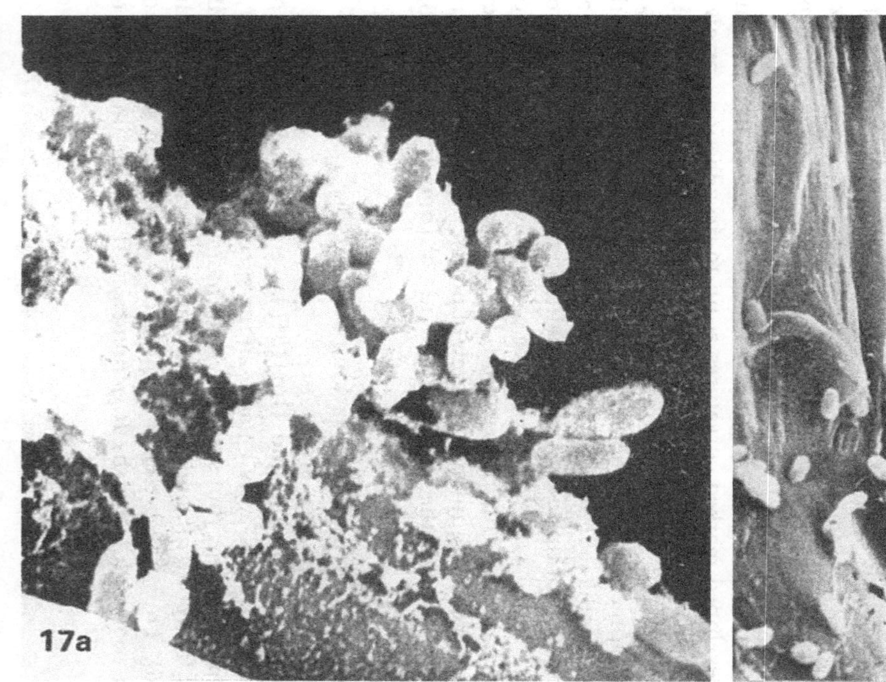

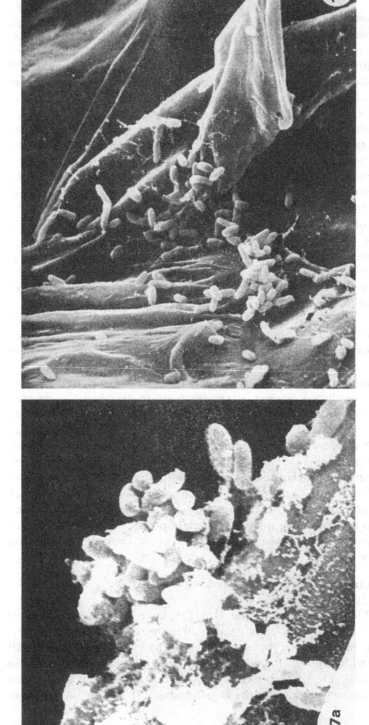

Fig. 17. Attachment of *Azospirillum brasilense* Sp7 to pearl millet roots in hydroponic culture. Scanning electron micrographs of bacteria attached to root hair surfaces *a*. 7,000×, and epidermis near lateral root emergence *b*. 3,000×. Note the particulate material in *a*, and the fibrillar material in *b*, associated with the adherent bacteria. From Umali-Garcia et al. [107], and courtesy of the American Society for Microbiology.

know whether the root hairs of these plants have this or other lectins which recognize the soybean rhizobia during infection. Another area worth investigating with these lines of soybean is to determine if the symbiont rhizobia affect *de novo* synthesis of root hair lectins.

Attachment of "associative" nitrogen-fixing bacteria to grass roots

The attachment of *Azospirillum brasilense* Sp7 to root hairs of pearl millet in hydroponic culture is similar in some respects to attachment of *R. trifolii* 0403 to clover root hairs [107]. Such attachment can be viewed as favouring the colonization of the rhizoplane by azospirilla and their subsequent entry into the grass root cortex [107]. Adherent azospirilla were associated with granular material on root hairs and fibrillar material on undifferentiated epidermal cells (Fig. 17 a–c). Significantly fewer cells of azospirilla attached to millet root hairs when the roots were grown in culture medium containing 5 mM nitrate (Fig. 18 a–b), whereas attachment to undifferentiated epidermal cells was unaffected. Root exudate from millet contained non-dialyzable and protease sensitive proteins which bound to azospirilla in the root environment and promoted their attachment to millet root hairs, but not to undifferentiated epidermal cells. Millet root hairs adsorbed more cells of azospirilla than of *R. trifolii, Pseudomonas* sp., *Azotobacter vinelandii, Klebsiella pneumoniae,* and *Escherichia coli.* It is clear from these studies that attachment of azospirilla to root hairs of millet is accomplished by mechanisms which differ from those that attach azospirilla to undifferentiated epidermal cell surfaces.

Korhonen et al. [108] have recently examined the *in vitro* adhesion of associative nitrogen-fixing *Klebsiella* spp. to grass roots. Klebsiellas were labelled with 3H-amino acids and incubated with the grass roots. Their attachment was dependent on inoculum density, incubation time and temperature, pH and ionic strength of the incubation buffer, and the growth phase of the bacterial cells. Type 1 (mannose-sensitive) and type 3 fimbriae (pili) were isolated from the labelled bacteria and examined for their possible role in root attachment. The binding of type 1 pili to roots was inhibited by the hapten a-methylmannoside and by Fab monovalent fragments directed against the purified pili. However, non-piliated mutant strains attached very well to the whole grass roots. More work on the role of pili in root attachment is needed, and will require detailed microscopy to sort out the mechanisms of attachment of these nitrogen-fixing organisms to grass root hairs and other epidermal cells.

CONCLUDING REMARKS

It seems unlikely that acceptance or rejection of the lectin-recognition hypothesis will ever become universal until the recognition code is deciphered and mutant strains of both symbionts—bacterium and plant—with

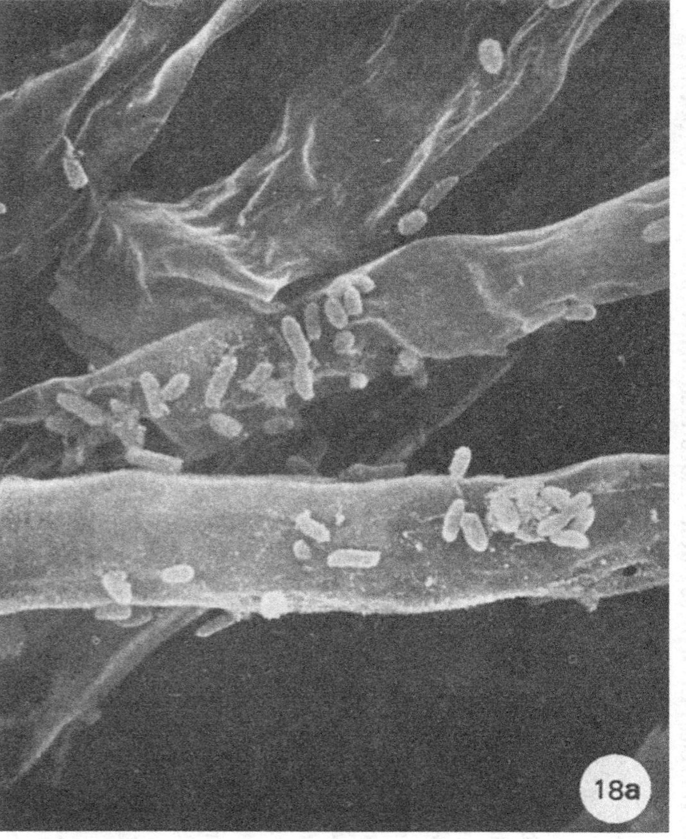

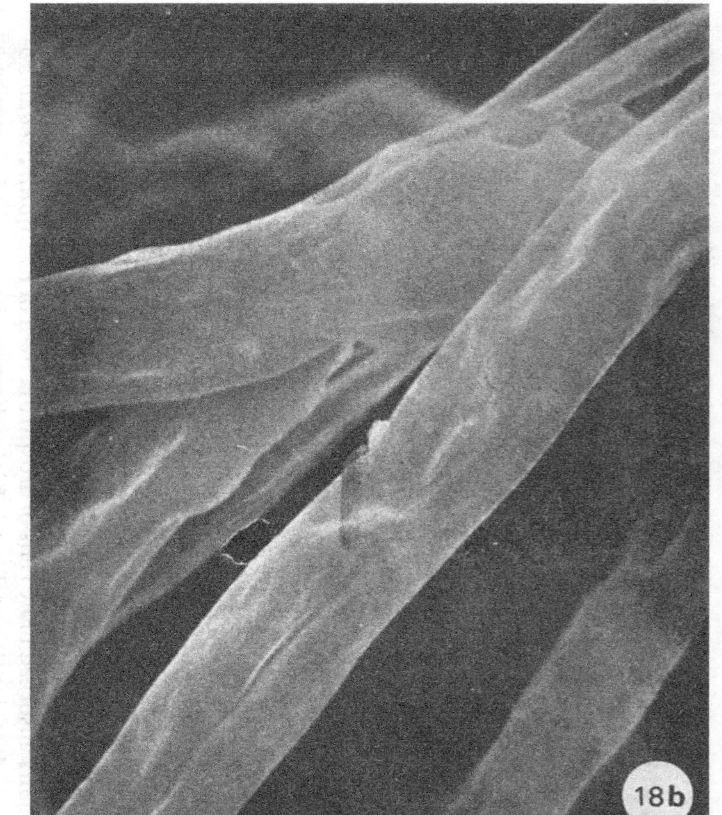

Fig. 18. Inhibition of adsorption of *A. brasilense* Sp7 to root hairs of pearl millet grown with nitrate. Root hairs were grown under nitrogen-free conditions in *a*, and with 5 mM KNO_3 in *b* (3,000×). From Umali-Garcia et al. [107], and courtesy of the American Society for Microbiology.

altered lectin/lectin receptors and no pleiotropic effects are available for analysis.

The *Rhizobium*-legume symbiosis can be viewed as a delicate balance of many cell-cell communications which, in coordination, culminate in the formation of root nodules that fix N_2 into ammonia fertilizer for the plant symbiont. In addition to attachment, there is curling and branching of the root hair [59], tip-to-tip adhesions [102], penetration of the root hair cell wall by the rhizobia without host cell lysis [26, 109], dome formation of the new infection thread [109], nucleus-directed growth and extension of the infection thread down the root hair shaft [101, 110], infection thread penetration of the cell wall at the base of the root hair, host cell proliferation in the inner cortex in front of the advancing infection thread [111], branching of the infection thread in the nodular cells, release and envelopment of the bacteria from the infection thread into peribacteroid membranes [112], development of the bacteroids [113–114], differentiation of nodular tissue [112, 115], leghemoglobin synthesis [116], nitrogenase synthesis and expression [117], and mechanisms for exchange of metabolites and energy between the legume and the respiring *Rhizobium* bacteroids [118]. Each of these events provides an excellent model to study the underlying biochemical mechanisms of plant-microorganism interactions.

ACKNOWLEDGEMENTS

Portions of this work were supported by Grants 78-59-2261-0-1-050-2 and 82-CRCR-1-1040 from the Competitive Research Grant Program of the United States Department of Agriculture, Grant No. 80-21906 from the National Science Foundation, Project No. 1314H from the Michigan Agricultural Experiment Station, and the European Molecular Biology Organization.

REFERENCES

1. Burnet, F.M. Self-recognition in colonial marine forms and flowering plants in relation to evaluation of immunity, *Nature* (London) *232*, 230–235 (1971).
2. Hamblin, J. and Kent, S.P. Possible role of phytohaemagglutinin in *Phaseolus vulgaris* L., *Nature New Biol. 245*, 28–29 (1973).
3. Bohlool, B.B. and Schmidt, E.L. Lectins: A possible basis for specificity in the *Rhizobium*-legume root nodule symbiosis, *Science 185*, 269–271 (1974).
4. Dazzo, F.B. and Hubbell, D.H. Cross-reactive antigens and lectin as determinants of symbiotic specificity in the *Rhizobium*-clover association, *Applied Microbiology 30*, 1017–1033 (1975).
5. Vincent, J.M. Factors controlling the legume-*Rhizobium* symbiosis, in *Nitrogen Fixation II* (Editors, W. Orme-Johnson and W.E. New-

ton), University Park Press, Baltimore, MD (1980).
6. Zurkowski, W. Specific adsorption of bacteria to clover root hairs, related to the presence of the plasmid pWZ2 in cells of *Rhizobium trifolii, Microbios 27*, 27–32 (1980).
7. Dazzo, F.B. Napoli, C.A. and Hubbell, D.H. Adsorption of bacteria to roots as related to host specificity in the *Rhizobium*-clover association, *Appl. Environ. Microbiol. 32*, 168–171 (1976).
8. Kato, G., Maruyama, Y. and Nakamura, M. Involvement of lectins in *Rhizobium*-pea recognition, *Plant Cell Physiol. 22*, 759–771 (1981).
9. Paau, A.S., Leps, W.T. and Brill, W.J. Agglutinin from alfalfa necessary for binding and nodulation by *Rhizobium meliloti, Science 213*, 1513–1515 (1981).
10. Broughton, W.J. A review: control of specificity in legume-*Rhizobium*-associations, *J. Appl. Bacteriol. 45*, 165–194 (1978).
11. Sequeira, L. Lectins and their role in host-pathogen specificity, *Ann. Rev. Phytopathol. 16*, 453–481 (1978).
12. Schmidt, E.L. Initiation of plant root-microbe interactions, *Ann. Rev. Microbiol. 33*, 355–376 (1979).
13. Dazzo, F.B. Adsorption of microorganisms to roots and other plant surfaces, pp. 253–316, in *Adsorption of Microorganisms to Surfaces* (Editors, G. Bitton and K. Marshall), J. Wiley, New York (1980a).
14. Dazzo, F.B. Infection processes in the *Rhizobium*-legume symbiosis, pp. 49–59, in *Advances in Legume Science* (Editors, R.J. Summerfield and A.H. Bunting), Crown Publishers, U.K. (1980b).
15. Dazzo, F.B. Lectins and their saccharide receptors as determinants of specificity in the *Rhizobium*-legume symbiosis, pp. 277, in *The Cell Surface: Mediator of Developmental Processes* (Editors, S. Subtelny and N. Wessells), Academic Press, New York (1980c).
16. Dazzo, F.B. Determinants of host-specificity in the *Rhizobium*-clover symbiosis, in *Nitrogen Fixation II* (Editors, W. Orme-Johnson and W.E. Newton), University Park Press, Baltimore, MD (1980d).
17. Kauss, H. Lectins and their physiological role in slime molds and in higher plants, in *Encyclopedia of Plant Physiology*, Vol. II, Plant Carbohydrates, Springer-Verlag, Berlin (1980).
18. Dazzo, F.B. Bacterial attachment as related to cellular recognition in the *Rhizobium*-legume symbiosis, *J. Supramol. Struct. Cell. Biochem. 16*, 29–41 (1981a).
19. Dazzo, F.B. Microbial adhesion to plant surfaces, pp. 311–328, in *Microbial Adhesion to Surfaces* (Editors, R. Berkeley, J. Lynch, J. Melling, P. Rutter and B. Vincent), Ellis Horwood Ltd., Chichester, U.K. (1981b).
20. Bauer, W.D. Infection of legumes by rhizobia, *Ann. Rev. Plant Physiol. 32*, 407–449 (1981).
21. Graham, T.L. Recognition in *Rhizobium*-legume symbioses, pp.

127–148, in *International Review of Cytology* (Editor, A. Atherly), Supplement 13, Academic Press, New York (1981).

22. Dazzo, F.B. Plant cell-cell interactions, in *Cell Interactions and Development: Molecular Mechanisms* (Editor, K. Yamada), John Wiley, New York.
23. Dazzo, F.B. and Hubbell, D.H. Control of root hair infection, pp. 274–310, in *Nitrogen Fixation II* (Editor, W. Broughton), Oxford University Press, Oxford, England.
24. Solheim, B. and Paxton, J. Recognition in *Rhizobium*-legume systems, in *Plant Disease Control: Resistance and Susceptibility*, John Wiley, New York (1981).
25. Dazzo, F.B. and Hubbell, D.H. Cross-reactive antigens and lectins as determinants of symbiotic specificity in the *Rhizobium*-clover association, *Appl. Microbiol. 30*, 1017–1033.
26. Napoli, C.A. and Hubbell, D.H. Ultrastructure of *Rhizobium*-induced infection threads in clover root hairs, *Appl. Microbiol. 30*, 1003–1009 (1975).
27. Kumarasinghe, M.K. and Nutman, P.S. *Rhizobium*-stimulated callose formation in clover root hairs and its relation to infection, *J. Exp. Bot. 28*, 961–967 (1977).
28. Stacey, G., Paau, A. and Brill, W.J. Host recognition in the *Rhizobium*-soybean symbiosis, *Plant Physiol. 66*, 609–614 (1980).
29. Kato, G., Maruyama, Y. and Nakamura, M. Role of bacterial polysaccharides in the adsorption process of the *Rhizobium*-pea symbiosis, *Agric. Biol. Chem. 44*, 2843–2855 (1980).
30. Reporter, M., Raveed, D. and Norris, G. Binding of *Rhizobium japonicum* to cultured soybean root cells: Morphological evidence, *Plant Sci. Lett. 5*, 73–76 (1975).
31. Chen, A.P. and Phillips, D.A. Attachment of *Rhizobium* to legume roots as the basis for specific associations, *Physiol. Plant. 38*, 83–88 (1976).
32. van Rensberg, H.J. and Strijdom, D.W. Root surface association in relation to nodulation of *Medicago sativa*, *Appl. Environ. Microbiol. 44*, 93–97 (1982).
33. Dazzo, F.B. and Brill, W.J. Bacterial polysaccharide which binds *Rhizobium trifolii* to clover root hairs, *J. Bacteriol. 137*, 1362–1371 (1979).
34. Dazzo, F.B., W.E. Yanke and Brill, W.J. Trifoliin: A *Rhizobium* recognition protein from white clover, *Biochim. Biophys. Acta 539*, 276–286 (1978).
35. Dazzo, F.B. and Brill, W.J. Receptor site on clover and alfalfa roots for *Rhizobium*, *Appl. Environ. Microbiol. 33*, 132–136 (1977).
36. Kijne, J.W., van der Schaal, I.A.M. and DeVries, G.E. Pea lectins and the recognition of *Rhizobium leguminosarum*, *Plant Sci. Lett. 18*, 65–74 (1980).

37. Kijne, J.W., van der Schaal, A.A., Diaz, C.L. and van Iren, F. Mannose-specific lectins and the recognition of pea roots by *Rhizobium leguminosarum*, pp. 521-530, in *Lectins*, Vol. III (Editors, Bog-Hansen and Spengler).
38. Gatehouse, J.A. and Boulter, D. Isolation and properties of lectin from the roots of *Pisum sativum. Physiol. Plant. 49*, 437-442 (1980).
39. van der Schaal, I.A.M. and Kijne, J.W. Pea lectins and surface carbohydrates of *Rhizobium leguminosarum*, p. 425, in *Current Perspectives in Nitrogen Fixation* (Editors, A.H. Gibson and W.E. Newton), Australian Academy of Science, Canberra (1981).
40. Gade, W., Jack, M.A., Dahl, J.B., Schmidt, E.L. and Wolf, F. The isolation and characterization of a root lectin from soybean (*Glycine max* L.) cultivar Chippewa, *J.Biol. Chem. 256*, 12905-12910 (1981).
41. Reeke, J.N., Becker, J.W., Cunningham, B.A., Wang, J.C., Yahara, I. and Edelman, G.M. Structure and function of concanavalin A, *Adv. Expt. Med. Biol. 55*, 13-33 (1975).
42. Hughes, T.A. and Elkan, G.H. Study of the *Rhizobium japonicum*-soybean symbiosis, *Plant Soil 61*, 87-91 (1981).
43. Hughes, T.A., Leece, J.C. and Elkan, G.H. Modified fluorescent technique using rhodamine for studies of *Rhizobium japonicum*-soybean symbiosis, *Appl. Environ. Microbiol. 37*, 1243-1244 (1979).
44. Hrabak, E.M., Urbano, M.R. and Dazzo, F.B. Growth-phase dependent immunodeterminants of *Rhizobium trifolii* lipopolysaccharide which bind trifoliin A, a white clover lectin, *J. Bacteriol. 148*, 697-711 (1981).
45. Bishop, P.E., Dazzo, F.B., Applebaum, E.R., Maier, R.J. and Brill, W.J. Intergeneric transformation of genes involved in the *Rhizobium*-legume symbiosis, *Science 198*, 938-939 (1977).
46. Saunders, R.E., Carlson, R.W. and Albersheim, P. A *Rhizobium* mutant incapable of nodulation and normal polysaccharide secretion, *Nature 271*, 240-242 (1978).
47. Napoli, C.A. and Albersheim, P. *Rhizobium leguminosarum* mutants incapable of normal extracellular polysaccharide production, *J. Bacteriol. 141*, 1454-1456 (1980).
48. Dazzo, F.B., Hrabak, E., Urbano, M.R., Sherwood, J. and Truchet, G. Regulation of Recognition in the *Rhizobium*-clover symbiosis, pp. 292-295, in *Current Perspectives in Nitrogen Fixation* (Editors, A. Gibson and W.E. Newton), Australian Academy Science, Canberra, Australia (1981).
49. Napoli, C.A., Dazzo, F.B. and Hubbell, D.H. Production of cellulose microfibrils by *Rhizobium, Appl. Microbiol. 30*, 123-131 (1975).
50. Deinema, M.H. and Zevenhuizen, L.P.T. Formation of cellulose fibrils by Gram negative bacteria and their role in bacterial flocculation, *Arch. Microbiol. 78*, 42-57 (1971).

51. Stemmer, P. and Sequeira, L. Pili of plant pathogenic bacteria, *Amer. Phytopathol. Soc. Ann. Mtg.*, Abstr. 328 (1981).
52. Bhuvaneswari, T.V., Turgeon, B.G. and Bauer, W.G. Early events in the infection of soybean (*Glycine max* L. Merr.) by *Rhizobium japonicum*, I. Location of infectible root cells, *Plant Physiol. 66*, 1027-1031 (1980).
53. Bhuvaneswari, T.V., Bhagwat, A.A. and Bauer, W.D. Transient susceptibility of root cells in four common legumes to nodulation by rhizobia, *Plant Physiol. 68*, 1144-1149 (1981).
54. Kamberger, W. Role of cell surface polysaccharides in the *Rhizobium*-pea symbiosis, *FEMS Microbiol. Lett. 6*, 361-365 (1979).
55. Hubbell, D.H., Morales, V.M. and Umali-Garcia, M. Pectolytic enzymes in *Rhizobium, Appl. Environ. Microbiol. 38*, 1186-1188 (1979).
56. Martinez-Molina, E. Morales, V.M. and Hubbell, D.H. Hydrolytic enzyme production by *Rhizobium, Appl. Environ. Microbiol. 35*, 210-213 (1979).
57. Ljunggren, H. and Fahraeus, G. Role of polygalacturonase in root hair invasion by nodule bacteria, *J. Gen. Microbiol. 26*, 521-528 (1961).
58. Palomares, A., Montoya, E. and Olivares, J. Induction of polygalacturonase production in legume roots as a consequence of extrachromosomal DNA carried by *Rhizobium meliloti, Microbios 21*, 33-39 (1978).
59. Yao, P.Y. and Vincent, J.M. Factors responsible for the curling and branching of clover root hairs by *Rhizobium, Plant Soil 45*, 1-16 (1976).
60. Higashi, S. and Abe, M. Promotion of infection thread formation by substances from *Rhizobium, Appl. Environ. Microbiol. 39*, 297-301 (1980).
61. Truchet, G.L. and Dazzo, F.B. Morphogenesis of lucerne root nodules incited by *Rhizobium meliloti* in the presence of combined nitrogen, *Planta 154*, 352-360 (1982).
62. Dazzo, F.B. and Brill, W.J. Regulation by fixed nitrogen of host-symbiont recognition in the *Rhizobium*-clover symbiosis, *Plant Physiol. 62*, 18-21 (1978).
63. Diaz, C., Kijne, J.W. and Quispel, A. Influence of nitrate on pea root cell wall composition, p. 426, in *Current Perspectives in Nitrogen Fixation* (Editors, A.H. Gibson and W.E. Newton), Australian Academy of Science, Cańberra (1981).
64. Dazzo, F.B. and Hrabak, E.M. Presence of trifoliin A, a *Rhizobium*-binding lectin, in clover root exudate, *J. Supramol. Struct. Cell. Biochem. 16*, 133-138 (1981).
65. Dazzo, F.B., Truchet, G.L., Sherwood, J.E., Hrabak, E.M. and Gar-

diol, A.E. Alteration of the trifoliin A-binding capsule of *Rhizobium trifolii* 0403 by enzymes released from clover roots, *Appl. Environ. Microbiol. 44*, 478-490 (1982).

66. Solheim, B. Possible role of lectin in the infection of legumes by *Rhizobium trifolii* and a model of the recognition reaction between *Rhizobium trifolii* and *Trifolium repens*, NATO Advanced Study Institute on Specificity in Plant Diseases, Sardinia (1975).
67. Dazzo, F.B., Urbano, M.R. and Brill, W.J. Transient appearance of lectin receptors on *Rhizobium trifolii*, *Curr. Microbiol. 2*, 15-20 (1979).
68. Bhuvaneswari, T.V., Pueppke, S.G. and Bauer, W.D. Role of lectins in plant-microorganism interactions, I. Binding of soybean lectin to rhizobia, *Plant Physiol. 60*, 486-491 (1977).
69. Mort, A.J. and Bauer, W.D. Composition of the capsular and extracellular polysaccharides of *Rhizobium japonicum*: changes with culture age and correlations with binding of soybean seed lectin to the bacteria, *Plant Physiol. 66*, 158-163 (1980).
70. Tsien, H.C. and Schmidt, E.L. Accumulation of soybean lectin-binding polysaccharide during growth of *Rhizobium japonicum* as determined by hemagglutination inhibition assay, *Appl. Environ. Microbiol. 39*, 1100-1104 (1980).
71. Wolpert, J.S. and Albersheim, P. Host-symbiont interactions, I. The lectins of legumes interact with the O-antigen containing lipopolysaccharides of their symbiont rhizobia, *Biochem. Biophys. Res. Commu. 170*, 729-737 (1976).
72. Planqué, H. and Kijne, J.W. Binding of pea lectins to a glycan type polysaccharide in the cell walls of *Rhizobium leguminosarum*, *FEBS Lett. 73*, 64-66 (1977).
73. Bal, A.K., Shantharam, S. and Ratnam, S. Ultrastructure of *Rhizobium japonicum* in relation to its attachment to root hairs, *J. Bacteriol. 133*, 1393-1400 (1978).
74. Calvert, H.E., Lalonde, M., Bhuvaneswari, T.V. and Bauer, W.D. Role of lectins in plant microorganism interactions, IV. Ultrastructural localization of soybean lectin binding sites on *Rhizobium japonicum*, *Can. J. Microbiol. 24*, 785-793 (1978).
75. Mort, A.J. and Bauer, W.D. Structure of the capsular and extracellular polysaccharides of *Rhizobium japonicum* that bind soybean lectin, Application of two new methods for cleavage of polysaccharides into specific oligosaccharide fragments, *J. Biol. Chem. 257*, 1870-1875 (1982).
76. Tsien, H.C. and Schmidt, E.L. Localization and partial characterization of soybean lectin binding polysaccharide of *Rhizobium japonicum*, *J. Bacteriol. 145*, 1063-1074 (1981).
77. Kamberger, W. An Ouchterlony double diffusion study on the inter-

action between legume lectins and rhizobial cell surface antigens, *Arch. Microbiol. 121,* 83-90 (1979).

78. Kato, G., Maruyama, Y. and Nakamura, M. Role of lectins and lipopolysaccharides in the recognition process of specific legume-*Rhizobium* symbiosis, *Agric. Biol. Chem. 43,* 1085-1092 (1979).
79. Bhagwat, A. and Thomas, J. Dual binding sites for peanut lectin on rhizobia, *J. Gen. Microbiol. 117,* 119-125 (1980).
80. Carlson, R.W., Saunders, R.E., Napoli, C.A. and Albersheim, P. Host-symbiont interactions, III. Isolation and partial characterization of lipopolysaccharides from *Rhizobium, Plant Physiol. 62,* 912-917 (1978).
81. Mutaftschiev, S., Vasse, J. and Truchet, G. Exostructures of *Rhizobium meliloti, FEMS Microb. Lett. 13,* 171-175 (1982).
82. Rolfe, B., Djordjevic, M., Scott, K.,Hughes, J.E. Badenoch-Jones, J., Gresshoff, P.M., Cen, Y., Dudman, W.F., Zurkowski, W. and Shine, J. Analysis of the nodule-forming ability of fast-growing *Rhizobium* strains, in *Current Perspectives in Nitrogen Fixation* (Editors, A. Gibson and W. Newton), Australian Academy of Science, Canberra.
83. Law, I.J., Yamamoto, Y., Mort, A.J. and Bauer, W.D. Nodulation of soybean by *Rhizobium japonicum* mutants with altered capsule synthesis, *Planta 154,* 150-156 (1982).
84. Johnston, A.W.B., Beyon, I.L., Buchanan-Wollaston, A., Setchell, S., Hirsch, P. and Beringer, J. High frequency transfer of nodulating ability between species and strains of *Rhizobium, Nature 276,* 635-638 (1978).
85. Zurkowski, W. and Lorkiewicz, Z. Plasmid-mediated control of nodulation in *Rhizobium trifolii, Arch. Microbiol. 123,* 195-201 (1979).
86. Prakash, R.K., Hooykaas, P., Ledeboer, R., Kijne, J.W., Schilperoort, R.A., Nuti, M.P., Lepidi, A.A., Casse, F., Boucher, C., Julliot, S. and Denarie, J. in *Detection, isolation, and characterization of large plasmids in Rhizobium* (Editors, W. Orme-Johnson and W. Newton), University Park Press, Baltimore, MD.
87. Zurkowski, W. Conjugal transfer of the nodulating-conferring plasmid pWZ2 in *Rhizobium trifolii, Mol. Gen. Genet. 181,* 522-524 (1981).
88. Zurkowski, W. Molecular mechanism for loss of nodulation properties of *Rhizobium trifolii, J. Bacteriol. 150,* 999-1007 (1982).
89. Hooykaas, P.J., van Brussell, A.A.N., den Hulk-Ras, H., van Slogteren, G.M. and Schilperoort, R.A. Syn plasmid of *Rhizobium trifolii* expressed in different rhizobia species and *Agrobacterium tumefaciens, Nature 291,* 351-353 (1981).
90. Bafalvi, Z., Sakanyan, V., Koncz, C., Kiss, A., Dusha, I. and Kondorosi, A. Location of nodulation and nitrogen fixing genes on a high molecular weight plasmid of *R. meliloti, Mol. Gen. Genet. 184,* 318-325 (1981).

91. Rosenberg, G., Boistard, P., Denarie, J. and Casse-Delbart, F. Genes controlling early and late functions in symbiosis are located on a megaplasmid in *Rhizobium meliloti, Mol. Gen. Genet. 184,* 326–333 (1981).
92. Russa, R.T., Urbanik A., Kowalczuk, E. and Lorkiewicz, Z. Correlation between occurrence of plasmid pUCS202 and lipopolysaccharide alterations in *Rhizobium, OKR Microb. Lett. 13,* 161–165 (1982).
93. Palomares, A., Montega, E. and Olivares, J. Induction of polygalacturonase production in legume roots as a consequence of extrachromosomal DNA carried by *Rhizobium meliloti, Microbios 21,* 33–39 (1978).
94. Brewin, N. Genetics of nitrogen fixation: Future prospects of gene manipulation, in *Proceedings of the Second National Symposium on Nitrogen Fixation* (Editor, P. Uomala), Helsinki, Finland (1982).
95. Osborne, M.J., Cynkin, M.A., Gilbert, J.M., Mueller, L. and Singh, M. Synthesis of bacterial O-antigens, *Meth. Enzymol. 28,* 583–601 (1972).
96. Gmeiner, J. and Schlecht, S. Molecular organization on the outer membrane of *Salmonella typhimurium, Eur. J. Biochem. 93,* 609–620 (1979).
97. Maier, R.J. and Brill, W.J. Involvement of O-antigen in nodulation of soybean by *Rhizobium japonicum, J. Bacteriol. 133,* 1295–1299 (1978).
98. Bhuvaneswari, T.V. and Bauer, W.D. The role of lectins in plant-microorganism interactions, III. Influence of rhizosphere/rhizoplane culture conditions on the soybean lectin-binding properties of rhizobia, *Plant Physiol. 62,* 71–74 (1978).
99. Shantharam, S. and Bal, A.K. The effect of growth medium on lectin binding by *Rhizobium japonicum, Plant Soil 62,* 327–330.
100. Fred, E.B., Baldwin, I.L. and McCoy, E. *Root Nodule Bacteria and Leguminous Plants,* University of Wisconsin Press, Madison (1932).
101. Fahraeus, G. The infection of clover root hairs by nodule bacteria studied by a simple glass slide technique, *J. Gen. Microbiol. 16,* 374–381 (1957).
102. Dazzo, F.B., Truchet, G.L. and Kijne, J.W. Lectin involvement in root hair tip adhesions as related to the *Rhizobium*-clover symbiosis, *Physiol. Plant 56,* 143–147.
103. Higashi, S. and Abe, M. Scanning electron microscopy of *Rhizobium trifolii* infection sites on root hairs of white clover, *Appl. Environ. Microbiol. 40,* 1094–1099 (1980b).
104. Ferrari, T.E., Lee, S.S. and Wallace, D.H. Biochemistry and physiology of recognition in pollen-stigma interactions, *Phytopathol. 71,* 752–755 (1981).
105. Pull, S.P., Pueppke, S.G., Hymowitz, T. and Orf, J.H. Soybean lines

lacking the 120,000 dalton seed lectin, *Science 200,* 1277–1279 (1978).

106. Pueppke, S.G. *Glycine* genotypes that lack soybean lectin: versatile tools for investing the microbial function of lectins, Interlectin-5, *Symposium, Abstr. 144* (1982).
107. Umali-Gracia, M., Hubbell, D.H., Gaskins, M.H. and Dazzo, F. Association of *Azospirillum* with grass roots, *Appl. Environ. Microbiol. 39,* 219–226 (1980).
108. Korhonen, T.K., Haahtela, K., Ahonen, A.M., Rhen, M., Pere, A. and Tarkka, E. *in vitro* adhesion of associative nitrogen-fixing klebsiellas to plant roots, in *Second National Symposium on Nitrogen Fixation,* Helsinki, Finland (Editor, P. Uomala), Abstr. 19 (1982).
109. Callaham, D.A. and Torrey, J.G. The structural basis for infection of root hairs of *Trifolium repens* by *Rhizobium, Can. J. Bot. 59,* 1647–1656 (1981).
110. Nutman, P., Doncaster, C.C. and Dart, P.J. *Infection of clover by root-nodule bacteria,* British Film Institute, London (1973).
111. Libbenga, K.R. and Boggers, R.R. Root nodule morphogenesis, pp. 430–472, in *The Biology of Nitrogen Fixation* (Editor, A. Quispel), North-Holland Publishing Co., Amsterdam (1974).
112. Newcomb, W. Nodule morphogenesis and differentitation, pp. 247–298, in *International Review of Cytology* (Editor, S. Atherly), Suppl. 13, Academic Press, New York (1981).
113. Sutton, W.D., Pankhurst, C.E. and Craig, A.S. The *Rhizobium* bacteriod state, pp. 149–177, in *International Review of Cytology* (Editor, A. Atherly), Suppl. 13, Academic Press, New York (1981).
114. Urban, J.E. and Dazzo, F.B. Succinate-induced morphology of *Rhizobium trifolii* 0403 resembles that of bacteriods in clover nodules, *Appl. Environ. Microbiol. 44,* 219–226 (1982).
115. Truchet, G., Michel, M. and Denarie, J. Sequential analysis of the organogenesis of lucerne (*Medicago sativa*) root nodules using symbiotically defective mutants of *Rhizobium meliloti, Differentiation 16,* 163–172 (1980).
116. Verma, D.P., Ball, S., Guerin, C. and Wasamaker, J. Leghemoglobin biosynthesis in soybean root nodules. Characterization of the nascent and released peptides and the relative rate of synthesis of the major leghemoglobins, *Biochem. 18,* 476–483 (1979).
117. Gresshoff, P.M., Carroll, B., Mohapatra, S.S., Reporter, M., Shine, J. and Rolfe, B.G. Host factor control of nitrogenase function, pp. 209–212, in *Current Perspectives in Nitrogen Fixation* (Editors, H. Gibson and W.E. Newton), Australian Academy of Science, Canberra.
118. Imsande, J. Exchange of metabolites and energy between legume and *Rhizobium*, pp. 179–189, in *International Review of Cytology* (Editor, A. Atherly), Suppl. 13, Academic Press, New York (1981).

4. Stem Nodules

N.S. Subba Rao and M. Yatazawa

The ability to form root nodules by *Rhizobium* is restricted to some plants of the large botanical family leguminosea [1] and to the genus *Parasponia* (*Trema*) of ulmaceae [2]. In several other diverse families of the plant kingdom, the root nodulation habit has also been recorded but the causative organism in these instances (Ex *Alnus* and *Casuarina*) has been designated as *Frankia,* a genus of actinomycetales [3]. Some of these nodule-bearing plants are important nitrogen-fixing species of temperate and tropical forest ecosystems.

Nodule-like structures are also found on leaves of plants such as *Psychotria* and *Pavetta* (Rubiaceae) and various types of bacteria have been isolated from such nodulated leaves although the involvement of leaf nodules in nitrogen fixation has not been conclusively demonstrated. Nevertheless, leaf nodules have been shown to produce cytokinin-like substances which may contribute to chlorophyll synthesis and hence better plant growth [4, 5].

Interestingly enough, within leguminosea certain aquatic legumes and excess water-tolerant legumes are known to bear nitrogen-fixing nodules on stems. These stem nodules are caused by *Rhizobium.* Only some species of *Aeschynomene* and *Sesbania* are known to bear stem nodules (Fig. 1) and have been investigated in this regard [6-16]. Recently, renewed interest on these nodulating plants has been evinced in view of the potential benefit that could be derived if similar stem nodules could be induced on cultivated plants. In a recent survey on stem nodulation of legumes, nitrogen-fixing stem nodules (Fig. 2) have been reported on ceasalpinaceous plants (Yatazawa et al., unpublished). Similarly, nitrogen fixation has been recorded in warty lenticellate barks of various trees inhabited by enterobacteria [17-19].

This chapter reviews the current status of knowledge in this expanding area, stresses the importance of the subject and creates an interest to understand the physiology and biochemistry of these plants in depth because stem nodules provide an ideal means to study both photosynthesis and nitrogen fixation at close proximity.

Fig. 1. Stem nodules of *Aeschynomene aspera* in potted plants.

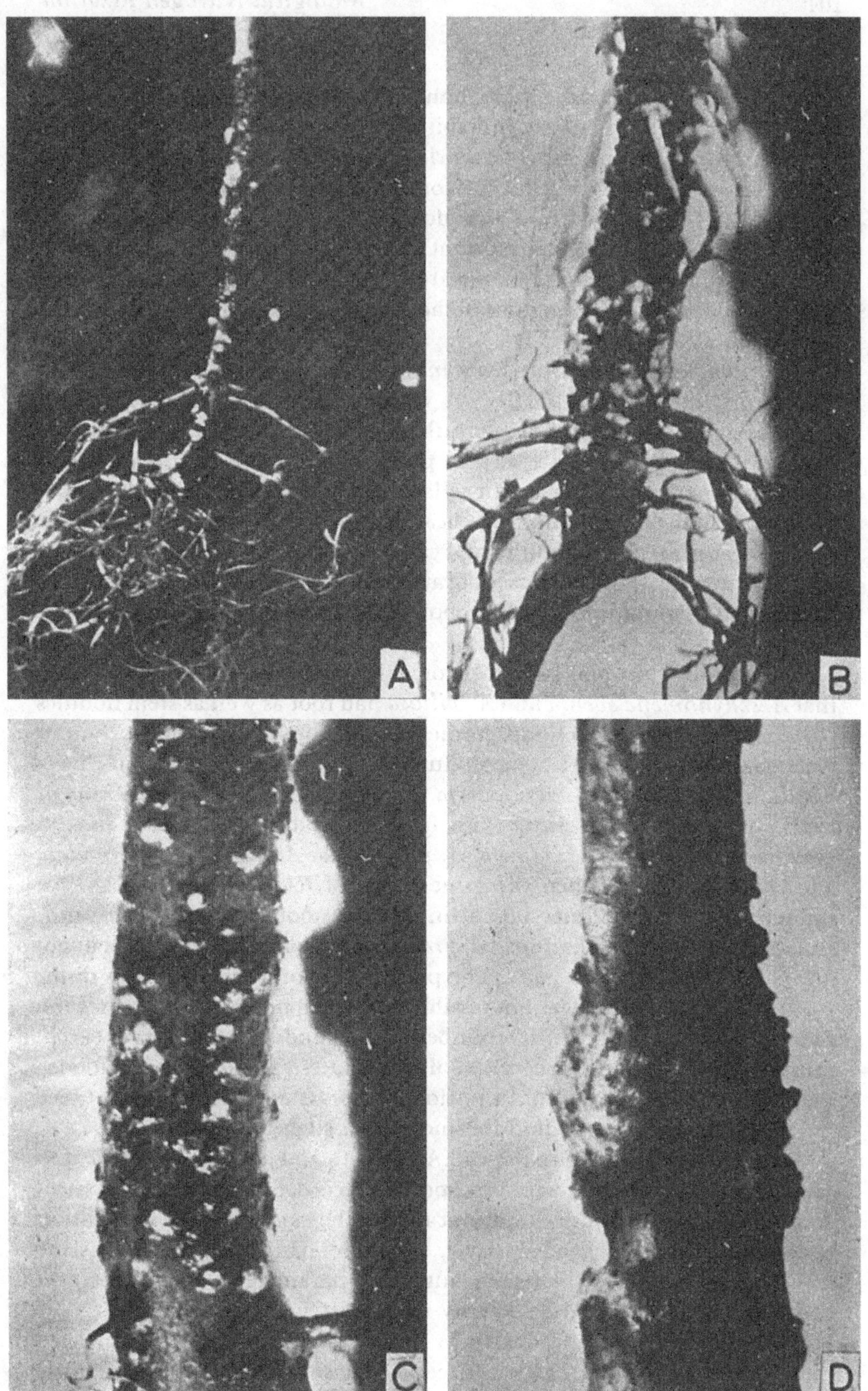

Fig. 2. Nodules on stem of *Cassia tora* observed by Dr. M. Yatazawa of Japan.

AESCHYNOMENE

The genus *Aeschynomene* has about 160 species which are primarily tropical in nature. They vary in habit from low growing herbs to tree-like shrubs growing up to 8 m. The world germplasm of *Aeschynomene* has been evaluated in Florida for performance and feasibility as a forage legume [20]. The evaluation report does not, however, mention the occurrence of stem nodules. One important forage legume in Florida has been *A. americana* and other potential ones are *A. brasiliana, A. elegans, A. evenia, A. falcata* and *A. villosa* but none of them have been described to possess the stem nodulation habit.

In India, two species of *Aeschynomene* have been encountered—*A. aspera* and *A. indica* [21]. They are widely distributed all over India. *A. aspera* is a relatively large bushy shrub whereas *A. indica* is herbaceous and relatively small. *A. aspera* is a 'pith plant' whose large stem is light and made up of parenchymatous pith cells and hence enables the plant to float on water. On the other hand, *A. indica* has a slender floating stem. Both the species bear green stem nodules and can withstand extreme drought when they are stunted. With the onset of rains, they grow profusely, flower and bear pods. Germination of seed is poor and the plants can be multiplied by cuttings.

A survey of rhizobial symbiosis on the Venezuelan Savannas revealed that *Aeschynomene evenia* and *A. villosa* had root as well as stem nodules [10]. In the U.S.A., green house studies revealed that a strain of *Rhizobium* isolated from submerged stem nodules of *A. indica* (BTA:1) effectively nodulated the roots of *A. denticulata, A. evenia, A. indica, A. pratensis, A. rudis, A. scabra, A. sensitiva,* but failed to nodulate *A. americana, A. brasiliana, A. elegans, A. falcata, A. fasicularia, A. histrix, A. paniculata* and *A. villosa* [20]. When the same strain of *Rhizobium* was used for induction of stem nodules on cuttings, stem nodules were visible on *A. scabra, A. denticulata, A. indica, A. pratensis, A. rudis* and *A. evenia* but not on *A. brasiliana* or *A. elegans*. The presence of combined nitrogen in the plant nutrient medium was not inhibitory to stem nodulation of *A. scabra*. Flooding had no effect on the number of stem nodules of *A. scabra* except for the size of the nodules because larger nodules appeared under flooded than non-flooded condition. In unflooded pots, the presence of combined nitrogen reduced root nodulation more than stem nodulation [22].

In the State of Florida in the U.S.A., growth and nitrogen fixation of *A. americana* was studied with no specific recorded observation of stem nodules. However, flooding did not reduce the extent of root nodulation but increased the nitrogenase activity of nodules. Under dry conditions, the dry weight of nodules increased but other parameters such as per cent nodulation and nitrogenase activity were drastically reduced [23].

In a study of root nodulation in *A. aspera* [12] from plants collected from Madurai, Southern India, fifteen isolates of *Rhizobium* were made from root and stem nodules and tested for nodulation on *A. aspera, Cajanus cajan* (pigeon pea), *Cicer arietinum* (chick pea), *Pisum sativum* (pea), *Trifolium repens* (clover), *Medicago sativa* (lucerne), *Lens culinaris* (lentil), *Glycine max* (soybean), *Vigna sinensis* (cowpea), *Vigna radiata* (mung bean), *Vigna mungo* (urd bean) and *Arachis hypogea* (pea nut). It was observed that *A. aspera* strains did not nodulate the roots of other plants tested and likewise none of the rhizobia from other cross-inoculation groups were able to nodulate *A. aspera.*

In plants collected from wet land rice fields in Tiruchirapally, southern India, both *A. aspera* and *A. indica* had stem as well as root nodules. Measurements of acetylene reduction activities (ARA) of field collected plants showed mean values (*n* moles C_2H_4/hr/dry weight of nodules) of 43.8 for stem nodules and 22.6 for root nodules of *A. aspera* while the values were 16.2 and 18.2, respectively for *A. indica* (Purushothaman and Subba Rao, unpublished).

Recently in a survey carried out in India on the occurrence and nature of stem nodules of *Aeschynomene* (Subba Rao, unpublished), both *A. aspera* and *A. indica* plants were examined on locations in Tamil Nadu (Madurai, Chidambaram), West Bengal (Haringhata) and New Delhi. In Chidambaram, the ecology of *A. aspera* was studied in detail during different seasons in a natural lake known as 'Veeranam lake'. During the dry season when the lake is dry, the shoot system of *A. aspera* plants gets telescoped with whorls of leaves truncated into small emergences on a rhizomatous stem which pushes an elongated tap root vertically down the soil in search of moisture. The plant survives the hot and humid summer under the shade of shrubs until the onset of monsoon when the plants develop into a profusely branched form with a spongy pith-like floating stem. In this particular ecosystem *A. aspera* did not bear root nodules. The plants collected from this place have been extensively grown in New Delhi in pots. In no instance, they bear root nodules but possess extensive stem nodules. On the other hand, *A. indica,* a relatively slender and smaller species, had root nodules only and was found to grow in wet land rice fields and in ditches or wherever it had access to moisture. Similar observations were made from plants in Delhi and West Bengal but the plants varied in the size of leaves, stem and pods. In a typical ecosystem in Madurai, *A. indica* plants developed both stem as well as root nodules and the plants exhibited wide variation in size of the shoot system and pod formation. There appeared to be wide variations in plant morphology, pod size and occurrence of nodules on stem and roots between locations. In general, the size and colour of seed (black to brown) varied between locations even within the same species and within the same plant. Hardly 20 per cent of the seed germinated and the establishment of plants from seed was difficult. Cuttings of *A. aspera* could however, be easily raised into plants in pots but only during the humid summer season with temperatures ranging from 26 to 42°C. The formation of green stem nodules could easily be observed within four to six weeks after planting of cuttings. The green cortex of nodules had chlorophyll whereas the inner central bacteroid zone developed pink to brown colouration probably due to leghaemoglobin. Isolation of *Rhizobium* from these nodules was fairly easy. Preliminary studies have revealed the existence of endosymbionts other than *Rhizobium* in stem nodules at some locations. This, however, needs confirmation.

In Japan, wetland rice fields support the growth of *A. indica* with large number of nitrogen fixing (acetylene reduction) stem and root nodules. *Rhizobium* isolates from stem nodules induced the formation of root nodules in fresh plants upon seed inoculation [25]. Further studies showed that the presence of combined nitrogen suppresses the development of stem nodules either in hydroponic culture solution or in soil. The formation of stem nodules in tap-water-fed cuttings was also poor, indicating other limiting factors in stem nodulation, possibly trace elements [25].

The developmental aspects of root and stem nodulation have been studied in detail in *A. indica.* The nodules occur singly and are widely distributed on the roots as well as stems, although they tend to cluster near the stem just above the water level. In size and shape both the root and stem nodules do not differ and they appear round and flat with an uneven surface and a broad basal attachment. The root nodules arise from the pericycle whereas the stem nodules arise from the outer cortex. Root hair development is scanty and no curling-effect can be seen. The root nodules arise from pericyclic cells in close proximity to the place of emergence of lateral roots. It is probable that the rupture of the tissue by the emergence of lateral roots creates points for entry of rhizobia. The formation of infection threads was not observed. In the submerged areas of the stem, lateral roots arise and provide niches for nodule primordia and there was no indication that nodule initials appeared in lenticels which are numerous on the aerial parts of the plant. There is, however, a need for intensive investigation on the mode of infection in these plants so as to throw more light on the initiation of stem nodules because it has been observed (Subba Rao, unpublished) that aerial stem nodules exist on regions where no indications were available about the origin of lateral roots.

SESBANIA

There are about 20 species of *Sesbania* distributed in the tropics. They are soft-wooded shrubs or herbs. The plants are grown as hedges, the wood is used for making gun-powder charcoal, the fibre made into ropes and the foliage used as cattle feed. *Sesbania rostrata* is an annual plant which grows in flooded soils of the Sahel region of West Africa during the rainy season, thriving in wetland rice fields. Stem nodulation in this plant has been reported [13-16]. Nodules occur at the sites of lenticels. They appear 2 m above water level and are usually speroidal and vary from 0.3 to 0.8 cm in size. The structural features of these nodules resemble other *Rhizobium*-induced nodules of soybean or cowpea, except that the nodule cortex is green due to chlorophyll. Exceptionally high ARA values (589 μm C_2H_4/h/plant) have been recorded in stem nodules. The root nodules which are smaller than their counterparts on stem exhibit less nitrogenase activity. A strain of *Rhizobium* isolated from *S. rostrata* nodulated its homologous host and produced effective nodules on roots and stem but induced ineffective nodules on *S. pachycarpa* and *S. aculeata* and did not nodulate roots of *Aeschynomene* sp. and *Macroptilium atropurpureum*. When a strain of *Rhizobium* isolated from nodules of *S. pachycarpa* was inoculated to *S. rostrata,* it produced ineffective nodules whereas strains of *Rhizobium* isolated from stem nodules of *Aeschynomene* sp. and cowpea did not induce nodulation on *S. rostrata*.

In another study [13], *S. rostrata* was grown in test tubes hydroponically in the presence and absence of combined nitrogen to study the effect of combined nitrogen on root and stem nodulation habit. Root nodulation was strongly inhibited by combined nitrogen while stem nodulation and acetylene reduction of stem nodules was enhanced. In *S. rostrata,* the sites of stem nodules have been designated as stem mamillae which are epidermal domes pierced by an incipient root. When the cells of the mamillae are infected by specific *Rhizobium,* the mamillae develop into root nodules [14]. In a comparative study on the mode of infection by *Rhizobium* in stem nodule calluses raised on a nutrient medium and in intact plants, it was observed that the infection ceased in calluses with the formation of pseudo-infection threads in the intercellular spaces of the callus cells whereas in intact plants, the proliferation of bacterial cells in the intercellular spaces of the cortex was observed followed by the protrusion of the typical infection threads accompanied by the liberation of bacteria into host cells [15].

WARTY LENTICELLATE BARKS

Based on ARA values, warty lenticellate barks of the several trees from various temperate and tropical areas have been shown to be of potential value for nitrogen accretion in the biosphere [17-19]. The ARA values (n mole C_2H_2 reduced per g (F.W./h)) for different species were as follows: *Acacia confusa* (0.3); *Albizia julibrissin* (7.0); *Amorpha fruticosa* (3.0); *Bauhinia purpurea* (0.3); *Ceasalpinia pulcherrima* (0.5); *Cassia siamea* (9.0); *Cyanomitra ramiflora* (17.0); *Inga laurina* (18.0); *Pterocarpus indicus* (0.2); *Robinia pseudo-acacia* (9.0); *Sophora tomentosa* (1.0); *Wisteria brachybotrys* (3.0); *Bruguiera gymorrhiza* (8.0); *Celtis boninensis* (2.0); *Clerodendron trichotomum* (5.0); *Ilex crenata* (13.0); *Ilex pedunculosa* (7.0); *Mallotus japonicus* (8.0); *Prunus donorium* (1.0); *Rhizophora mucronata* (8.0); *Weigela decora* (1.0).

Bacterial colonies isolated from the barks of *Bruguiera* and *Rhizophora* showed reasonably high nitrogenase activities on a nitrogen-free medium.

The nitrogen-fixing activities were confirmed by means of $^{15}N_2$ techniques. Based on routine tests, the bacteria were identified as *Enterobacter cloacae, E. aerogenes* and *Klebsiella pneumoniae* [16].

Potentialities: As green manure crops, *Sesbania* and *Aeschynomene* appear to be good candidates in wetland rice cultivation. In Senegal, it has been estimated that *S. rostrata* contributes 267 kg N/ha of which one-third is transferred to the rice crop and the remainder is left behind in the soil for subsequent utilization. The grain yield of rice was estimated at 3.72 t/ha by the application of *S. rostrata* [26]. *Aeschynomene* spp. have also a similar potential in supplementing fertilizer nitrogen application in the production technology of rice cultivation. The practice of green manuring using other well-known leguminous crops has been recognized but those legumes which bear stem nodules possess additional advantages over others in their twin abilities to use both the nodules on stem as well as on roots to harness atmospheric nitrogen. A more recent observation (Yatazawa et al. unpublished) on the occurrence of stem nodules on *Cassia tora* plants is a case in point to demonstrate the need to survey and explore new habitats for genotypes of plants which bear stem nodules.

In inter- and multiple-cropping, legumes are often used under irrigated conditions. There is a distinct possibility for selecting genotypes of legume species which possess a tendency to bear nodules on the hypocotyl or the crown region of the shoot system so that effective utilization of exposed plant surface is done to maximize biological nitrogen fixation.

At the level of fundamental research, stem nodules offer a challenging and convenient system to investigate the inter-relationship between photosynthesis and biological nitrogen fixation. The cortex of stem nodules is green due to chlorophyll and the bacteroid tissue is pink possibly due to leghaemoglobin. The transfer of carbon skeletons from the cortex into the bacteroid tissue for the process of nitrogen fixation can easily be traced and measured with accuracy by the use of labelled carbon. The stem nodules can easily be peeled off at the base for laboratory experiments.

The genetic basis for the stem nodulation habit may have to be understood in great detail so as to induce stem nodules in cultivated crops to obtain better fixation of atmospheric nitrogen.

REFERENCES

1. Fred, E.B., Baldwin, I.L. and McCoy. *Root Nodule Bacteria and Leguminous Plants,* University of Wisconsin, Wisconsin, U.S.A. (1932).
2. Trinick, M.J. Structure of nitrogen fixing root nodules formed by *Rhizobium* on root of *Parasponia andersonii, Canadian Journal of Microbiology 25,* 565-578 (1979).

3. Becking, J.H. Key to family Frankiaceae and genus *Frankia,* pp. 702-706, 871-872, in *Bergey's Manual of Determinative Bacteriology* (8th edition, Editors, R.E. Buchanan and N.E. Gibbons), Williams and Wilkins, Baltimore, U.S.A. (1974).
4. Becking, J.H. The physiological significance of the leaf nodules of *Psychotria, Plant and Soil, Special Volume* 361-374 (1971).
5. Lersten, N.R. and Horner, H.T. Jr. Bacterial leaf nodule symbiosis in angiosperms with emphasis on Rubiaceae and Myrsinaceae, *The Botanical Review 42,* 145-214 (1976).
6. Hagerup, O. En hygrofil Baelgplante (*A. aspera* L.). med Bakteriek-nolde paa Staengelen. *Dan. Bot. Ark. 15* (14), 1-9 (1928).
7. Suessenguth, K. von and Beyerle, R. Über Bakterienknollchen am Spross von *A. Paniculata* Willd., *Hedwigia 75,* 234-237 (1936).
8. Arora, N. Morphological development of the root and stem nodules of *Aeschynomene indica, Phytomorphology 4,* 211-216 (1954).
9. Jenik, J. and Kublikova, J. Root system of tropical trees, 3. The heterorhizis of *Aeschynomene elaphoroxylon* (Guill et Perr) Taub, *Preslia* (Praha) *41,* 220-226 (1969).
10. Barrios, S. and Gonzalez, V. Rhizobial symbiosis on Venezuelan Savannas, *Plant and Soil 34,* 707-719 (1971).
11. Yatazawa, M. and Yoshida, S. Stem nodules in *Aeschynomene indica* and their capacity to fix nitrogen, *Physiologia Plantarum 45,* 293-295 (1979).
12. Subba Rao, N.S., Tilak, K.V.B.R. and Singh, C.S. Root nodulation studies in *Aeschynomene aspera, Plant and Soil 56,* 491-494 (1980).
13. Dreyfus, B.L. and Dommergues, Y.R. Non-inhibition de la fixation d'azote atmospherique par l'azote combine' chez une legumineuse a nodules caulinaires, *Sesbania rostrata, C.R. Acad. Sci.* (Paris) *291,* 767-770 (1980).
14. Duhoux, E. and Dreyfus, B. Nature des sites d'infection par le *Rhizobium* de la tige de la legumineuse, *Sesbania rostrata, C.R. Acad. Sci.* (Paris) *294,* 407-411 (1982).
15. Duhoux, E. and Alazard, D. Culture *in vitro* de nodules de *Sesbania rostrata:* mode d'infection des tissus neoformes et comparaison avec l'infection chez la plante, *C.R. Acad. Sci.* (Paris) *269,* 93-100 (1983).
16. Dreyfus, B.L. and Dommergues, Y.R. Nitrogen-fixing nodules induced by *Rhizobium* on the stem of the tropical legume *Sesbania rostrata, FEMS Microbiology Letters 10,* 313-317 (1981).
17. Yatazawa, M., Hambali, G.G. and Uchino, F. Nitrogen fixing activity in warty lenticellate tree barks, *Soil Science and Plant Nutrition* (in press).
18. Yatazawa, M., Yoshida, S., Maeda, E., Sasakawa, H., Uchino, F. and Hambali, G.G. N-fixing activities in aerial part of plant stems, in *Symposium on Nitrogen Mobility in Ecosystems,* XIII International

Botanical Congress, University of Sydney (1981).

19. Yatazawa, M., Hambali, G.G. and Uchino, F. Partial oxygen pressure and corresponding nitrogen fixing activity in warty tree barks, *Soil Science and Plant Nutrition* (in press).
20. Quesenberry, K.H. and Ocumpaugh, W.R. Forage potential of *Aeschynomene* species in north central Florida, *Soil and Crop Science Society of Florida, Proceedings 40,* 160-162 (1981).
21. Roxburghw, *Flora Indica,* Today & Tomorrow Printers and Publishers, New Delhi (1971).
22. Eaglesham, A.R.J. and Szalay, A.A. Aerial stem nodules of *Aeschynomene* spp., *Plant Science Letters 29,* 265-272 (1983).
23. Albert, S.L., Bennett, J.M. and Quesenberry, K.H. Growth and nitrogen fixation of *Aeschynomene* under water stressed conditions, *Plant and Soil 60,* 309-315 (1981).
24. Yatazawa, M. and Yoshida, S. Stem nodules in *Aeschynomene indica* and their capacity of nitrogen fixation, *Physiologia Plantarum 45* (2), 293-295 (1979).
25. Yatazawa, M. and Haryanto, S. Development of upper stem nodules in *Aeschynomene indica* under experimental conditions, *Soil Science and Plant Nutrition* (Japan) *26* (2), 317-319 (1980).
26. Rinaudo, G., Dreyfus, B. and Dommergues, Y. *Sesbania rostrata* green manure and the nitrogen content of rice crop and soil, *Soil Biology and Biochemistry 15,* 111-113 (1983).

5. Oxygen Control Mechanisms in Nitrogen-Fixing Systems

B.D. Shaw

INTRODUCTION

Molecular nitrogen (N_2, dinitrogen) is a particularly stable molecule [1, 2]. Accordingly, nitrogen fixation, whether oxidative during lightning storms and aurorae [3], or reductive as in the industrial synthesis of ammonia and biological nitrogen fixation, requires a high input of energy. This energy is variously obtained. Excluding cyanobacteria (blue-green algae) at least 30 genera from 12 families of phototrophic and chemotrophic bacteria have been reported to fix nitrogen [4].

Biological nitrogen fixation is extremely susceptible to oxygen 'poisoning' and loss of function. Nevertheless, several nitrogen-fixing bacteria obtain energy by respiration, with oxygen as the final oxidant. Others may have to contend with the oxygen produced during photosynthesis.

This chapter covers the effects exerted by oxygen upon dinitrogen reduction, and the control of oxygen level to permit nitrogen fixation to proceed. The literature surveyed covered the mid-1970s until January 1983, but older references have been given wherever considered appropriate. Only a small number of the nitrogen-fixing genera have been mentioned: *Azospirillum* (formerly *Spirillum* [5]), *Azotobacter, Chromatium, Derxia, Frankia, Klebsiella, Rhizobium* (now *Rhizobium* and *Bradyrhizobium* [6]), *Rhodopseudomonas* and *Rhodospirillum*.

SOME PROPERTIES OF OXYGEN

Chemical forms

The reduction of a molecule of oxygen (O_2) to water requires four electrons. The stepwise addition of these electrons is chemically the most favourable sequence [7, 8, 9]. The superoxide radical $O_2^{-\cdot}$ (usually written as $O_2^{-\cdot}$), hydrogen peroxide H_2O_2, and the hydroxyl radical $OH^{\cdot}$. are intermediates in this reduction. They are all toxic, the latter particularly so [8, 10]. Superoxide radicals are removed enzymatically by superoxide dismutases, and hydrogen peroxide by catalases and peroxidases. The removal of these species prevents the formation of the hydroxyl radical, which is extremely reactive.

Some oxidative enzymes transfer two or four electrons to oxygen without releasing intermediates. The semi-reduced oxygen species however, can be produced by certain flavoproteins [7] and by autoxidations of thiols, hydro-quinones, reduced ferrodoxins, ferrous [Fe(II)] salts [10, 11] and a number of other compounds [7]. These chemicals are found within nitrogen-fixing organisms, or in their immediate environment. A review of hydrogen peroxide generation in membranes [9], while mentioning no examples of studies made with nitrogen-fixing bacteria, indicates that H_2O_2 production is a widespread property of membrane systems, particularly of microsomes and the plasmalemma. Lignin formation in plant cell walls requires H_2O_2 [9], and $OH^{\cdot}$ may also be required for lignin degradation [12].

Oxygen may also be involved in the production of reactive species besides its reduction products. In animal systems, if chloride ions are oxidized by H_2O_2 and peroxidase, hypochlorous acid and other active, toxic species of chlorine are formed [8, 10].

Solubility

The solubility of oxygen depends on the partial pressure of oxygen in the gas phase (the pO_2) and hence on the atmospheric pressure [13, 14]. The solubility of oxygen in aqueous media is very dependent on the temperature and solute composition [13, 14, 15]. It decreases with increasing temperature and increasing ionic strength. Oxygen is more soluble in lipids than in water [16, 17, 18]. Within the cell dissolved oxygen would be expected to partition between cell water, membranes and lipid deposits, perhaps including poly-β-hydroxybutyrate. The peribacteroid membranes in lupin nodules are richer in lipids than the other cell membranes, and their possible effect on gas transport has been discussed [19]. The solubility of oxygen in lipid increases with temperature [18]. Nitrogen is less soluble than oxygen in both water and lipid, but its partition coefficient into lipid from water is slightly greater [16].

A variety of units have been used in literature to describe pO_2, and they may be interconverted as follows. 101.325 kPa (S.I. units) = 1013.25 mbar = 1 atmosphere = 760 mm Hg. In this chapter, values have been expressed as kilopascals, kPa.

Transport

The solubility of oxygen is defined as the equilibrium concentration reached when oxygen movement to and from the solvent is equal. The flux density of oxygen (the rate of movement of a given amount through a specified area) to a population of cells is usually more relevant. Many experiments with nitrogen-fixing bacteria have used suspensions of cells shaken in gas. The flux of oxygen across the gas-liquid interface depends not only on solubility but also on mixing in the liquid near the interface, the ratio of interfacial area to liquid volume, and the removal rate by the cells [14, 15, 20, 21]. In the absence of measurements of the oxygen concentration in solution, the volumetric proportion of oxygen in the gas phase provides only an approximate indication of the possible concentration of dissolved oxygen. Air-saturated water at 25°C, 101 kPa pressure, is 258 μM in oxygen.

In calculating oxygen flux into tissues, Fick's first law has often been employed [36–38]. This states that

$$\text{flux density} = D_0 \frac{\partial c}{\partial x} \qquad (1)$$

with D_0 being the molecular diffusion coefficient of oxygen (0.18 $cm^2 s^{-1}$ in air [34, 37, 38] and 1.8×10^{-5} cm^2 s^{-1} in water [37] at 25°C), t is time, x distance. To describe the movement of oxygen in soils, a modified form of Fick's second law has been used [22, 23] with the terms as above, and α to account for the oxygen removed by other sites in the soil:

$$\frac{\partial c}{\partial t} = D\left(\frac{\partial^2 c}{\partial x^2}\right) = \alpha \qquad (2)$$

This removal of oxygen may be a result of microbial or root respiration, or by chemical oxidations such as the conversion of Fe (II) to Fe (III) ions. D does not have the usual meaning of gaseous oxygen molecular diffusion coefficient. Instead D is called the 'apparent diffusivity' because it is not universally constant for all soils or boundary conditions. Here D is a function both of the porosity of the soil and the twisted pathways of the airspaces (the tortuosity) through which the oxygen diffuses [22, 24]. These factors may vary for any particular soil, depending on the water content [26–28], the tillage methods used [29], the activity of plants and burrowing creatures [30, 31] and microorganisms [32]. The Penman expression for diffusivity [33] is $D = 0.6\ SD_0$, where S is the air-filled porosity and D_0 as above, is discussed elsewhere [22, 34].

It is evident that particular local soil conditions play a very important role in determining oxygen movement to microorganisms and plant roots, and this makes prediction difficult. Measurement is not easy either, because oxygen electrodes measure uniformly only in rapidly-stirred fluids [35, 154], whereas the distribution of soil oxygen may be extremely heterogeneous.

CONTROL BY OXYGEN

Exposure of nitrogen-fixing bacteria or of extracted components of the nitrogen-fixing system to unphysiological concentrations of oxygen has shown that oxygen affects a number of biochemical systems. For the purpose of this section, the promotion of such effects will be termed 'control'.

Control of enzyme activity

Nitrogenase

In all bacteria where the enzyme has been prepared from cell extracts, dinitrogen is reduced to ammonia by the joint action of two proteins, collectively called nitrogenase. The dinitrogen binding site is contained in the MoFe protein. This is a tetramer of about 200,000 molecular weight. It is a molybdenum-iron-sulphur protein. The dimeric Fe protein of molecular weight of about 60,000 is an iron-sulphur protein. Both proteins are irreversibly inactivated by oxygen. At 30°C the half-lives in air for MoFe proteins from *Bradyrhizobium japonicum, Klebsiella pneumoniae* and *Azotobacter chroococcum* were 4.5, 8 and 10 minutes, respectively. The Fe proteins of *K. pneumoniae* and *A. chroococcum* had half-lives of approximately 45 seconds in air, and even briefer with ATP present [39]. The isolated iron-molybdenum cofactor from nitrogenase was inactivated within a minute in air [40]. Robson and Postgate concluded that oxygen inactivated nitrogenase by attacking the metal-sulphur groups [21].

Ancillary proteins

The *nif* gene cluster of *K. pneumoniae* consists of 17 linked genes. Of its gene products, four proteins, including nitrogenase, have been characterized. The *nifJ* gene codes for an iron sulphur protein [41]. This protein is involved in the acetylene reduction supported by pyruvate: it may transfer electrons from a pyruvate metabolite to the *nifF* protein. It loses activity irreversibly after exposure to oxygen, with a half-life of several hours. The *nifF* protein has been purified from a *nifJ*⁻strain [42]. It is a flavoprotein containing FMN which is stable to oxygen. Strains which are *nifF* lack nitrogenase activity, but it can be activated by ATP and dithionite.

The Fe-protein of *Rhodospirillum rubrum* has an inactive form which can be activated by an Fe-protein activating enzyme. This enzyme is oxygen sensitive [43, 44]. It may also be mentioned that the uptake hydrogenases of both *Anabaena* and *Bradyrhizobium japonicum* are inactivated by oxygen *in vitro* [45]. These enzymes recycle hydrogen produced by nitrogenase, and are induced under nitrogen-fixing conditions.

Metabolic support

It has been observed that nitrogen-fixing cultures of *Bradyrhizobium japonicum* at pO_2 of 0.2 kPa had one form of glutamine synthetase, but two forms in aerobic cultures [46]. The low oxygen form, GSI, has been implicated in the control of nitrogen fixation in *Bradyrhizobium* sp. strain 32H1 [47]. Data on the effect of oxygen on metabolic pathways are sparse, but two studies may be noted. Haem and tetrapyrrole production by *Bradyrhizobium japonicum* cultures increased at low pO_2 as did the concentrations of some relevant enzymes [48]. When rhizobial haem is produced in legume nodules similar conditions may exist. In *Azotobacter beijerinckii* the

Entner Doudoroff pathway was the major route of glucose metabolism at all oxygen concentrations tested [49]. Other patterns of oxygen regulation of metabolism are possible [50] but studies of more nitrogen-fixing bacteria are required to discover whether they occur.

Control of protein synthesis

Nitrogenase

Since St. John et al. [51] reported that *Klebsiella pneumoniae* did not form nitrogenase antigens under aerobic conditions there have been more detailed studies of oxygen repression in this organism. Eady et al. [52] showed that exposure of nitrogen-fixing cultures to 5 kPa oxygen caused repression of nitrogenase synthesis. The synthesis of MoFe protein subunits, of which there are two types, was repressed with a half-life of 33 min; the half-life of Fe protein synthesis was 30 min. In air the repression was more rapid, and synthesis of some Fe protein was prolonged. Degradation of pre-formed nitrogenase occurred slowly with chloramphenicol present [52], although other workers stated it was more rapid without the inhibitor [53]. Examination of the gene regulation of nitrogenase and other oxygen-sensitive proteins has been facilitated by the use of gene fusions. In a study of *nif* gene regulation made by this means, the *nifHDK* operon (nitrogenase genes) was sensitive to 1 μM oxygen, in both *nif+* and *nif* deletion backgrounds. It was concluded that the nitrogenase structural genes were unlikely to mediate oxygen repression.

Because gene fusions made with phage Mu are unstable, Hill et al. [55] used revertants of *nifL*mutants to show that *nifL* was involved in the oxygen control of nitrogenase genes. While the synthesis of *nifH*, *nifD* and *nifK* gene products (the polypeptides of nitrogenase) in the wild type declined steadily at 3 and 6 μM oxygen, in a *nif*$^+$ revertant of a *nifL* mutant the synthesis of these proteins first declined, then climbed to initial levels. Transduction from this mutant allowed colonies carrying *nifH*::*lacZ* to express β galactosidase in air. Oxygen repression could be restored by a Nif$^+$ plasmid [55]. Concentrations of oxygen below those of previous studies (i.e. 0.02–3 μM) inhibited nitrogenase derepression in a strain containing a chromosomal *nifH*::*lacZ* fusion and a Nif$^+$ plasmid [56]. Initially derepression was more rapid in samples with very low oxygen concentrations, but after several hours the anaerobic treatments had the greatest derepressed activity. The oxygen concentration for 50 per cent inhibition was similar to the binding constant of the main respiratory oxidase. It was suggested that this oxidase, or an oxidized component of the electron transport pathway, triggered *nifL* repression of the nitrogenase proteins [56].

Oxygen repression of MoFe protein synthesis has been observed in *Bradyrhizobium japonicum* [57], and of both nitrogenase proteins in *Azotobacter chroococcum* [58] and *Bradyrhizobium* sp. (*Lupinus*) [59] but not in *Rhizobium leguminosarum* [60]. The synthesis of Fe protein (but not MoFe protein) appeared to be repressed in *R. leguminosarum* bacteroids from waterlogged pea nodules [61], perhaps due to insufficient oxygen. In *Azotobacter chroococcum* the synthesis of all three nitrogenase polypeptides was abruptly halted by 50 kPa oxygen and resumed at a higher rate when the cells were returned to 10 kPa oxygen [58]. In the *Bradyrhizobium* strains examined MoFe protein synthesis was slowly repressed after exposure of free-living cultures [57] and bacteroids [59] to air. Fe protein synthesis was repressed more rapidly under these conditions [59]. This non-coordinate repression by oxygen contrasts with the coordinate repression observed in the other bacteria discussed. How the abrupt changes in oxygen concentration used in these studies interact with cellular protective mechanisms is unknown but may be relevant.

It has been stated that oxygen repression prevents the wasteful synthesis of nitrogenase [21, 51]. The half-life of nitrogenase in *Rhizobium leguminosarum* bacteroids is about two days [62]. It may be noted that in a few minutes of nitrogenase activity as much ATP is used as that required for its gene transcription and translation. Transcription will cost the least energy because the half-life of *nif* mRNA is long. It has been measured at 18 min in *Klebsiella pneumoniae,* although it was less in air [63]. In *Rhizobium leguminosarum* bacteroids it was estimated as 1.6 minutes by rifampicin-inhibited translation [60]. Aerobic oxidation of rifampicin can generate superoxide [64], but whether that occurs in rhizobia is unknown.

Ancillary proteins

In *Klebsiella pneumoniae* gene fusion strains, synthesis of β galactosidase from all the *nif* promoters was inhibited by six-hour aeration. The *nifL A* operon was much less sensitive to oxygen repression than the *nifHDK* operon [54]. Results with *nifL* revertants led to the hypothesis that the *nif L* gene product was involved in the oxygen regulation of nitrogen fixation as a negative effector [55]. Mutations which relieved the oxygen repression of *nifHDK* did the same for *nifJ* and *nifEN* [55]. MacNeil et al. [65] suggested that factors besides the *nifL* product might be involved in oxygen regulation. They reported that a *nifL* :: *lacZ* fusion strain (a Mu lysogen) was still partially repressed in aerobic media, and this was discussed further in a review [66]. Preformed proteins from *nif* genes were degraded under aerobic conditions, some more rapidly than others [53].

Transfer of *Bradyrhizobium japonicum* cells from 1 kPa oxygen to air inhibited the induction of hydrogenase activity in a similar fashion to protein synthesis inhibitors [67].

CONTROL OF OXYGEN

While some of the proteins required for nitrogen fixation require an anoxic location, only some nitrogen-fixing bacteria are anaerobes. The various means by which this anoxic location is provided have been discussed in detail in other reviews [21, 68-70]. Gallon categorized these as: avoidance of oxygen, physical barriers to oxygen, metabolic removal of oxygen, conformational protection and nitrogenase synthesis [70]. Only selected aspects will be discussed here.

Control by physical location

Free-living and associative bacteria

The temporal and spatial segregation of nitrogenase in cyanobacteria have been discussed elsewhere [71], and provide good examples of control by physical barriers to diffusion as well. The nitrogen fixation of bacteria is adapted to some unusual environments [72-74], the pO_2 being one factor which will determine what organisms may compete successfully. In the case of *Azospirillum brasilense* the bacteria fix nitrogen at 0.5-0.7 kPa oxygen [75, 76] and actively move to a suitable oxygen environment [77]. This can be seen in cultures as the movement of a pellicle of bacteria [78, 79]. Free-living nitrogen fixation in rhizobia occurs at a defined depth in soft-agar layers which delimits a suitable oxygen flux [80].

Plant symbioses

The plant symbioses which are considered in this section almost all involve the root zone. Let us first consider some aspects of oxygen movement there. Stolzy and Letey [35] concluded that oxygen fluxes of around 1×10^{-10} mol $cm^{-2}s^{-1}$ were required for root growth in a number of plants, with greater rates being optimal. However, these rates may not be reached in a number of unsaturated soil conditions, including moderate rates of irrigation [26, 27]. Nitrogen-fixing nodules are likely to be even more affected, at least in the short term. Soybean acetylene reduction in pot-grown plants does adapt to low pO_2 [82], but comparison of these results with oxygen fluxes under field conditions needs to be made as diffusivities in the field may be much lower. Other reviews [69, 83, 84] have emphasized the deleterious effects of waterlogging on some nitrogen-fixing symbioses. Less extreme water potentials may still depress potential nitrogen fixation by restricting oxygen diffusion.

It has been known for over 40 years [85, 86] that respiration in legume nodules is limited by oxygen supply. More recently, several authors have emphasized this conclusion [36, 87, 88]. Bergersen [87] found that oxygen uptake of excised soybean nodules increased with pO_2 in two stages. He suggested that this resulted from progressive saturation of plant and bacteroid respiration, with a diffusion barrier between the two. Tjepkema and Yocum [36] concluded that nodules were diffusion-limited for oxygen, in order to explain their biphasic nodule respiration curves. They observed that the Q_{10} for soybean nodule respiration was 1.07 (between 13°C and 23°C). This is characteristic of a non-enzymatic process, which they proposed was diffusion. Similar Q_{10} values for these conditions were found in nodulated French beans as well as soybeans [89], but Arrhenius plots of respiration rates were biphasic. Below about 10°C, a Q_{10} of 2 applied. This Q_{10} is typical of enzymatic limitation.

While aquatic plants including rice [90] can transport oxygen to the roots, anoxic soybean nodules showed no acetylene reduction when the roots were exposed to extra oxygen [91, 92]. Approximately 200 $cm^3 h^{-1} g^{-1}$ of air-saturated water would have to flow through the tissue to support a typical acetylene reduction rate [93] of 15 μ mol $h^{-1} g^{-1}$. The translocation of fluid in phloem is likely to be very much slower [34, 94], given the area of nodule vascular tissue [95].

There are different hypotheses concerning the site of diffusion resistance in nodules [36, 87, 96]. Tjepkema and Yocum [36] argued that the cortex was the site of resistance, assuming that it contained insufficient air spaces, and that gaseous oxygen had to diffuse through a liquid barrier. Measurements of the spatial distribution of oxygen within the nodule, made with oxygen microelectrodes, showed a sharp change between the cortex and the bacteroid zone [97]. Bergersen and Goodchild [96] described detailed studies of the network of intercellular spaces within soybean nodules. These spaces were filled with gas. They were fewer in number in the cortex, with a wider range of areas (0.2-9 μm^2) than in the zone containing bacteroids. There the spaces were 1-5 μm^2 in area and more numerous, occupying 2.4 per cent of the zone. It proved difficult to trace spaces across the scleroid layer in the cortex. Free diffusion does not occur in very small pores [34] because the average free path of the molecules is comparable to the pore diameter. The ratio of these two lengths, termed the Knudsen length [98, 99], varies from about 0.1 to 0.02 for the areas mentioned. However, the corrections due to this effect may be small.

Published tests of the diffusion barrier hypothesis have assumed free diffusion [37, 38]. Fick's law was used, with the molecular diffusivity modified to account for the fraction of nodule space occupied by intercellular spaces. Tortuosity was not considered. However, because the spaces are predominantly radial in soybean [96], tortuosity may be less important than in the soil.

A number of parameters needed for constructing a mathematical model of oxygen flux in nodules have been measured: bacteroid respiration rate in relation to acetylene reduction [100, 101], tissue respiration rate [36, 102], nodule acetylene reduction rate [93, 103], oxygen concentration at the nodule centre [88], nodule size [36, 91, 102] and intercellular space dimensions [96]. Bergersen [102, 104] has integrated the biochemical data into a description of nodule function which repays detailed study. In later papers, Bergersen and Turner [100, 105] suggested that the free oxygen concentration in the tissue increased with pO_2 switching the bacteroids to a less efficient oxidase system. Tests of this hypothesis require inclusion of leghaemoglobin in the model [96].

The corralloid nodules formed by rhizobia on the roots of *Parasponia* trees [106] have central vascular tissue surrounded by infected cells in which the rhizobia remain within infection threads [107]. Impeded oxygen diffusion in the inner cortex has been observed [38], as predicted by Trinick [107]. Indeed the major features of oxygen control in legume nodules are shared by *Parasponia* nodules. These are: diffusion-limited respiration in the infected zone caused by a diffusion barrier in the cortex, and possession of protein which reversibly binds oxygen. In the most active part of *Parasponia* nodules, nitrogen fixation is comparable to the activity in legumes. It decreases from tip to base [108]. Acetylene reduction in the nodules had a narrow pO_2 optimum [108]. It was suggested that *Parasponia* nodules had a less efficient buffering system for oxygen than legumes.

Actinorhizal nodules resemble lateral roots morphologically [109] and possess continuous intercellular airspaces from the atmosphere to the infected tissue [38]. In culture the endophytes form vesicles and fix nitrogen at pO_2's of 5 to 20 kPa [110–112]. The oxygen response of intact nodules is similar to this [110, 111]. Measured as the ratio of respiration rate to acetylene reduction, the metabolic cost for actinorhizal nodules was similar to that for legume nodules [109]. This indicates that respiratory protection of nitrogen fixation is unlikely. It does not establish firmly that the energy cost in terms of reduced carbon used is the same, because non-aerobic energy metabolism [113] was not considered.

As sites for oxygen effects, it is the cells filled with rhizobia which have received the greatest attention. They contain a haemoglobin in legume and *Parasponia* nodules. In legumes, this protein is called leghaemoglobin. It has been the subject of recent reviews [95, 104, 114–116]. Leghaemoglobin at 15–25 mg ml^{-1} is present in the infected soybean nodule cells [93]. This high concentration has caused difficulties in studies of its localization [25, 115]. The average oxygenation of leghaemoglobin was 20 per cent in soybean nodules [117], which would leave 10 nM oxygen in solution [118]. *Bradyrhizobium japonicum* bacteroids have an efficiently-coupled respiratory oxidase functioning at this oxygen concentration [100, 105]. It supports high rates of acetylene reduction with oxyleghaemoglobin [101]. The flux of oxygen by free diffusion of 10 nM oxygen would be less than 4×10^{-20} mol s^{-1} per bacteroid (assuming equation 1, a bacteroid area of 2×10^{-8} cm^2 [96], a diffusion length of 1×10^{-4} cm [96] and $c = 0$ at the bacteroid surface). This could support <50 per cent of the observed respiration rate [102]. It should be noted that the value is approximate because the calculation takes no account of the boundary layer effects around bacteroids [119]. Wittenberg et al. [120] used the Wyman equation to describe the transport of oxygen by the diffusion of oxyleghaemoglobin itself. There are now sufficient values in the literature to make the calculation [96, 117, 120]. The transported oxygen flux would be about 1×10^{-17} mol s^{-1} per bacteroid (assuming leghaemoglobin was completely deoxygenated at the bacteroid surface and with the dimensions as above). Bergersen [102] concluded that the respiration rate of bacteroids in the nodule was $3\text{-}7\times10^{-20}$ mol s^{-1} per bacteroid. Thus the calculated flux of oxygen transported by the diffusion of oxyleghaemoglobin would appear sufficient to account for the observed respiration rate. However, boundary layer effects may render this 'facilitated diffusion' less efficient than these simple calculations would indicate [104, 114, 119]. It has been emphasized that leghaemoglobin will stabilize the oxygen gradient across the nodule [114, 119, 120] and the concentration at which oxygen is supplied to the bacteroids. The spatial separation of the bacteroids may also be important in maintaining diffusion paths (F.J. Bergersen, personal communication).

A haemoglobin has been extracted from nodules of *Parasponia andersonii* under anaerobic conditions, with polyphenol oxidation prevented by soluble polyvinylpyrrollidone [121]. It is notable that the haemoglobin was not detected in *Parasponia rugosa* [106] nodules when insoluble polyvinylpyrrollidone was used in the extraction medium [122]. The yield of haemoglobin in the *Parasponia* nodules was about one-third that of leghaemoglobin elicited by the same rhizobial strain in *Vigna unguiculata*. The kinetic properties of the *Parasponia* haemoglobin seem sufficient to allow oxygen transport to the rhizobia [121]. The necessary conditions for oxygen transport and stabilization appear to be satisfied, but their sufficiency has yet to be established. The diffusivity of the haemoglobin will be less than that for leghaemoglobin because *Parasponia* globin is larger.

Haemoglobin has also been extracted from nodules of *Casuarina cunninghamiana* [121], which has an actinomycete endophyte [112]. From the reports available (see above), actinorhizal nodules seem likely to provide a more oxygenous environment than rhizobial nodules. It may be expected that the *Casuarina* haemoglobin might stabilize the oxygen concentration within the nodules, but would bind oxygen less tenaciously than leghaemoglobin. Stem nodules form another nitrogen-fixing tissue in certain plants (see Chapter 4). A *Rhizobium* strain isolated from *Sesbania rostrata* stem nodules showed maximum nitrogenase activity at a pO_2 of 3 kPa in free-living culture [123]. It would require a low oxygen environment within the nodules.

Earlier beliefs that haemoglobins were unique to legumes led some authors to conclude that they fulfilled a dispensable function. The evidence available now allows the opposite conclusion to be contemplated: that oxygen-transport proteins are necessary for nitrogen fixation in plants. It should not be assumed at this stage that oxygen transport and stabilization are the only functions of haemoglobins. There are 12 nmol of leghaemoglobin in a 30-day soybean nodule [102]. The 2.3×10^5 infected cells containing $3-4\times10^4$ bacteroids each [102] will have a total of about 0.4 nmol of nitrogenase (from data in Refs. 59, 124). Nitrogenase contains 24–30 atoms of iron per tetramer [39]. Therefore similar amounts of iron are complexed in the plant cytoplasm in leghaemoglobin and within the bacteroids in nitrogenase. We may wonder if haemoglobin also functions to prevent sequestration of iron by the endophyte as suggested by W.D. Sutton in 1977. Further investigation of the role of rhizobial siderophores and their reductases, and of iron metabolism in general, may shed light on this question.

Control by catabolism

The respiratory protection of *Azotobacter* species has been discussed in detail elsewhere [21]. This may also operate in *Bradyrhizobium japonicum* bacteroids which have a low affinity oxidase which operates above 1 μm oxygen with inefficient ATP production [100, 105]. The respiratory activity of the photosynthetic bacteria *Rhodopseudomonas capsulata, Rhodospirillum rubrum* and *Chromatium vinosum* scavenges oxygen over a limited concentration range and keeps the interior anoxic. Acetylene reduction could be completely inhibited by oxygen with the nitrogenase still active if tested *in vitro*. It was suggested that oxygen competed with nitrogenase for electrons [125].

Uptake hydrogenase [126] is produced by a number of bacteria when they are fixing nitrogen, at activities sufficient to recycle the hydrogen produced by nitrogenase [127–131]. This enzyme would be expected to scavenge oxygen but may not be able to provide respiratory protection, as in *Azospirillum brasilense,* because of inhibition by oxygen [132]. *Derxia,* known for producing gum which restricts oxygen diffusion [21], grows autotrophically with hydrogen and oxygen [133]; the hydrogenase will provide some oxygen removal. Different substrates may vary in ability to support acetylene reduction at a particular pO_2, as observed for rhizobial bacteroids [134, 135].

Control by cellular protectants

Information about oxygen toxicity has increased since the review of Yates [68] in 1977. Superoxide dismutases have been reported in cyanobacteria [71]. The enzyme is inducible in shaken aerobic cultures of *Bradyrhizobium japonicum* and other rhizobial strains [137]. Low activity was detected at low pO_2. All rhizobial superoxide dismutases had the characteristics of an Fe (III) enzyme. The superoxide dismutase activity increased with increasing oxygen concentration in *Azospirillum brasilense* cultures [138] but only slightly in nitrogen-fixing *Azotobacter chroococcum* [139] from which an Mn (III) superoxide dismutase was purified [140]. Cultures of the latter organism responded with increased yield to very high rates of aeration [20]. Although oxygen in leghaemoglobin has superoxide character, it appears to be transferred as oxygen [120]. By the use of exogenous superoxides and free radical scavengers, Buchanan implicated superoxide in the inhibition of nitrogen fixation in *Azotobacter chroococcum* [139]. The inclusion of superoxide dismutase and catalase lessened inactivation of the *Clostridium pasteurianum* Fe protein during oxidation [141], encouraging the hypothesis that nitrogenase is the target for oxygen toxicity mediated by superoxide. However, information about the effect of oxygen species on other components required for nitrogen fixation is lacking. So too are data about the intracellular species of oxygen produced during habitual exposure to oxygen, and during oxygen shock. Superoxide dismutases are also present in legume tissue but no correlation with nitrogen fixation has been sought [142, 143].

In *Azotobacter chroococcum*, nitrogenase can be prepared in an oxygen-stable form, associated with an iron-sulphur protein which affords it protection [144]. Yet another form of cellular protection has been proposed for *Azospirillum brasilense* which shows an optimum pO_2 of 0.5–0.7 kPa oxygen for nitrogen fixation [75, 76]. Under aerobic conditions some strains produce carotenoid pigments which may protect nitrogenase from oxygen [138, 145]. A variety of agents, then, are used by nitrogen-fixing bacteria to control the adverse effects of oxygen.

CONCLUSION

Nitrogen reduction requires considerable energy, as already noted. However, this can also be true of the assimilation of nitrates [146, 147], often the predominant form of inorganic nitrogenous material in aerated soil [148]. In studies of nitrogen nutrition in general, and of nitrogen fixation in particular, optimum conditions have often been sought [148–150], although stress situations have also been studied [83, 84, 151] to some extent. In nature nitrogen fixation may not reach its potential because conditions are suboptimal. It has been suggested in this chapter that oxygen flux may impose such a limitation, and merits further attention. When cellular improvements to nitrogen fixation have been contemplated, discussion has often involved a few major protein species such as nitrogenase and haemoglobin. In legume nodules nitrogen fixation is accomplished by very high concentrations of these proteins: leghaemoglobin comprises 8–12 per cent of soluble plant protein in nodules [136], nitrogenase 7 per cent of bacteroid protein [124]. Calculations based on quantity and *in vitro* turnover rate [81] show nitrogenase is more than adequate for the observed ammonia synthesis, and so too is leghaemoglobin. A prime constraint seems rather to be the diffusion limitation on respiration. Relief of this constraint, within narrow limits, would involve changes to nodule morphology. Plant breeding for larger numbers of smaller nodules [147, 152] (to optimize surface to volume ratio) [83] combined with selection in a suboptimal environment might be of value. The considerable morphological variation arising in tissue culture might also prove useful if changes in nodule respiration were sought. The role of non-respiratory metabolism [69, 113] also requires further exploration in a range of nitrogen-fixing systems because it provides an alternative source of energy and reductant.

In flooded soils NH_4^+ ions accumulate, as do other reduced species such as Fe (II) and sulphide [153]. Oxygen fluxes in the rhizosphere may influence not only the rate of nitrogen fixation but also the local balance of ions in the microenvironment, and hence its suitability for bacterial colonization [90, 154].

ACKNOWLEDGEMENTS

I would like to thank Sugirthamani Shaw, Dr B.E. Clothier and colleagues at the DSIR, Palmerston North, for helpful discussions. Drs C.A. Appleby, F.J. Bergersen and R.C. van den Bos kindly supplied useful material, some in advance of publication.

REFERENCES

1. Cotton, F.A. and Wilkinson, G. *Advanced Inorganic Chemistry*, 3rd edition, Wiley-Interscience, New York (1972).
2. Olive, G.H. and Olivé, S. The dinitrogen molecule, pp. 3–29, in *A Treatise on Dinitrogen Fixation*, Sections I and II (Editors, R.W.F. Hardy, F. Bottomley and R.C. Burns), John Wiley & Sons, New York (1979).
3. Friedmann, E.I. and Kibler, A.P. Nitrogen economy of endolithic microbial communities in hot and cold deserts, *Microbial Ecology 6*, 95–108 (1980).
4. Gordon, J.K. Introduction to the nitrogen-fixing prokaryotes, pp. 781–794, in *The Prokaryotes*, Volume I (Editors, M.P. Starr, H. Stolp, H.G. Trüper, A. Balows and H.G. Schlegel), Springer-Verlag, Berlin (1981).
5. Tarrand, J.J., Krieg, N.R. and Döbereiner, J. A taxonomic study of the *Spirillum lipoferum* group, with descriptions of a new genus, *Azospirillum* gen. nov. and two species, *Azospirillum lipoferum* (Beijerinck) comb. nov. and *Azospirillum brasilense* sp. nov., *Canadian Journal of Microbiology 24*, 967–980 (1978).
6. Jordan, D.C. Trasfer of *Rhizobium japonicum* 1980 to *Bradyrhizobium* gen. nov., a genus of slow-growing, root nodule bacteria from leguminous plants, *International Journal of Systematic Bacteriology 32*, 136–139 (1982).
7. Fridovich, I. The biology of oxygen radicals, *Science 201*, 875–879 (1978).
8. Badwey, J.A. and Karnovsky, M.L. Active oxygen species and the functions of phagocytic leukocytes, *Annual Review of Biochemistry 49*, 695–726 (1980).
9. Ramasarma, T. Generation of H_2O_2 in biomembranes, *Biochimica et Biophysica Acta 694*, 69–93 (1982).
10. Tauber, A.I. The human neutrophil oxygen armory, *Trends in Biochemical Sciences* 7, 411–414 (1982).

11. Gutteridge, J.M.C. The role of superoxide and hydroxyl radicals in phospholipid peroxidation catalysed by iron salts, *FEBS Letters 150*, 454-458 (1982).
12. Kutsuki, H. and Gold, M.H. Generation of hydroxyl radical and its involvement in lignin degradation by *Phanaerochaete chrysoporium*, *Biochemical and Biophysical Research Communications 109*, 320-327 (1982).
13. Umbreit, W.W., Burris, R.H. and Stauffer, J.F. *Manometric Techniques*, Burgess, Minneapolis (1957).
14. Lessler, M.A. and Brierley, G.P. Oxygen electrode measurements in biochemical analysis, *Methods of Biochemical Analysis 17*, 1-29 (1969).
15. Degn, H., Lundsgaard, J.S., Petersen, L.C. and Ormicki, A. Polarographic measurement of steady state kinetics of oxygen uptake by biochemical samples, *Methods of Biochemical Analysis 26*, 47-77 (1980).
16. Battino, R., Evans, F.D. and Danforth, W.F. The solubilities of seven gases in olive oil with reference to theories of transport through the cell membrane, *Journal of the American Oil Chemists' Society 45*, 830-833 (1968).
17. Fischkoff, S. and Vanderkooi, J.M. Oxygen diffusion in biological and artificial membranes determined by the fluorochrome pyrene, *Journal of General Physiology 65*, 663-676 (1975).
18. Windrem, D.A. and Plachy, W.Z. The diffusion-solubility of oxygen in lipid bilayers, *Biochimica et Biophysica Acta 600*, 655-665 (1980).
19. Robertson, J.G., Warburton, M.P., Lyttleton, P., Fordyce, A.M. and Bullivant, S. Membranes in lupin root nodules. II. Preparation and properties of peribacteroid membranes and bacteroid envelope inner membranes from developing lupin nodules, *Journal of Cell Science 30*, 151-174 (1978).
20. Hine, P.W. and Lees, H. The growth of nitrogen-fixing *Azotobacter chroococcum* in continuous culture under intense aeration, *Canadian Journal of Microbiology 22*, 611-618 (1976).
21. Robson, R.L. and Postgate, J.R. Oxygen and hydrogen in biological nitrogen fixation, *Annual Review of Microbiology 34*, 183-207 (1980).
22. Kirkham, D. and Powers, W.L. *Advanced Soil Physics*, Wiley-Interscience, New York (1972).
23. Kowalik, P., Barnes, C.J. and Smiles, D.E. Oxidation of liquid animal wastes in soil, *Soil Science Society of America Journal 43*, 255-260 (1979).
24. Currie, J.A. Rothamsted studies of soil structure. IV. Porosity, gas diffusion and pore complexity in dry soil crumbs, *Journal of Soil Science 30*. 441-452 (1979).

25. Robertson, J.G. and Farnden, K.J.F. Ultrastructure and metabolism of the developing legume root nodule, pp. 65-113, in *The Biochemistry of Plants,* Volume 5 (Editor, B.J. Miflin), Academic Press, New York (1980).
26. Gornat, B., Enoch, H. and Goldberg, D. The effect of sprinkling intensity and soil type on oxygen flux during irrigation and drainage, *Soil Science Society of America Proceedings 35*, 668-670 (1971).
27. Rankin, J.M. and Sumner, M.E. Oxygen flux measurement in unsaturated soils, *Soil Science Society of America Journal 42*, 869-871 (1978).
28. Bridge, B.J. and Rixon, A.J. Oxygen uptake and respiratory quotient of field soil cores in relation to their air-filled pore space, *Journal of Soil Science 27*, 279-286 (1976).
29. Dowdell, R.J., Crees, R., Burford, J.R. and Cannel, R.Q. Oxygen concentrations in a clay soil after ploughing or direct drilling, *Journal of Soil Science 30*, 239-245 (1979).
30. Clothier, B.E. and White, I. Water diffusivity of a field soil, *Soil Science Society of America Journal 46*, 155-158 (1982).
31. Edwards, C.A. and Lofty, J.R. Nitrogenous fertilizers and earthworm populations in agricultural soils, *Soil Biology & Biochemistry 14*, 515-521 (1982).
32. Campbell, R. and Rovira, A.D. The study of the rhizosphere by scanning electron microscopy, *Soil Biology & Biochemistry 5*, 747-752 (1973).
33. Penman, H.L. Gas and vapour movements in the soil. I. Diffusion of vapours through porous solids, *Journal of Agricultural Research 30*, 437-462 (1940).
34. Milthorpe, F.L. and Moorby, J. *An Introduction to Crop Physiology*, 2nd edition, Cambridge University Press, Cambridge (1979).
35. Stolzy, L.H. and Letey, J. Characterizing soil oxygen conditions with a platinum microelectrode, *Advances in Agronomy 16*, 249-279 (1964).
36. Tjepkema, J.D. and Yocum, C.S. Respiration and oxygen transport in soybean nodules, *Planta 115*, 59-72 (1973).
37. Sinclair, T.R. and Goudriaan, J. Physical and morphological constraints on transport in nodules, *Plant Physiology 67*, 143-145 (1981).
38. Tjepkema, J.D. and Cartica, R.J. Diffusion limitation of oxygen uptake and nitrogenase activity in the root nodules of *Parasponia rigida* Merr. and Perry, *Plant Physiology 69*, 728-733 (1982).
39. Eady, R.R. and Smith, B.E. Physico-chemical properties of nitrogenase and its components, pp. 399-490, in *A Treatise on Dinitrogen Fixation*, Sections I and II (Editors, R.W.F. Hardy, F. Bottomley and R.C. Burns), John Wiley & Sons, New York (1979).
40. Shah, V.K. and Brill, W.J. Isolation of an iron-molybdenum cofactor

from nitrogenase, *Proceedings of the National Academy of Sciences (USA) 74*, 3249-3253 (1977).

41. Bogusz, D., Houmard, J. and Aubert, J.-P. Electron transport to nitrogenase in *Klebsiella pneumoniae*. Purification and properties of the *nifJ* protein, *European Journal of Biochemistry 120*, 421-426 (1981).
42. Nieva-Gómez, D., Roberts, G.P., Klevickis, S. and Brill, W.J. Electron transport to nitrogenase in *Klebsiella pneumoniae*, *Proceedings of the National Academy of Sciences (USA)* 77, 2555-2558 (1980).
43. Gotto, J.W. and Yoch, D.C. Purification and Mn^{2+} activation of *Rhodospirillum rubrum* nitrogenase activating enzyme, *Journal of Bacteriology 152*, 714-721 (1982).
44. Triplett, E.W., Wall, J.D. and Ludden, P.W. Expression of the activating enzyme and Fe protein of nitrogenase from *Rhodospirillum rubrum*, *Journal of Bacteriology 152*, 786-791 (1982).
45. Burris, R.H., Arp, D.J., Hageman, R.V., Houchins, J.P., Sweet, W.J. and Tso, M. Mechanism of nitrogenase action, pp. 56-66, in *Current Perspectives in Nitrogen Fixation* (Editors, A.H. Gibson and W.E. Newton), Australian Academy of Science, Canberra (1981).
46. Ranga Rao, V., Darrow, R.A. and Keister, D.L. Effect of oxygen tension on nitrogenase and on glutamine synthetases I and II in *Rhizobium japonicum* 61A76, *Biochemical and Biophysical Research Communications 81*, 224-231 (1978).
47. Ludwig, R.A. Regulation of *Rhizobium* nitrogen fixation by the unadenylylated glutamine synthetase I system, *Proceedings of the National Academy of Sciences (USA)* 77, 5817-5821 (1980).
48. Avissar, Y.J. and Nadler, K.D. Stimulation of tetrapyrrole formation in *Rhizobium japonicum* by restricted aeration, *Journal of Bacteriology 135*, 782-789 (1978).
49. Carter, I.S. and Dawes, E.A. Effect of oxygen concentration and growth rate on glucose metabolism, poly-β-hydroxybutyrate biosynthesis and respiration of *Azotobacter beijerinckii*, *Journal of General Microbiology 110*, 393-400 (1979).
50. Mitchell, C.G. and Dawes, E.A. The role of oxygen in the regulation of glucose metabolism, transport and the tricarboxylic acid cycle in *Pseudomonas aeruginosa*, *Journal of General Microbiology 128*, 49-59 (1982).
51. St John, R.T., Shah, V.K. and Brill, W.J. Regulation of nitrogenase synthesis by oxygen in *Klebsiella pneumoniae*, *Journal of Bacteriology 119*, 226-269 (1974).
52. Eady, R.R., Issack, R., Kennedy, C., Postgate, J.R. and Ratcliffe, H.D. Nitrogenase synthesis in *Klebsiella pneumoniae:* Comparison of ammonium and oxygen regulation, *Journal of General Microbiology 104*, 277-285 (1978).

53. Roberts, G.P. and Brill, W.J. Gene-product relationships of the *nif* regulon of *Klebsiella pneumoniae, Journal of Bacteriology 144,* 210-216 (1980).
54. Dixon, R., Eady, R.R., Espin, G., Hill, S., Iaccarino, M., Kahn, D. and Merrick, M. Analysis of regulation of *Klebsiella pneumoniae* nitrogen fixation (*nif*) gene cluster with gene fusions, *Nature 286,* 128-132 (1980).
55. Hill, S., Kennedy, C., Kavanagh, E., Golberg, R.B. and Hanau, R. Nitrogen fixation gene (*nifL*) involved in oxygen regulation of nitrogenase synthesis in *K. pneumoniae, Nature 290,* 424-426 (1981).
56. Bergersen, F.J., Kennedy, C. and Hill, S. Influence of low oxygen concentration on derepression of nitrogenase in *Klebsiella pneumoniae, Journal of General Microbiology 128,* 909-915 (1982).
57. Scott, D.B., Hennecke, H. and Lim, S.T. The biosynthesis of nitrogenase MoFe protein polypeptides in free-living cultures of *Rhizobium japonicum, Biochimica et Biophysica Acta 565,* 365-378 (1979).
58. Robson, R.L. O_2-repression of nitrogenase synthesis in *Azotobacter chroococcum, FEMS Microbiology Letters 5,* 259-262 (1979).
59. Shaw, B.D. Non-coordinate regulation of *Rhizobium* nitrogenase synthesis by oxygen: studies with bacteroids from nodulated *Lupinus angustifolius, Journal of General Microbiology 129,* 849-857 (1983).
60. van den Bos, R.C., Schots, A., Hontelez, J. and van Kammen, A. Nitrogenase synthesis in isolated *Rhizobium leguminosarum* bacteroids: constitutive synthesis from *de novo* transcribed mRNA (in press).
61. Bisseling, T., van Steveren, W. and van Kammen, A. The effect of waterlogging on the synthesis of the nitrogenase components in bacteroids of *Rhizobium leguminosarum* in root nodules of *Pisum sativum, Biochemical and Biophysical Research Communications 93,* 687-693 (1980).
62. Bisseling, T., van Straten, J. and Houwaard, F. Turnover of nitrogenase and leghaemoglobin in root nodules of *Pisum sativum, Biochimica et Biophysica Acta 610,* 360-370 (1980).
63. Kaluza, K. and Hennecke, H. Regulation of nitrogenase messenger RNA synthesis and stability in *Klebsiella pneumoniae, Archives of Microbiology 130,* 38-43 (1981).
64. Kono, Y. Oxygen enhancement of bactericidal activity of rifamycin SV on *Escherichia coli* and aerobic oxidation of rifamycin SV to rifamycin S catalysed by manganous ions: the role of superoxide, *Journal of Biochemistry 91,* 381-395 (1982).
65. MacNeil, D., Zhu, J. and Brill, W.J. Regulation of *Klebsiella pneumoniae:* isolation and characterization of strains with *nif-lac* fusions, *Journal of Bacteriology 145,* 348-357 (1981).
66. Roberts, G.P. and Brill, W.J. Genetics and regulation of nitrogen

fixation, *Annual Review of Microbiology 35,* 207-235 (1981).
67. Maier, R.J., Hanus, F.J. and Evans, H.J. Regulation of hydrogenase in *Rhizobium japonicum, Journal of Bacteriology 137,* 824-829 (1979).
68. Yates, M.G. Physiological aspects of nitrogen fixation, pp. 219-270, in *Recent Developments in Nitrogen Fixation* (Editors, W. Newton, J.R. Postgate and C. Rodriguez-Barrueco), Academic Press, London (1977).
69. Sprent, J.I. *The Biology of Nitrogen-fixing Organisms,* McGraw-Hill, London (1979).
70. Gallon, J.R. The oxygen sensitivity of nitrogenase: A problem for biochemists and micro-organisms, *Trends in Biochemical Sciences 6,* 19-23 (1981).
71. Stewart, W.D.P. Some aspects of structure and function in N_2-fixing cyanobacteria, *Annual Review of Microbiology 34,* 497-536 (1980).
72. Neilson, A.H. and Sparell, L. Acetylene reduction (nitrogen fixation) by *Enterobacteriaceae* isolated from paper mill process waters, *Applied and Environmental Microbiology 32,* 197-205 (1976).
73. Guerinot, M.L. and Patriquin, D.G. N_2-fixing vibrios isolated from the gastrointestinal tract of sea urchins, *Canadian Journal of Microbiology 27,* 311-317 (1981).
74. Potrikus, C.J. and Breznak, J.A. Nitrogen-fixing *Enterobacter agglomerans* isolated from guts of wood-eating termites, *Applied and Environmental Microbiology 33,* 392-399 (1977).
75. Okon, Y., Houchins, J.P., Albrecht, S.L. and Burris, R.H. Growth of *Spirillum lipoferum* at constant partial pressures of oxygen, and the properties of its nitrogenase in cell-free systems, *Journal of General Microbiology 98,* 87-93 (1977).
76. Nelson, L.M. and Knowles, R. Effect of oxygen and nitrate on nitrogen fixation and denitrification by *Azospirillum brasilense* grown in continuous culture, *Canadian Journal of Microbiology 24,* 1395-1403 (1978).
77. Barak, R., Nur, I., Okon, Y. and Henis, Y. Aerotactic response of *Azospirillum brasilense, Journal of Bacteriology 152,* 643-649 (1982).
78. Rao, A.V. and Venkateswarlu, B. Nitrogen fixation by *Azospirillum* isolated from tropical grasses native to Indian desert, *Indian Journal of Experimental Biology 20,* 316-318 (1982).
79. Das, A. and Mishra, A.K. Effect of yeast extract, casamino acids, peptone and various L-amino acids on growth and acetylene reduction in *Azospirillum brasilense, Indian Journal of Experimental Biology 20,* 751-755 (1982).
80. Pankhurst, C.E. and Craig, A.S. Effect of oxygen concentration, temperature and combined nitrogen on the morphology and nitrogenase activity of *Rhizobium* sp. strain 32H1 in agar culture, *Journal of*

General Microbiology 106, 207-219 (1978).

81. Emerich, D.W., Hageman, R.V. and Burris, R.H. Interactions of dinitrogenase and dinitrogenase reductase, *Advances in Enzymology 52*, 1-22 (1981).
82. Criswell, J.G., Havelka, U.D., Quebedeaux, B. and Hardy, R.W.F. Adaptation of nitrogen fixation by intact soybean nodules to altered rhizosphere pO_2, *Plant Physiology 58*, 622-625 (1976).
83. Sprent, J. Nitrogen fixation by legumes subjected to water and light stresses, pp. 405-420, in *Symbiotic Nitrogen Fixation in Plants* (Editor, P.S. Nutman), Cambridge University Press, Cambridge (1976).
84. Gibson, A.H. The influence of the environment and managerial practices on the legume-*Rhizobium* symbiosis, pp. 393-450, in *A Treatise on Dinitrogen Fixation*, Section IV, Agronomy and Ecology (Editors, R.W.F. Hardy and A.H. Gibson), John Wiley, New York (1977).
85. Allison, F.E., Ludwig, C.A., Hoover, S.R. and Minor, F.W. Biochemical nitrogen fixation studies. I. Evidence for limited oxygen supply within the nodule, *Botanical Gazette 101*, 513-533 (1940).
86. Smith, J.D. Haemoglobin and the oxygen uptake of leguminous root nodules, *Biochemical Journal 44*, 591-598 (1949).
87. Bergersen, F.J. The effects of partial pressure of oxygen upon respiration and nitrogen fixation by soybean root nodules, *Journal of General Microbiology 29*, 113-125 (1962).
88. Wittenberg, J.B., Appleby, C.A. and Wittenberg, B.A. The kinetics of the reactions of leghaemoglobin with oxygen and carbon monoxide, *Journal of Biological Chemistry 247*, 527-531 (1972).
89. Pankhurst, C.E. and Sprent, J.I. Effects of temperature and oxygen tension on the nitrogenase and respiratory activities of turgid and water-stressed soybean and French bean root nodules, *Journal of Experimental Botany 27*, 1-9 (1976).
90. van Berkum, P. and Sloger, C. Physiology of root-associated nitrogenase activity in *Oryza sativa*, *Plant Physiology 69*, 1161-1164 (1982).
91. Ralston, E.J. and Imsande, J. Entry of oxygen and nitrogen into intact soybean nodules, *Journal of Experimental Botany 33*, 208-214 (1982).
92. Sprent, J.I. The effects of water stress on nitrogen-fixing root nodules. II. Effects on the fine structure of detached soybean nodules, *New Phytologist 71*, 443-450 (1972).
93. Bergersen, F.J. and Goodchild, D.J. Cellular location and concentration of leghaemoglobin in soybean root nodules, *Australian Journal of Biological Science 26*, 741-756 (1973).
94. Ho, L.C. The relationship between the rates of carbon transport and of photosynthesis in tomato leaves, *Journal of Experimental Botany 27*, 87-97 (1976).

95. Goodchild, D.J. The ultrastructure of root nodules in relation to nitrogen fixation, *International Review of Cytology* Supplement 6, 235–288 (1977).
96. Bergersen, F.J. and Goodchild, D.J. Aeration pathways in soybean root nodules, *Australian Journal of Biological Science 26,* 729–740 (1973).
97. Tjepkema, J.D. and Yocum, C.S. Measurement of oxygen partial pressure within soybean nodules by oxygen microelectrodes, *Planta 119,* 351–360 (1974).
98. John, J.E.A. *Gas Dynamics,* pp. 316–327, Allynn and Bacon, Boston (1969).
99. Dullien, F.A.L. *Porous Media. Fluid Transport and Pore Structure,* pp. 200–234, Academic Press, New York (1979).
100. Bergersen, F.J. and Turner, G.L. Properties of terminal oxidase systems of bacteroids from root nodules of soybean and cowpea and of N_2-fixing bacteria grown in continuous culture, *Journal of General Microbiology 118,* 235–252 (1980).
101. Bergersen, F.J. and Turner, G.L. Leghaemoglobin and the supply of O_2 to nitrogen-fixing root nodule bacteroids: Studies of an experimental system with no gas phase, *Journal of General Microbiology 89,* 31–47 (1975).
102. Bergersen, F.J. Rhizobium-legume symbiosis. The intracellular environment of *Rhizobium japonicum* in soybean root nodules, pp. 376–383, in *Microbial Ecology* (Editors, M.W. Loutit and J.A.R. Miles), Springer-Verlag, Berlin (1978).
103. Pankhurst, C.E. and Sprent, J.I. Effects of water stress on the respiratory and nitrogen-fixing activity of soybean root nodules. *Journal of Experimental Botany 26,* 287–304 (1975).
104. Bergersen, F.J. *Root Nodules of Legumes: Structure and Functions,* Research Studies Press, Chichester (1982).
105. Bergersen, F.J. and Turner, G.L. Leghaemoglobin and the supply of O_2 to nitrogen-fixing root nodule bacteroids: Presence of two oxidase systems and ATP production at low free O_2 concentration, *Journal of General Microbiology 91,* 345–354 (1975).
106. Akkermans, H.D.L., Abdulkadir, S. and Trinick, M.J. N_2-fixing root nodules in Ulmaceae: *Parasponia* or (and) *Trema* spp.? *Plant and Soil 49,* 711–716 (1978).
107. Trinick, M.J. Structure of nitrogen-fixing nodules formed by *Rhizobium* on roots of *Parasponia andersonii* Planch, *Canadian Journal of Microbiology 25,* 565–578 (1979).
108. Trinick, M.J. Effects of oxygen, temperature and other factors on the reduction of acetylene by root nodules formed by *Rhizobium* on *Parasponia andersonii* Planch, *New Phytologist 86,* 27–38 (1980).
109. Tjepkema, J.D. and Winship, L.J. Energy requirement for nitrogen

fixation in actinorhizal and legume root nodules, *Science 209*, 279-281 (1980).

110. Tjepkema, J.D., Ormerod, W. and Torrey, J.G. Vesicle formation and acetylene reduction activity in *Frankia* sp. CPI1 cultured in defined nutrient media, *Nature 287*, 633-635 (1980).
111. Tjepkema, J.D., Ormerod, W. and Torrey, J.G. Factors affecting vesicle formation and acetylene reduction (nitrogenase activity) in *Frankia* sp. CPI1, *Canadian Journal of Microbiology 27*, 815-823 (1981).
112. Gauthier, D., Diem, H.G. and Dommergues, Y. *In vitro* nitrogen fixation by two actinomycete strains isolated from *Casuarina* nodules, *Applied and Environmental Microbiology 41*, 306-308 (1981).
113. Tajima, S. and La Rue, T.A. Enzymes for acetaldehyde and ethanol formation in legume nodules, *Plant Physiology 70*, 388-392 (1982).
114. Appleby, C.A. Leghaemoglobin, pp. 521-554, in *The Biology of Nitrogen Fixation* (Editor, A. Quispel), North Holland, Amsterdam (1974).
115. Bergersen, F.J. Leghaemoglobin, oxygen supply and nitrogen fixation: Studies with soybean nodules, pp. 247-261, in *Limitations and Potentials for Biological Nitrogen Fixation in the Tropics* (Editors, J. Döbereiner, R.H. Burris and A. Hollaender), Plenum Press, New York (1978).
116. Dilworth, M.J. and Appleby, C.A. Leghemoglobin and *Rhizobium* hemoproteins, pp. 691-764, in *A Treatise on Dinitrogen Fixation*, Sections I and II, Inorganic and Physical Chemistry and Biochemistry (Editors, R.W.F. Hardy, F. Bottomley and R.C. Burns), John Wiley and Sons, New York (1979).
117. Appleby, C.A. Properties of leghaemoglobin *in vivo*, and its isolation as ferrous oxyleghaemoglobin, *Biochimica et Biophysica Acta 188*, 222-229 (1969).
118. Appleby, C.A. The oxygen equilibrium of leghaemoglobin, *Biochimica et Biophysica Acta 60*, 226-235 (1962).
119. Stokes, A.N. Facilitated diffusion: The elasticity of oxygen supply, *Journal of Theoretical Biology 52*, 285-297 (1975).
120. Wittenberg, J.B., Bergersen, F.J., Appleby, C.A. and Turner, G.L. Facilitated oxygen diffusion. The role of leghaemoglobin in nitrogen fixation by bacteroids isolated from soybean root nodules, *Journal of Biological Chemistry 249*, 4057-4066 (1974).
121. Appleby, C.A., Tjepkema, J.D. and Trinick, M.J. Hemoglobin in a non-legume plant, *Parasponia:* Possible genetic origin and function in nitrogen fixation, *Science 220*, 951-953 (1983).
122. Coventry, D.R., Trinick, M.J. and Appleby, C.A. A search for a leghaemoglobin-like compound in root nodules of *Trema cannabina*

Lour., *Biochimica et Biophysica Acta 420,* 105-111 (1976).

123. Dreyfus, B.L. and Dommergues, Y.R. Stem nodules on the tropical legume, *Sesbania rostrata,* p. 471, in *Current Perspectives in Nitrogen Fixation* (Editors, A.H. Gibson and W.E. Newton), Australian Academy of Science, Canberra (1981).
124. Whiting, M.J. and Dilworth, M.J. Legume root nodule nitrogenase: Purification, properties and studies on its genetic control, *Biochimica et Biophysica Acta 371,* 337-351 (1974).
125. Hochman, A. and Burris, R.H. Effect of oxygen on acetylene reduction by photosynthetic bacteria, *Journal of Bacteriology 147,* 492-499 (1981).
126. Adams, M.W.W., Mortenson, L.E. and Chen, J.-S. Hydrogenase, *Biochimica et Biophysica Acta 594,* 105-176 (1981).
127. Partridge, C.D.P., Walker, C.C., Yates, M.G. and Postgate, J.R. The relationship between hydrogenase and nitrogenase in *Azotobacter chroococcum:* Effect of nitrogen sources on hydrogenase activity. *Journal of General Microbiology 119,* 313-319 (1980).
128. Chan, Y.K., Nelson, L.M. and Knowles, R. Hydrogen metabolism of *Azospirillum brasilense* in nitrogen-free medium, *Canadian Journal of Microbiology 26,* 1126-1131 (1980).
129. Benson, D.R., Arp, D.J. and Burris, R.H. Cell-free nitrogenase and hydrogenase from actinorhizal root nodules, *Science 205,* 688-689 (1979).
130. Eisbrenner, G. and Evans, H.J. Carriers in electron transport from molecular hydrogen to oxygen in *Rhizobium japonicum* bacteroids, *Journal of Bacteriology 149,* 1005-1012 (1982).
131. O'Brian, M.R. and Maier, R.J. Electron transport components involved in hydrogen oxidation in free-living *Rhizobium japonicum, Journal of Bacteriology 152,* 422-430 (1982).
132. Pedrosa, F.O., Stephan, M., Döbereiner, J. and Yates, M.G. Hydrogen-uptake hydrogenase activity in nitrogen-fixing *Azospirillum brasilense, Journal of General Microbiology 128,* 161-166 (1982).
133. Pedrosa, F.O., Döbereiner, J. and Yates, M.G. Hydrogen-dependent growth and autotrophic carbon dioxide fixation in *Derxia, Journal of General Microbiology 119,* 547-551 (1980).
134. Trinchant, J.C. and Rigaud, J. Sur les substrats énergétiques utilisés, lors de la réduction de C_2H_2, par les bactéroides extraits des nodosités de *Phaseolus vulgaris* L. *Physiologie Végétale 17,* 547-556 (1979).
135. Trinchant, J.C., Birot, A.M. and Rigaud, J. Oxygen supply and energy-yielding substrates for nitrogen fixation (acetylene reduction) by bacteroid preparations, *Journal of General Microbiology 125,* 159-165 (1981).
136. Dilworth, M.J. and Coventry, D.R. Stability of leghaemoglobin in

yellow lupin nodules, pp. 431-442, in *Recent Developments in Nitrogen Fixation* (Editors, W. Newton, J.R. Postgate and C. Rodriguez-Barrueco), Academic Press, London (1977).

137. Stowers, M.D. and Elkan, G.H. An inducible iron-containing superoxide dismutase in *Rhizobium japonicum*, *Canadian Journal of Microbiology 27*, 1202-1208 (1981).

138. Nur, I., Okon, Y. and Henis, Y. Effect of dissolved oxygen tension on production of carotenoids, poly-β-hydroxybutyrate, succinate oxidase and superoxide dismutase by *Azospirillum brasilense* Col grown in continuous culture, *Journal of General Microbiology 128*, 2937-2943 (1982).

139. Buchanan, A.G. The response of *Azotobacter chroococcum* to oxygen: Superoxide-mediated effects, *Canadian Journal of Microbiology 23*, 1548-1558 (1977).

140. Buchanan, A.G. and Less, H. Superoxide dismutase from nitrogen-fixing *Azotobacter chroococcum*: Purification, characterization, and intracellular location, *Canadian Journal of Microbiology 26*, 441-447 (1980).

141. Mortenson, L.E., Walker, N.M. and Walker, G.A. In *Proceedings of the 1st International Symposium on Nitrogen Fixation*, Volume 1, pp. 117-149 (Editors, W.E. Newton and C.J. Nyman), Washington State University Press, Washington (1976).

142. Lee, E.H. and Bennett, J.H. Superoxide dismutase. A possible protective enzyme against ozone injury in snap beans (*Phaseolus vulgaris* L.), *Plant Physiology 69*, 1444-1449 (1982).

143. Sevilla, F., Lopéz-Gorgé, J. and del Rio, L.A. Characterization of a manganese superoxide dismutase from the higher plant *Pisum sativum*, *Plant Physiology 70*, 1321-1326 (1982).

144. Robson, R.L. Characterization of an oxygen-stable nitrogenase complex isolated from *Azotobacter chroococcum*, *Biochemical Journal 181*, 569-575 (1979).

145. Nur, I., Steinitz, Y.L., Okon, Y. and Henis, Y. Carotenoid composition and function in nitrogen-fixing bacteria of the genus *Azospirillum*, *Journal of General Microbiology 122*, 27-32 (1981).

146. Imsande, J. Exchange of metabolites and energy between legume and *Rhizobium*, pp. 179-190, in *International Review of Cytology*, supplement 13 (Editors, K.L. Giles, and A.G. Atherly), Academic Press, New York (1981).

147. Phillips, D.A. Efficiency of symbiotic nitrogen fixation in legumes, *Annual Review of Plant Physiology 31*, 29-49 (1980).

148. Pate, J.S. Transport and partitioning of nitrogenous solutes, *Annual Review of Plant Physiology 31*, 313-340 (1980).

149. Pate, J.S. and Minchin, F.R. Comparative studies of carbon and nitrogen nutrition of selected grain legumes, pp. 105-114, in *Advan-*

ces in Legume Science, Volume 1 (Editors, R.J. Summerfield and A.H. Bunting), Royal Botanic Gardens, Kew (1980).

150. Broughton, W.J. Effect of light intensity on net assimilation rates of nitrate-supplied or nitrogen-fixing legumes, pp. 285–299, in *Photosynthesis and Plant Development* (Editors, R. Marcelle, H. Clijsters and M. Van Poucke), W. Junk, The Hague (1979).

151. De Jong, T.M. and Phillips, D.A. Water stress effects on nitrogen assimilation and growth of *Trifolium subterraneum* L. using dinitrogen or ammonium nitrate, *Plant Physiology 69*, 416–420 (1982).

152. Nutman, P.S. Varietal differences in the nodulation of subterranean clover, *Australian Journal of Agricultural Research 18*, 381–425 (1967).

153. Reddy, K.R., Rao, P.S.C. and Patrick, W.H. Factors influencing oxygen consumption rates in flooded soils, *Soil Science Society of America Journal 44*, 741–744 (1980).

154. Armstrong, W. Aeration in higher plants, *Advances in Botanical Research 7*, 225–332 (1979).

6. Plasmids Governing Symbiotic Nitrogen Fixation

A. Kondorosi, G.B. Kiss and I. Dusha

INTRODUCTION

One of the most beneficial interactions between bacteria and plants is symbiotic nitrogen fixation. Bacteria of the genus *Rhizobium* invade the root cells of different leguminous plants resulting in nodule formation and eventually nitrogen fixation. The mutual advantages of the symbiosis for the plants and bacteria are obvious. Plants are supplied by fixed nitrogen which is one of the most serious limiting nutrients. On the other hand, bacteria are under protected environment inside the nodule cells and are supplied with the product of photosynthesis (photosynthate) for the requirement of their carbon and energy.

According to quantitative analyses, biological nitrogen fixation contributes about 170×10^6 tonnes of combined nitrogen per year to the ecosystem [1]. Leguminous plants in symbiosis with rhizobia can fix nitrogen in a range of 52-300 kg per hectare per year [2]. These amounts of nitrogen are sufficient to support good plant growth and yield. It is unnecessary to apply expensive man-made nitrogen fertilizers which in most cases render the ecosystem off its balance and spoil the human environment, if greater reliance on biological nitrogen fixation is made. These facts stimulated scientists to become acquainted with the biological process of nitrogen fixation.

In the past two decades basic experimental conditions were established to study symbiotic genes in the genus *Rhizobium*. The exchange of genetic material by conjugation and transduction was worked out enabling the construction of linkage maps of different *Rhizobium* species [3]. Using various mutagenic treatments many mutants altered in symbiotic properties were generated and characterized [4, 5]. The finding of interspecific hybridization of the cloned *nif H* gene of *Klebsiella pneumoniae* accelerated the dissection of symbiotic genes by molecular cloning and DNA sequence analysis [6, 7, 8]. New agarose gel electrophoresis techniques made possible the discovery and analysis of large plasmids in rhizobia [9].

The recent reviews on rhizobia [3, 5, 10, 11] presenting experimental data brought out that functions of plasmids play a key role in symbiotic nitrogen fixation. In this review these new discoveries have been summarised and a comprehensive survey on symbiotic genes of rhizobia has been given.

PRESENCE OF PLASMIDS IN RHIZOBIA

The presence of large indigenous plasmids is a common feature of different *Rhizobium* species [9, 12]. More than one resident plasmid was detected in most of the species, especially in the so-called fast-growing rhizobia. In different *R. leguminosarum* strains the number of plasmids ranges from two [13, 14] to eight (W. Lotz, personal communication); similarly in *R. trifolii* two to eight plasmids have been described so far [15-17],(L.K. Dunican, personal communication); in *R. phaseoli* two to three plasmids have been identified [16-18]. With a few exceptions, a wide variety of *R. meliloti* strains also carry two to three or more plasmids [19-20] (W. Broughton, personal communication). Slow-growing rhizobia also do contain plasmids, although data are still rather scanty [21-24].

In earlier studies detection and isolation of *Rhizobium* plasmids were hindered by the fact that no appropriate methods were available for studying plasmids of such high molecular weight. New procedures, developed originally for the isolation of *Agrobacterium* plasmids [25], were successfully applied in many cases for the purification of *Rhizobium* plasmids [21, 26].

Moreover the increased interest in the symbiotic nitrogen fixation has also contributed to the development of techniques suitable for detection and isolation of large (up to 300 Md) plasmids [27-29]. The slight modification of the method of Eckhardt [30] for rhizobia [19, 20] allowed the detection of extremely large (300-600 Md) plasmids (megaplasmids). These studies revealed that in several cases more than 50 per cent of the total DNA was extrachromosomal. It is not excluded that plasmids with even higher molecular weight are present in some bacteria.

USE OF Tn5 IN LABELLING OF PLASMIDS AND MUTAGENESIS OF SYMBIOTIC GENES

Apart from a few cases (bacteriocinogenic plasmids, pigment production, etc.) no selectable markers were localized on most of the indigenous *Rhizobium* plasmids. To facilitate genetic and biochemical work, a method was developed for labelling plasmids with transposons [31]. The antibiotic resistance marker coded by the transposon could easily be followed and this method allowed simple identification of a given plasmid.

On the other hand, insertion of transposons can be used as an efficient technique to mutagenize symbiotic genes. Host specificity, nodule formation and nodule functions are usually not expressed in a free-living state.

Therefore a plant inoculation test of individual colonies is always necessary for screening mutants. Transposon-induced mutations have many advantages: insertion into a gene leads to a complete loss of its function; antibiotic resistance marker of the transposon is a positively selectable marker which makes genetic mapping of the affected gene simple; molecular cloning of the transposon-marked gene is easy.

Tn5 is the most frequently used transposon for insertion mutagenesis. It is 5.7 kb in size, its integration into the genome is random, it has no *Eco* RI restriction site and carries an easily selectable marker, kanamycin. Recently many papers have appeared on the isolation of symbiotic mutants with Tn5 [19, 32–35].

Transposon Tn7 which codes for Sp Sm Tp resistance was also used for mutagenesis of the *sym* plasmid of *R. meliloti* 2011 [36, 124].

PLASMID-BORNE SYMBIOTIC GENES

Genes controlling symbiotic development

The development of nodules and effective nitrogen fixation in symbiosis is a multistep process depending on the coordinate expression of specific genes acting at different stages in both partners. In bacteria several genes have been attributed to be involved exclusively in this process. According to the apparent pheno-type, mutants incapable to *fix* nitrogen in symbiosis can be classified into two main groups. Nod^- bacteria are not able to form nodule on plants, while Fix^- bacteria form nodules but are impaired in nitrogen fixation and/or nodule persistence. Plants inoculated with either of the two mutant types show chlorosis, a typical symptom of nitrogen deficiency.

In the next sections, the localization and structural analysis of the genes involved in nodule formation and nitrogen fixation are summarized.

GENETIC DETERMINANTS FOR NODULATION

Rhizobium species are classified by their ability to nodulate related leguminous plants. Members of one group (cross inoculation group) form nitrogen-fixing nodules on one set of legumes, but not on others [37]. Genes determining host specificity of nodulation (*hsn*) are functioning at the very beginning of the infection process and are probably involved in the recognition of the plant host. Mutations in these genes result in non-nodulating, Nod^- phenotype. Besides *hsn* genes, other gene(s), not specific to one cross-inoculation group, are also involved in nodule formation. These genes may be considered as "common" *nod* genes. Unlike mutations in *hsn* genes, "common" *nod*$^-$ mutations can be complemented by definition with genes present in other rhizobia, belonging to another cross-inoculation group. Since mutations in "common" *nod* genes have the same phenotype as mutations in *hsn* genes, a Nod^- character has to be critically considered.

Both hsn and "common" nod genes are plasmid-borne

Early genetic data had suggested that plasmids detected in different *Rhizobium* strains [21, 38-40] might carry nodulation functions in rhizobia [41, 42]. Johnston et al. [43] presented the first convincing evidence for the location of genes for nodulation in a self-transmissible plasmid, pRLIJI, found in a *R. leguminosarum* field isolate (strain 248). In this and the subsequent experiments no distinction was made between *hsn* and "common" *nod* genes. (This distinction however is made in the following section.) In some experiments, however, circumstantial evidence could be drawn for the location of either *hsn* and/or "common" *nod* genes on plasmids. For example, if a *R. leguminosarum* plasmid was transferred to other *Rhizobium* species and the transconjugants also acquired the nodulation ability on peas the natural host of *R. leguminosarum, hsn* gene must have been transferred. In these cases the transfer of the "common" *nod* genes cannot be concluded because their possible presence in the genome of the recipient. The location of both *hsn* and "common" *nod* genes could be deduced if plasmids were transferred, for instance, to an appropriate *Agrobacterium tumefaciens* strain and the transconjugants formed nodules on the natural host of the donor strains. In the cases reported [44, 45] the nodules were ineffective suggesting expression problems and/or the presence of essential genes for symbiotic nitrogen fixations on the chromosome or on other resident plasmid (s). In the following section those experiments have been summarised where the plasmid location of the nodulation ability was demonstrated. When conclusive, distinction was made between *hsn* and "common" *nod* genes. Johnston et al. [43] introduced plasmid pJB5JI (a Tn5 derivative of pRL1JI) into other *Rhizobium* species and the symbiotic properties of the transconjugants were tested on peas, the natural host of *R. leguminosarum*. All transconjugants containing pJB5JI of *R. trifolii* 6661 and 6710, *R. phaseoli* 1233, and *R. "species"* 9009 induced nodules on peas but they nodulated more slowly and reduced acetylene less efficiently than *R. leguminosarum*. This result shows that the ability to nodulate peas could be transferred to other species of *Rhizobium*. Interspecific transconjugants were also tested for nodulation on their normal hosts. All were capable of nodulating their respective hosts but nodule formation was delayed and the nodules were smaller and fewer in number. This kind of response was attributed to physiological incompatibility between two plasmids of different origin [17].

R. phaseoli 1233 derivatives containing pJB5JI were analysed in more detail by Beynon et al. [17]. The majority of the transconjugants (97%) nodulated both peas and *Phaseolus* beans poorly. They contained pJB5JI and both indigenous plasmids of *R. phaseoli* 1233. These transconjugants lost their nodulation ability on *Phaseolus* beans following passage through pea nodules. Bacteria reisolated from these nodules had lost or suffered deletion on the smaller plasmid. Interestingly, passing through on

Phaseolus nodules no phenotypic change had been observed. Three per cent of the transconjugants failed to produce nitrogen-fixing nodules on *Phaseolus* beans but nodulated peas normally. They were phenotypically stable following passage through both pea and *Phaseolus* nodules. Plasmid analysis of these transconjugants revealed that they lost pRP1JI, the smaller of the two plasmids of this strain, suggesting the localization of *hsn* genes for *Phaseolus* on this plasmid. Spontaneous deletion derivatives of pRP1JI are Nod$^-$, providing further evidence that pRP1JI carries symbiotic functions.

The *hsn* genes for pea could be transferred on pJB5JI into *R. trifolii* ANU 794 [15]. Nine out of ten transconjugants nodulated peas and plasmid analysis of the transconjugants showed that all retained the three indigenous plasmids of *R. trifolii* ANU 794 and acquired pJB5JI. The Nod$^-$ transconjugant, however, contained a deleted derivative of pJB5JI suggesting the loss of *hsn* genes for peas. The Nod+ character both on peas and white clover, was stable after passing through on pea nodules, whereas bacteria isolated from white clover nodules suffered deletion in pJB5JI. These derivatives were not able to nodulate peas while the effective nodulation of clover was always retained.

In *R. leguminosarum* RCC1001, the smaller of the two plasmids (pSyml) carries *hsn* genes for pea [14]. This was demonstrated by the mobilization of a Tn5 derivative of pSyml (pSyml::Tn5) into *R. trifolii* (LPR 5001). Transconjugants were able to form effective nitrogen-fixing nodules both on clover and vetches, the member of the pea vetch cross-inoculation group. On the other hand, *A. tumefaciens* and *R. meliloti* transconjugants carrying pSyml::Tn5 induced ineffective nodules on vetches.

Plasmid pRL6JI was found to carry *nod* genes in an *R. leguminosarum* field isolate. If plasmid pRP2JI::Tn5, a derivative of the indigenous plasmid in *R. phaseoli* strain 8002 was introduced into a pRL6JI cured derivative of *R. leguminosarum,* transconjugant nodulated only *Phaseolus* beans. The lack of nodulation on peas indicated *hsn* genes on pRL6JI [18, 46]. In *R. trifolii* LPR5001, among other plasmids, pRtr5a (=p*Sym*5) was identified by its self-transfer ability [44]. A Tn5 derivative of this plasmid, pRtr5a::Tn5 was eliminated from Nod$^+$ Fix$^+$ *R. trifolii.* The cured, Km^s colonies had lost their nodulation properties but following reintroducing pRtr5a::Tn5, transconjugants recovered their nodulation ability on clover. When p*Sym*1::Tn5, the *sym* plasmid of *R. leguminosarum* RCC1001 was transferred into the pRtr5a::Tn5 cured derivative of *R. trifolii,* transconjugants acquired effective nodulation ability only on peas. This result suggests that pRtr5a::Tn5 carries *hsn* determinants for clover [14]. This was further supported by introducing clover nodulation ability by pRtr5a::Tn5 into *R. leguminosarum* and *A. tumefaciens. R. leguminosarum* RCR1001 transconjugants were capable of nodulating not only their normal symbiotic partners (peas and vetches) but also clover.

A Nod⁻ *R. leguminosarum* (LPR1802) carrying pRtr5a::Tn5 nodulated clover but not peas or vetches. A Ti plasmid-cured *A. tumefaciens* derivative carrying pRtr5a::Tn5 was Nod+on clover, but the small, non-nitrogen-fixing root nodules appeared after a long period of incubation.

Plasmid pRtr514a from *R. trifolii* strain NZP514 is not self-transmissible but could be mobilized by the broad host range P-group plasmid R 68.45 to a pRtr 514a-cured Nod⁻ derivative of strain NZP 514. The Nod+ recipients had received a cointegrate plasmid (pPNI) comprising pRtr514a an R 68.45 [47]. The ability to nodulate clover could be transferred on pPNI into *R. leguminosarum, R. phaseoli* and *R. meliloti,* but only *R. leguminosarum* transconjugants formed effective nodules. *R. meliloti* (pPNI) transconjugants did not nodulate clover. Testing the transconjugants on their original host revealed that *R. leguminosarum* strain with pPNI lost the ability to nodulate pea plants as a consequence of plasmid loss by incompatibility. The lack of or ineffective nodulation by *R. meliloti* on lucerne and clover was due to functional interference [17].

In *R. phaseoli* strain 8002 three plasmids can be detected (pRP2JI, pRP3JI, pRP5JI). When pRP2JI was labelled with Tn5, Km^R could be transferred into a non-nodulating *R. leguminosarum* strain. Transconjugants induced nitrogen-fixing nodules on *Phaseolus* beans and all failed to nodulate peas [18].

Host specificity genes for lucerne was aemonstrated on pRme41b, the *Sym* plasmid in *R. meliloti* 41 [45]. Since plasmid pRme41*b* could not be mobilized by self transfer, it was made susceptible for mobilization. For this the mobilization region of RP4, (mob_{RP4}) was inserted by *in vitro* techniques into the *nif* genes from Rm41 carried on pIDI [19]. The resulting plasmid, pAK11, was then integrated into pRme41*b* by homologous recombination *in vivo.* The resulting plasmid pRme41*b*::pAK11 was then mobilized into *Rhizobium* strain PN4003, a fast growing strain nodulating *Lotus pedunculatus,* into *Rhizobium* strain MPIK3030, a fast growing strain nodulating a number of tropical legumes, and into *A. tumefaciens,* strain GV3101, a Ti plasmid-cured derivative of strain C58. Interspecific transconjugants were able to form small white nodules on lucerne, but were not able to *fix* nitrogen. Electron and light microscopic analysis of these nodules demonstrated that root hair curling and infection thread formation did occur but the infection threads aborted early. Moreover, no bacteria or bacteroids were found in the plant cells [48]. The above results clearly demonstrate that pRme41*b* carries *hsn* and "common" *nod* determinants. All *nod* genes required for the early steps of nodulation were localized on an R-prime carrying 80 kb region of pRme41*b* [49].

As discussed above in many *Rhizobium* strains, Nod⁻ phenotype was concomitant to the alteration of indigenous plasmids.

Fine structural analysis of genes involved in nodulation

Characterization of Nod$^-$ mutants by light and/or electron microscopy revealed at least 10 stages in nodule development. Some of these steps have been described [10] and it seems that a deficiency (mutation) in any step blocks the further nodule development.

Some *R. meliloti* Nod$^-$ mutants for example are unable to evoke root hair curling (Hac$^-$, or "nonreactive" mutants; [35, 50]). These mutants do not penetrate the plant cells and there is no sign of any bacterium-plant interaction, except that bacteria may bind to the root hair surface. Such Hac$^-$ mutants were isolated also from *R. trifolii* [51].

Hac$^-$ mutants were used to identify genes coding for root hair curling located on the *sym* megaplasmid of *R. meliloti* [52, 53]. Long et al. [52] have cloned this gene by direct complementation of Nod mutants. A gene bank of *R. meliloti* 2011 was prepared into the wide host range vector pALFRI [54] and a population of the recombinant plasmids was mass-conjugated into Nod$^-$mutants. By testing the transconjugants for their nodulation ability on *Medicago,* recombinant plasmids carrying the wild type *nod* allele were identified. The recombinant plasmids restored the ability of two Nod$^-$mutants to evoke root hair curling and to form nodules on alfalfa. Both Nod$^-$mutants contained an insertion in a 8.7 kb *Eco*RI fragment, which mapped 20 kb away from the *nif* structural genes (Fig. 1)

In *R. meliloti* 41 the same *nod* region was identified by deletion mapping and complementation analysis of Nod$^-$point, insertion and deletion mutants [53, 55]. Cosmid clones overlapping a 135 kb region of pRme41*b* including the *nif* structural genes (Fig. 1) were used as hybridization probes against restriction endonuclease digested DNA from various Nod$^-$mutants. This approach allowed to identify two *nod* gene clusters on the megaplasmid. One *nod* cluster was located on a 8.5 kb *Eco*RI fragment [53] which is very likely identical with the 8.7 kb *nod* fragment of *R. meliloti* 2011 (Fig. 1).

The 8.5 kb *Eco*RI was recloned into pRK 290 and then transferred into various Nod$^-$recipients. The Nod+ Fix+ phenotype was restored in several Nod$^-$deletion and point mutants [53]. Interestingly, Banfalvi et al. [19] have found, that these Nod$^-$ mutations were also suppressed upon the introduction of pJB5JI, a *sym* plasmid of *R. leguminosarum* [31]. This latter result indicated that the (se) gene (s) code for such nodulation functions which are coded by other *Rhizobium* species as well ("common" *nod* genes).

It seems that this nodulation function is common for various rhizobia. When the 8.5 kb fragment was used as hybridization probe against DNS from several other *Rhizobium* species [56] (E. Kondorosi, A. Kondorosi, A. Szalay and R. Hadley, unpublished), hybridization was observed nearly in all cases. Recently, the exact location of *nod* genes on the 8.5 kb fragment was determined by directed Tn5 mutagenesis (E. Kondorosi and A. Kondo-

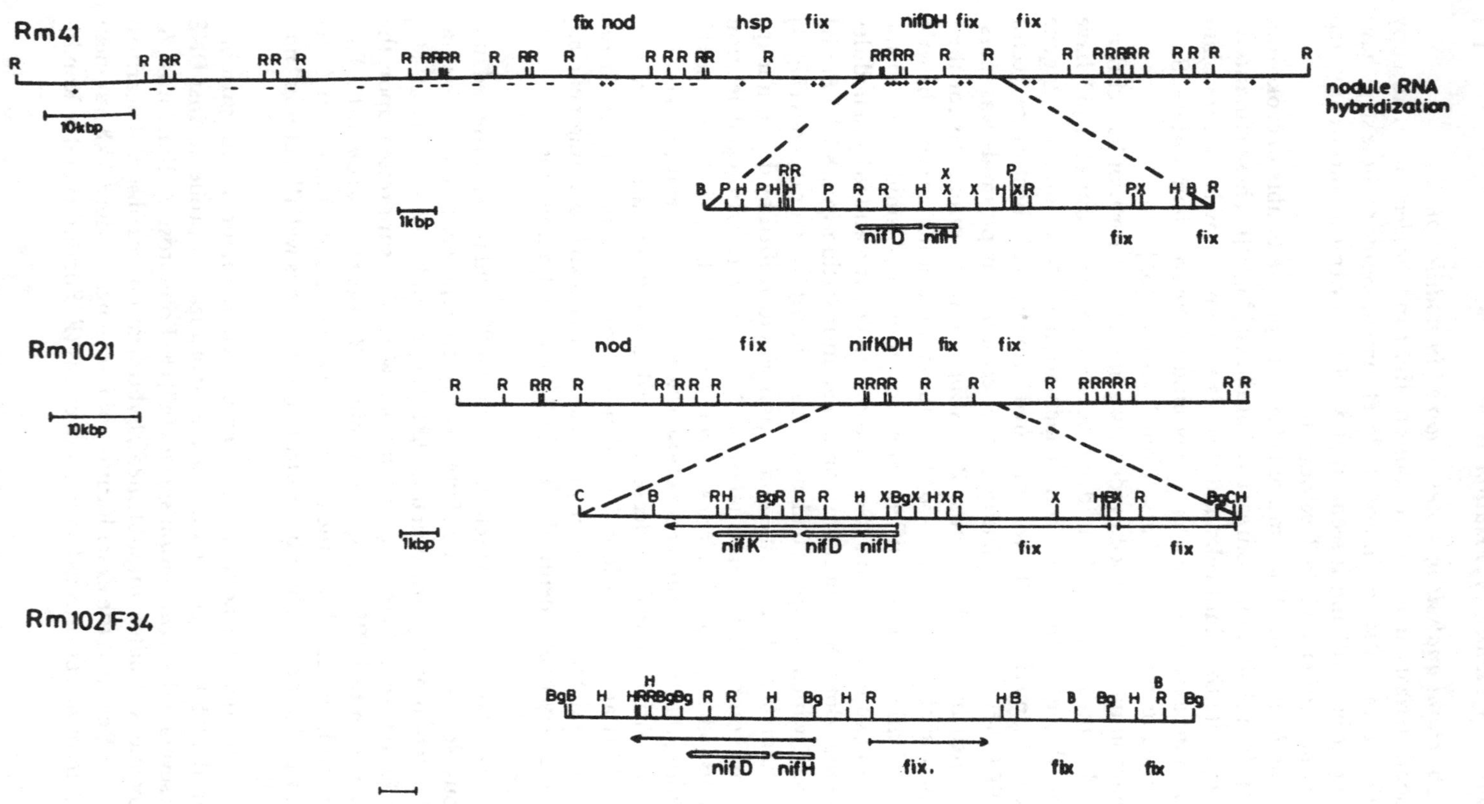
Rm41
fix nod
hsp
fix
nifDH
fix
fix
nodule RNA hybridization
10kbp
1kbp
nif D
nifH
fix
fix
Rm1021
nod
fix
nifKDH
fix
fix
10kbp
1kbp
nifK
nifD
nifH
fix
fix
Rm102F34
nif D
nifH
fix.
fix
fix

rosi, unpublished). Using about a 2 kb *nod* subfragment, the interspecies homology of *nod* region was further supported (C. Bachem, unpublished). The fact that these *nod* genes are conserved in many *Rhizobium* species and can express in other *Rhizobium* species, greatly facilitates the identification and cloning the corresponding *nod* genes from other *Rhizobium* species.

The nature and function of these *nod* gene products is not known. Indole-3-acetic acid (IAA) has been suggested to play a role in root hair curling [57]. Measurements of IAA production by wild type and Hac⁻ mutants of *R. trifolii*, however, do not support this idea [58]. Moreover, the genetic information coding for the enzymes required in the pathways of IAA biosynthesis is not carried by the *sym* plasmid of *R. trifolii* or of *R. leguminosarum* [58, 59]. According to Wang et al. [60] the cytokinin content of the nodules is rather low, in contrast to earlier reports [61, 62]. The second *nod* cluster was identified between the "common" *nod* genes and the *nif* genes in *R. meliloti* 41 [53, 55]. In the course of the genetic analysis of Nod⁻ deletion mutants it was found that the 8.5 kb *Eco*RI *nod* fragment did not complement Nod⁻ mutants carrying large deletions in pRme41*b*. On the other hand, a *nod-nif* R-prime, pGR3, containing about an 80 kb insert, restored nodulation in these mutants [49]. *A. tumefaciens* carrying pGR3 induced nodules on alfalfa, indicating that all essential *nod* genes, coding for early nodulation functions are present on this region. From these results a second *nod* region was suggested to be present on pGR3. Since the nodulation ability coded by pGR3 is specific for the plant host alfalfa, it was reasonable to suggest that host specificity gene (s) of nodulation (*hsn*) are carried by pGR3.

In line with the above results, a Nod⁻ mutant, obtained after random Tn5 mutagenesis carried an insertion in a 6.8 kb *Eco*RI fragment located between the *nod* and *nif* genes (Fig. 1). Introduction of the wild type 6.8 kb *Eco*RI cloned into pRK 290 into the mutant resulted in the restoration of the Nod+ phenotype. Directed Tn5 mutagenesis of this DNA fragment showed that about a 2 kb region codes for nodulation functions. This mutant was not complementable by the *sym* plasmid of *R. leguminosarum* (A. Kondorosi and E. Kondorosi, unpublished). Therefore, this *nod* gene may be involved in the control of host specificity of nodulation. A third class of *nod* mutations were not located on the *nod⁻ nif* region [63] shown in Fig. 2. These Nod⁻ mutants of *R. meliloti* were able to evoke root hair curling (Hac+ or "reactive" [50]). After infection of alfalfa with such mutants bacteria were observed inside the epidermal cells, although no

Fig. 1. Physical-genetic maps of the *nod⁻ nif* region for different *R. meliloti* strains. Maps for *R. meliloti* 41 (Rm 41)[55], for Rm1021[63] and for Rm102F34[95] show only *Eco*RI restriction sites. Hybridization of cloned fragments of Rm41 was performed with total nodule RNA as hybridization probe[55].
+ hybridization; — no hybridization above background.

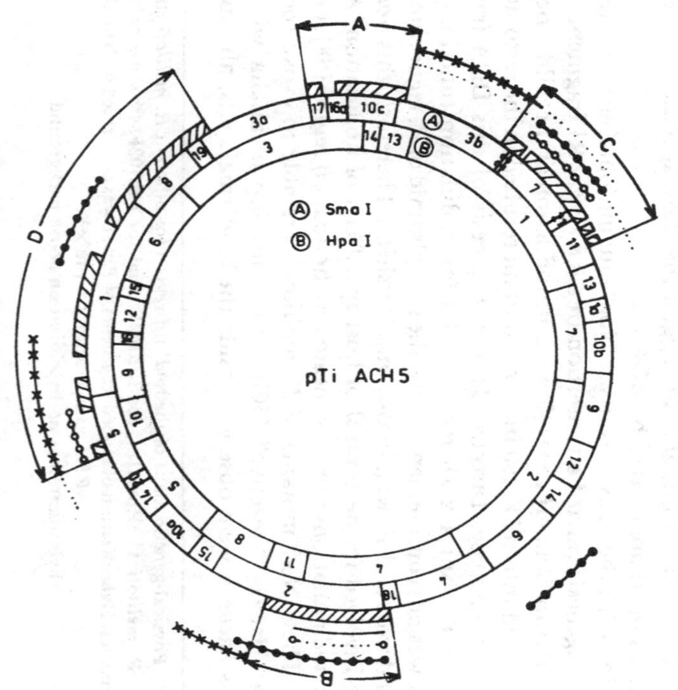
pTi ACH5
Ⓐ Sma I
Ⓑ Hpa I
A
B
C
D

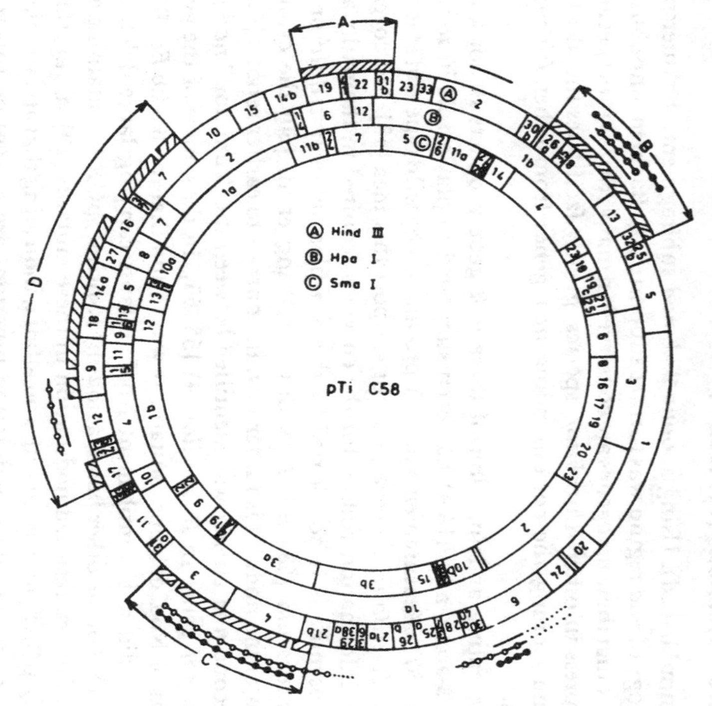
pTi C58
Ⓐ Hind III
Ⓑ Hpa I
Ⓒ Sma I
A
B
C
D

infection threads were found. The mode of entry of these bacteria into plant cells was apparently different from that of the wild type *R. meliloti,* since necrotic symptoms on the plant root tissue were seen [50]. These mutants did not attach to root hairs and did not bind purified alfalfa agglutinin [64].

It is possible to classify Nod⁻mutants by using mixtures of different mutants to inoculate plants [34, 65]; (G.B. Kiss et al., unpublished). Some Nod⁻mutants could be "actively helped" into nodules formed by Nod+ strains, whilst others could not. Interestingly Nod⁻strains, mutated in the common *nod* genes, were those which could be helped (the authors' laboratory).

Studies on the involvement of extracellular polisaccharides suggested that these cell components might be implicated in the nodulation process [66, 67]. Analysis of a number of *R. trifolii* and *R. leguminosarum* mutants did not give an absolute correlation between the nodulation ability and the ability to synthesize extracellular polysaccharides in the form of microfibrils or capsules [51].

NIF GENES ARE ALSO PLASMID BORNE

The reduction of dinitrogen to ammonia takes place in the bacteroids located within the plant cells in the nodule. Dinitrogen is reduced by an enzyme called nitrogenase consisting of two subunits, coded by *nif*K, *nif*D genes. Nitrogenase reductase, the *nif*H gene product also participates in the enzymatic reduction of N_2 These three structural genes transcribed in one transcriptional unit were cloned on an *Eco*RI fragment from *K. pneumoniae* [68]. The resulting plasmid, pSA30 was used as a probe in DNA-DNA hybridization experiment to show homology with *Eco*RI fragments with different nitrogen-fixing procaryotes [7]. Six out of 19 organisms tested were *Rhizobium* species. Similar hybridization experiment clearly demonstrated the conservation of the *nif* genes (see below) in unrelated bacteria and in blue-green algae [8]. From these results, however the location of the *nif* genes could not be deduced. Nuti et al. [6] demonstrated, that *nif* genes are on plasmids isolated from three different *R. leguminosarum* strains. The modification of the agarose gelelectrophoresis of Eckhardt [30] made possible the detection of very large plasmids (megaplasmids>300 Md) in agarose gel from different bacteria [19, 69]. Transferring the plasmid bands to nitrocellulose filter following hybridization with specific probe, the *nif* genes could be located on the appropriate plasmid in strains carrying more than one plasmids [19, 69, 70, 71].

Fig. 2. Map of Ti-plasmids and regions of homology with *Rhizobium* plasmids. The lines outside the map indicate homologous regions of Ti-plasmids to pRle100la———; pRtr5a o-o-o ; pRph3622b •-•-•; pRmeL5-30×-×-×; The weak homology is represented by dotted lines. Common regions A, B, C and D of Ti-plasmids are indicated by cross-hatched areas (data reprinted from Prakash and Schilperoort [138] with the kind permission of the publisher).

In strains of *R. leguminosarum, R. phaseoli,* and *R. trifolii* the *nif* sequences were found on plasmids with different molecular weight. This is in contrast with all *R. meliloti* strains tested in which a megaplasmid (molecular weight more than 300 Md) carried the *nif* genes.

There are several reports on the failure to demonstrate the presence of plasmid DNA in effective *Rhizobium* strains carrying *nif* sequences [23, 56]. Since plasmid identification might be a technical problem, no strong conclusion can be drawn for the chromosomal location of the *nif* genes.

FINE STRUCTURAL ANALYSIS OF *NIF* GENES

As discussed above, the *nif* structural genes in fast-growing rhizobia are plasmid-borne. In slow-growers this has not been substantiated, but it cannot be ruled out. Our knowledge on the structure and regulation of *nif* genes in various *Rhizobium* species is summarized below. The structural genes of enzyme nitrogenase are highly conserved in various nitrogen-fixing microorganisms.

It was found that the *nif*H and *nif*D genes hybridized strongly to DNA from many *Rhizobium* species, such as *R. meliloti* [72, 19], *R. leguminosarum* [73], and *R. phaseoli* [74]. In hybridization experiments with *R. japonicum* strong homology was found only with the *nif*D sequences [75]. Homology for the *nif*K gene was not detected in the various *Rhizobium* species tested. Under less stringent hybridization conditions, however, weak hybridization with the *R. meliloti nif*K gene was observed [76]. Other *K. pneumoniae nif* genes did not exhibit detectable homology with *R. meliloti* DNA, but the *nif*A hybridized to the *R. leguminosarum* DNA sequences [77] (see below).

Based on the conservation of *nif*H and *nif*D genes, the *nif* structural genes were identified and cloned from *R. meliloti* [7, 78, 19], *R. phaseoli* [74], *R. trifolii* [79], *R. leguminosarum* [73], *R. parasponia* [56, 79] and from *R. japonicum* [80]. Detailed restriction maps for the *nif* regions have been reported so far for three different *R. meliloti* strains (Fig. 1). These maps are fairly similar, as expected. The entire nucleotide sequence of *nif*H and a partial sequence of *nif*D have been determined for *R. meliloti* 41 [81]. Partial *nif*H and *nif*D sequences for *R. meliloti* 2011 have also been published [76]. Comparison of available *nif*H and *D* sequence data for the two strains indicated that the structural genes have essentially the same nucleotide sequence.

By comparing these sequence data with nucleotide and amino acid sequences of *nif*H and *nif*D from *K. pneumoniae* [82, 83], *Anabaena* 7120 [84], *Clostridium pasteurianum* [85, 86] and *Azotobacter vinelandii* [87], the amino acid sequences of the *R. meliloti nif*H [81] and *nif*D [88] gene products could be deduced. The *R. meliloti nif*H product (nitrogenase reductase) consists of 297 amino acid residues and has a molecular weight of 32,740 daltons. Approximately the same size of protein was detected in *E. coli* minicells, when the *nif*H was placed after a strong *E. coli* promoter

[89]. The *R. meliloti* nitrogenase reductase shares 70, 67 and 60 per cent amino acid homology with nitrogenase reductase from *Anabaena* 7120, *K. pneumoniae* and *C. pasteurianum*, respectively. At triplet codon level, however, the homology with the *nif*H of *Anabaena* 7120 and of *K. pneumoniae* is only 27 per cent and 34 per cent, respectively. This indicates that the amino acid sequences are more conserved than the nucleotide sequences [81].

A late evolution of the nitrogenase genes has been suggested by Postgate [90]. The fact that *nif* genes are located on plasmids in many *Rhizobium* species and in some other nitrogen-fixing organisms seems to give support to the recent appearance and lateral distribution of the *nif* genes. On the other hand, however, the structural conservation of nitrogenase reductase is most likely related to the function and not to the late evolution of the *nif* genes [81, 82].

Analysis of the nucleotide sequence upstream from the traslation initiation codon of *nif*H, revealed a potential ribosome-binding site [91] at positions —6 to —10 [81]. Although the 3′ end of the *Rhizobium* 16S RNA has not been determined yet, it is likely that these sequences for enterobacteria and rhizobia are similar. This is supported by studies on the expression of *Rhizobium* genes in *E. coli* [75, 89].

The organization of the *nif* structural genes in some fast-growing rhizobia is strikingly similar to that in *K. pneumoniae*. The *nif*HDK genes form one transcriptional unit in *K. pneumoniae* [92, 93, 94] and also in *R. meliloti* [76], in *R. leguminosarum* [70], and in *R. trifolii* [79]. In *R. meliloti* a 5–6 kb transcript from the *nif* structural genes was identified by Northern hybridization [95] (E. Kondorosi and A. Kondorosi, unpublished). Nucleotide sequencing [81], genetic complementation [76] and S_1 nuclease mapping [95] revealed the same polarity (*nif*HDK) in *R. meliloti* as in *K. pneumoniae* [93].

In *R. leguminosarum* a somewhat smaller transcript from the *nif* structural genes was detected by Krol et al. [70]. Despite this difference in size, it was suggested that *nif*HD and K form one operon also in this species. On the other hand, studies on the regulation of the synthesis of nitrogenase components indicated that the synthesis of the two components (MoFe and Fe proteins) are regulated independently in *R. leguminosarum* [70, 96]. Krol et al. [97] suggested that this is probably due to the regulation of the *nif* messenger RNA at the level of translation. Obviously, further experiments are needed to support this hypothesis.

In *R. parasponia* (the same strain as *Rhizobium* sp. NGR 234 [56, 79, 98]) and in *R. japonicum* [75, 98], however, the *nif*H and D genes occur in separate operons. In *R. japonicum nif*D and K are organized in one transcriptional unit [75].

The structure and control of the *nif*HDK promoter regions of *R. meliloti* and *K. pneumoniae* also show similarities. By mapping the *in vivo* transcription initiation sites, the *nif*H promoter regions of both organisms have been determined [99]. Analysis of their nucleotide sequences revealed that they do not exhibit strong homology with consensus promoter sequences of *E. coli* [100]. Using *E. coli* RNA polymerase and purified *R. meliloti* or *K. pneumoniae nif* DNA fragments as templates, no significant transcription *in vitro* could be detected [99] (I. Török, unpublished). Comparison of the two DNA sequences upstream from the transcriptional start site revealed a fairly high homology between the two promoters. A 40 bp region upstream from the transcription point exhibited 50 per cent homology in the two species. Three regions of exact homology were found: at –70 (6 bp), –32 (8 bp) and –14 (5 bp). The most interesting is the conservation of eight base pairs (ACGGCTGG) from –32 to –40, since the –30 to –40 region upstream from the transcription start site is generally involved in the positive regulation of transcription [100].

From the sequence homology of the two *nif*H promoters, it was plausible to suppose that the two genes are under the same type of control [99]. To test this, both promoters were fused to the *lac*Z gene of *E. coli*. By measuring the level of β-galactosidase, both promoters were shown to be activated by the *K. pneumoniae nif*A gene. Moreover, the presence of multicopy plasmids containing either the *K. pneumoniae* or *R. meliloti nif*H promoter inhibited N_2 fixation in *K. pneumoniae* [99, 101]. These results suggested that the *K. pneumoniae nif*A product could bind to both *K. pneumoniae* and *R. meliloti nif*H promoters and the higher copy number of *nif*H promoter titrated the *nif*A product in both cases. Furthermore, a regulatory protein similar to *K. pneumoniae nif*A should be present in *R. meliloti*. Since *R. meliloti nif* genes are not expressed under free-living conditions, this regulatory protein must be under some kind of symbiotic control.

In *K. pneumoniae* the other 14 *nif* genes are located in the vicinity of the *nif* structural genes on a 24 kb contiguous chromosomal region, forming 7 or 8 operons [102]. In *R. meliloti* several *fix* genes have been localized to the right of the *nif* structural genes (Fig. 1) [53, 76, 103].

Directed Tn5 mutagenesis was performed on a 18 kb region adjacent to and surrounding the *nif* structural genes of *R. meliloti* 2011 and the Tn5 insertions were assayed for Fix^- phenotype [76]. Fix^- mutations were found in two clusters: one is the 6.3 kb *nif*HDK operon and the other is at least 5 kb in size. The two clusters are separated by a 1.6 kb region which is non-essential for N_2 fixation. Complementation analysis indicated that the second *fix* cluster forms at least two transcriptional units. This region carries gene (s) controlling the expression of *nif* structural genes (W. Szeto and L. Zimmermann, personal communication).

In *R. meliloti* 102F34 directed Tn5 mutagenesis revealed a 15 kb *fix* region comprising the *nif* structural genes [95]. An unessential 1.9 kb region to the right of the *nif*H gene was found also in this strain. No symbiotic genes were found on the neighbouring 10 kb regions on both sides of the *fix* region. It is likely that this 15 kb *fix* gene cluster corresponds to the *nif* cluster of *K. pneumoniae*. The number of *nif* genes, however, is either lower in *R. meliloti* or several *nif* genes are located somewhere else in the *R. meliloti* DNA.

In *R. leguminosarum* two *nif* clusters separated by a *nod* region were identified by Tn5 insertion mutagenesis and molecular cloning on plasmid pRL1JI [77]. The first region carried genes hybridizing to the nitrogenase genes from *K. pneumoniae* [104]. The other region was homologous to the *nif*A gene of *K. pneumoniae*. No homology was found with the cloned *nif*L sequences [77].

In *K. pneumoniae* not only the *nif*H promoter but other *nif* promoters are also under the control of *nif*A gene product [105]. It is likely that these promoter sequences are homologous to each other. Preliminary characterization of *nif* sequences in *R. phaseoli* revealed reiteration of *nif* sequences. At least some of these repeated sequences seem to be located on a large plasmid [74]. It is possible that some of these reiterated sequences are other *nif* promoters. If this is true, other symbiotic genes with the same type of promoter could be easily identified and cloned.

Studies on the transcription of *nif* genes from several fast-growing *Rhizobium* species indicated that these genes are strongly expressed in the nodules [70, 103, 106, 107, 108]. On the other hand, the *nif* genes were not transcribed under free-living conditions. In these experiments *sym* plasmid DNA or cloned *nif* fragments were hybridized with ^{32}P-labelled RNA, isolated either from the nodules or from vegetatively grown bacteria.

Using the same approach, Paau and Brill [64] have found that *nif*-specific transcripts appeared at a very early stage of symbiotic association between *R. meliloti* and alfalfa when bacteria were still in the infection thread. Mature bacteroids of young nodules contained the highest amount of *nif* transcripts. In senescent nodules, however, DNA-RNA hybridization experiments indicated that no selective amplification of the *nif* genes occurs during symbiosis [64, 70]. Moreover, there is no evidence for rearrangements of DNA regions carrying the *nif* genes [64].

In *K. pneumoniae* positive correlation (direct or indirect) between derepression of *nif* genes and the accumulation of guanosine tetraphosphate, ppGpp, was found [109]. In *R. meliloti*, however, ppGpp, is not accumulated during amino acid starvation and the metabolism of ppGpp is different from that observed in *Enterobacteriaceae* [110]. One can speculate that the inability of most *Rhizobium* species to *fix* nitrogen *ex planta* may be related to their unusual ppGpp metabolism.

Other genes involved in symbiotic nitrogen fixation

Genes involved in the later steps of the nodulation process, except *nif* genes, have not been identified yet. Since mutations in these genes result in Nod$^+$ Fix$^-$ phenotypes, at present such mutants are classified as Fix$^-$.

Genetic analyses on various *Rhizobium* species indicate that the number of *fix* genes is much higher than the number of *nod* genes [33, 35, 111, 112]. In many cases, however, these *fix* genes are not specifically involved in symbiotic nitrogen fixation. Many mutations with recognizable phenotype in the free-living state, lead to the loss of symbiotic activity [113, 114]. In a number of reports on such mutants, however, the possibility of two independent mutations has not been ruled out. Therefore, in many cases the reported correlations between biochemical and symbiotic lesions need to be reinvestigated.

Another difficulty arises from the lack of expression of symbiotic genes under free-living conditions. Therefore, to isolate mutants defective in these genes, direct screening of mutagenized populations for symbiotic mutants is needed. In fact, mutants isolated after such procedure are now available for several *Rhizobium* species such as *R. japonicum* [111], *R. leguminosarum* [112], *R. trifolii* [34] and *R. meliloti* [33, 35]. Recently many papers appeared on the isolation of symbiotic mutants with Tn5 [18, 19, 32–35, 83, 104]. Tn5 directed mutagenesis of cloned DNA fragments [72] from the *sym* plasmid [53, 76, 95] resulted also in obtaining large number of symbiotic mutants. Scott et al. [115] demonstrated that about 0.3–0.5 per cent of Tn5-induced mutants of *R. trifolii* were affected in symbiotic nitrogen fixation. These results suggested that at least 15–20 genes are involved in establishing symbiosis. The phenotype of the Tn5 induced mutants varied from the complete loss of nodulation to the production of ineffective nodules with different morphology. The presence of Tn5 in the genome was demonstrated by hybridization experiments using Tn5 specific probes.

In *R. phaseoli* 187 Tn5 derivatives of pRP2JI were isolated of which 7 showed Fix^- phenotype. Since melanine non-producing Tn5 derivatives were Fix+, melanine production is not necessary for the induction of nitrogen-fixing nodules [18].

Several *fix* genes are present on the *sym* plasmids outside the *nif* region in *R. leguminosarum*. Plasmid pRL1JI was tagged with Tn5 and Fix^- mutations were mapped about 30 kb away from the *nif* genes [104]. Eight independent Tn5 derivatives formed nodules on pea but failed to reduce acetylene. Two derivatives produced low amounts of leghemoglobine while the remaining part induced green or yellow nodules. These bacteria did not differentiate to bacteroids in nodule cells. Electronmicroscopy of the nodules induced by these mutants indicated that bacteria were not released from infection threads, indicating that the (se) gene (s) controls a later step of symbiotic development.

In *R. meliloti fix* genes are present at several regions of the *sym* megaplasmid. Hybridization of nodule RNA to R-primes carrying various segments of the megaplasmid gave some support to this [108]. Analysis of Fix^- deletion mutants in *R. meliloti* 41 clearly indicated the presence of *fix* genes outside the region mapped [53](Fig. 1). In some Fix^- Tn5 mutants of *R. meliloti* 41 the mutation mapped on the chromosome while the others on pRme41*b* [35].

Some *fix* genes are probably involved in the maintenance and functioning of the nitrogen-fixing nodules. It is possible that several of these genes are carried by indigenous plasmids, as was shown for the *hup* genes [13].

Hydrogen gas is always produced during nitrogen fixation. The reaction is catalyzed by nitrogenase, consuming one-third of the energy available for the enzyme. Another enzyme called uptake hydrogenase is able to recycle H_2 with energy production. Some nitrogen-fixing bacteria possess active hydrogen uptake (*hup*) system resulting in more effective nitrogen-fixation [116]. Brewin et al. [117] demonstrated that in *R. leguminosarum* strain 128C53 the determinants for hydrogenase activity are genetically linked to plasmid pRL6JI. This plasmid can be mobilized by a plasmid making it possible to introduce uptake hydrogenase activity to other species lacking these determinants.

There are a number of properties coded by various *Rhizobium* strains which are important with respect to field inoculation. For instance, the ability of a *Rhizobium* strain to compete with another strain is an agriculturally important property. Unfortunately, we know very little about the basis of competitiveness and other related properties. On the other hand, *Rhizobium* strain improvement by plasmid transfer has been reported [118]. Its genetic determinants have not been identified yet. Behavioural mutants altered in mobility are now available in *R. meliloti* [119, 120]. The mobile strains have some selective advantage over non-mobile ones. The genes responsible for this property have not been analysed yet.

PLASMID-CODED NON-SYMBIOTIC FUNCTIONS IN RHIZOBIA

The present knowledge about plasmid-coded symbiotic functions was discussed in the earlier sections. Here is summarized data concerning other non-symbiotic functions which were also shown to be located on indigenous plasmids. Such functions used as selectable markers of plasmids greatly facilitated genetic work.

Table 1. Plasmid-coded functions in rhizobia

Species	*nod*	*nif*	*fix*	Bacteriocin production	Melanin production	*hup*	Self-transfer
R. leguminosarum	+	+	+	+		+	+
R. phaseoli	+	+	+		+		
R. trifolii	+	+	+				+
R. meliloti	+	+	+				+
R. species NGR 234	+	+					

Pigment production

Production of a dark pigment is a characteristic feature of nearly all *R. phaseoli* strains. Pigment production was not observed on minimal medium, unless tyrosine was added. Beynon et al. [17] found a direct correlation between loss of pigment production and the loss or deletion of pRP1JI, one of the plasmids of *R. phaseoli*. Therefore, Pig^{+} phenotype was used as a marker for the presence of that plasmid. No pigment was produced by 20 different strains of both the *R. leguminosarum* and *R. trifolii* examined.

Bacteriocin production

One of the best known non-symbiotic property of various *R. leguminosarum* isolates is bacteriocin production. 97 different *R. leguminosarum* isolates were tested by Hirsch [121] and another 52 strains were studied by Wijffelman et al. [122] for their ability to produce bacteriocin. The majority of the strains were characterized by the production of small molecular weight bacteriocin (e.g. strain 300 and derivatives). Another group was found to produce medium size bacteriocin, like strain 248, 306, 309, etc. Even in those strains which do not excrete small bacteriocin, the genes responsible for small bacteriocin production were demonstrated. However, their function was repressed in wild type strain. A correlation between the presence of repression of small bacteriocin production (Rsp) and the presence of highly self-transmissible plasmids carrying also medium bacteriocin production (Med) was found. These plasmid-coded functions (Rsp, Med) were expressed also in other *R. leguminosarum, R. trifolii, R. phaseoli* and *A. tumefaciens* strains [27, 122].

Transfer properties

Due to the number and size of the resident plasmids in rhizobia genetical and biochemical work to study plasmid-coded functions is difficult. Therefore the property of some of the *Rhizobium* plasmids that they are able to transfer into a new host with a high frequency is extremely valuable. Identification of symbiotic functions on such plasmids, their expression in new hosts, the ability to complement symbiotic defects, etc. are extensively studied (see previous sections). In addition, the transfer of one of the resident plasmids into a new host (e.g. Ti plasmid-cured *Agrobacterium tumefaciens*) allows the purification of a single plasmid [16]. Furthermore, the selective and more efficient mutagenesis of genes coded by one specific plasmid is possible.

SELF-TRANSMISSIBLE PLASMIDS

Some of the indigenous plasmids of various *Rhizobium* species carry all informations necessary for conjugal transfer. Self-transmissibility of such plasmids has been demonstrated by following the transfer of either symbiotic or non-symbiotic markers located on these plasmids.

R. leguminosarum

Field isolate of *R. leguminosarum* strains 248, 306 and 309 were shown to harbour self-transmissible plasmids [121]. From strain 248 the transfer of the second smallest plasmid pRL1JI (mol. weight 130×10^6) was observed [27]. From strain 306 the second smallest plasmid pRL3JI, (mol. weight 180×10^6, [71]), and from strain 309 the third smallest plasmid pRL4JI (mol. weight 160×10^6) were transferred [27]. In each case transconjugants gained the ability to produce medium molecular weight bacteriocin, and the small bacteriocin production characteristic of the recipient was repressed (see previous section). pRL1JI, in addition, was able to transfer nodulation properties [46]. Plasmids pRL1JI, pRL3JI and pRL4JI transferred between strains of *R. leguminosarum* at a high frequency: 10^{-1}-10^{-2}/recipient [46]. Although pRL3JI and pRL4JI did not transfer symbiotic genes at high frequency, some of the transconjugants acquired nodulation ability (frequency: 10^{-6}/recipient). Data suggest that mobilization of nodulation functions occurred by formation of a cointegrate plasmid which consisted of bacteriocin plasmid and nodulation genes. These self-transmissible plasmids are also known to mobilize chromosomal markers at a frequency of 10^{-7}-10^{-8}/recipient.

Transfer properties in interspecific matings with different *Rhizobium* species were also studied [15, 17, 71]. pJB5JI, a derivative of pRL1JI, into which the Km^R transposon Tn5 had been inserted [43] was transferred into *R. phaseoli* or *R. trifolii.* Transconjugants gained the ability to nodulate peas.

The self-transfer of the two smallest of the six indigenous plasmids of *R. leguminosarum* strain 300, pRL7JI and pRL8JI was also demonstrated [123]. pIJ1001::Tn5, a deleted and Tn5-marked derivative of pRL7JI, transferred at a frequency of 10^{-8}/recipient, while transconjugants carrying pRL8JI::Tn5 arose with a frequency of 10^{-6}/recipient. Neither pRL8JI or pIJ1001 carry genes for nodulation or nitrogen fixation.

One of the three plasmids of *R. leguminosarum* strain TOM, pRL5JI (mol. weight 160×10^6) was able to transfer into other *R. leguminosarum* strains [13]. Analysis of transconjugants revealed that pRL5JI carried not only nodulation functions but also host range specificity of TOM. Several characteristics of pRL5JI different from that of pRL1JI, pRL3JI and pRL4JI were found: its transfer frequency was lower (10^{-5}-10^{-6}/recipient), and, in addition, it did not transfer the ability of medium bacteriocin production, although the donor strain TOM was able to produce it.

R. trifolii

A self-transmissible plasmid was identified in *R. trifolii* strain LPR5001 [44]. pRtr5a (mol. weight 180×10^6) was transferred at a frequency of 10^{-4}-10^{-6} into other *R. trifolii* strains. Introduction of pRtr5a restored the nodulation and nitrogen fixation ability of a plasmid-cured *R. trifolii.*

This *sym* plasmid was also transferred into *R. leguminosarum* and *A. tumefaciens* and the expression of plasmid coded functions was observed.

R. meliloti

Relatively little is known about the transfer properties of *R. meliloti* plasmids. Self-transmissibility of the megaplasmid pRme41*b* of *R. meliloti* 41 was observed only at a barely detectable frequency, which did not allow interspecies transfer of the megaplasmid (Z. Banfalvi and A. Kondorosi, unpublished).

R. loti

Transfer of an indigenous plasmid of *R. loti* strain NZP2037 to other rhizobia was studied by Pankhurst et al. [124]. The plasmid pRL2037a (mol. weight 240×10^6) was labelled with Tn5 and the transfer of Km^R was followed. The frequency of transfer varied between 10^{-4} and 5×10^{-7} depending on the recipient species used.

MOBILIZATION OF NON-SELF-TRANSMISSIBLE PLASMIDS

Use of P-1 type plasmids

For a large number of *Rhizobium* plasmids self-transmissibility was not demonstrated, and some of them proved to be Tra^-. In a number of cases, however, such plasmids could be mobilized by the help of broad host range P-1 type plasmids. When an R plasmid pRL180 was introduced into *R. leguminosarum* RCC1001 mobilization of the *sym* plasmid was observed at a frequency of 10^{-5}-10^{-6}/recipient [14].

Similarly pWZ2, the nodulation-conferring plasmid of *R. trifolii* 24 was mobilized by RP4[17].

In other cases first a cointegrate was formed between an R plasmid and a *Rhizobium* plasmid. The construction of one of such hybrid pPN1 consisting of R68.45 and *sym* plasmid of *R. trifolii* [47], has been described earlier. This cointegrate could be transferred into *R. leguminosarum*, *R. phaseoli*, *E. coli* and *P. aeruginosa* (Ronson, unpublished). A somewhat different approach was used when plasmid pAK11 carrying the *mob* region of RP4 was inserted into the indigenous megaplasmid [45] pRme41*b* of *R. meliloti* 41. By using pJB3JI (KmS derivative of R68.45) the hybrid was transferred into *R. meliloti* 41 at a very low frequency: 10^{-7}-10^{-8}/recipient. However, when recipients with a deletion in the *nif-nod* region of pRme41*b* were used, transconjugants appeared at a higher frequency (10^{-4}-10^{-5}/recipient), suggesting that some genes coding for entry exclusion of the same type of plasmid might have also been deleted.

Mobilization by Rhizobium Plasmids

Self-transmissible plasmids of *R. leguminosarum* strains could also be used for mobilization of other *R. leguminosarum* plasmids in intra- and interspecific matings. Brewin [117] used two self-transmissible *R. legumin-*

osarum plasmids for mobilization of pRL6JI, a plasmid of strain 128C53 which was a likely carrier of Nod^+ Hup^+ determinants. In addition to the genetic markers of the transmissible plasmids, transconjugants acquired also nodulation ability and hydrogenase activity. As it was mentioned earlier, the two smallest plasmids of *R. leguminosarum* strain 300 are able to transfer at a low frequency. Introduction of pRL1JI, a highly transmissible plasmid, into this strain increased the transfer of the two resident plasmids by a factor 10^2-10^5, compared to the frequency of their self-transfer [17, 123].

R-primes

It was possible to obtain shortened and transmissible derivatives of the large *Rhizobium* plasmids, by constructing P-1 type R-prime derivatives [9, 36, 49, 125, 126].

Incompatibility properties of *Rhizobium* plasmids

As it has been shown, *Rhizobium* plasmids could be transferred between strains of either the same or different species by conjugal transfer or by mobilization with transmissible plasmids. Therefore data about incompatibility properties of a number of plasmids are known.

Detailed investigations confirmed that self-transmissible bacteriocinogenic plasmids of *R. leguminosarum* field isolates were incompatible [46, 127]. Introduction of the Tn5 marked derivatives of any of these plasmids resulted in the elimination of medium bacteriocin production of the recipient strain. The presence of the resident incompatible plasmid decreased the transfer of the incoming plasmid 10-100 fold.

On the contrary, when any of the medium bacteriocinogenic plasmids was introduced into *R. leguminosarum* strain 300, no loss of the resident plasmids could be observed by gel electrophoresis [27, 128]. Although small bacteriocin production was eliminated from the transconjugants, that was due to repression coded by the entering plasmid [122].

It is worthwhile to note that a self-transmissible plasmid of *R. leguminosarum* strain TOM, which carries host specificity and *nod* functions, was also shown to be compatible with another nodulation-conferring plasmid of *R. leguminosarum* strain 248 [13]. The two plasmids could coexist either in *R. leguminosarum* strain 300 or TOM.

Similar observations were found when compatibility properties of *sym* plasmids of different strains were studied [14]. All of the indigenous plasmids of the recipient strains were present in the transconjugants which had received *pSym1* of *R. leguminosarum* strain 1001. Therefore unidentified *sym* plasmids of these strains seem to be compatible with *pSym1*. The symbiotic plasmid of *R. trifolii*, *pSym5*, could also coexist with *pSym1* either in *R. trifolii* [14] or in *R. leguminosarum* [44], and plasmid-coded functions could be expressed from both of them. This stable coexistence of *sym* plasmids of different species is surprising however, since they carry

DNA regions of extensive homology as discussed later.

Transferring the Tn5 labelled derivative of pRL1JI, of *R. leguminosarum* into other *Rhizobium* species, it was found to be compatible with the resident plasmids of *R. trifolii* [15] and with the plasmids of *R. phaseoli* [17]. In *R. phaseoli* transconjugants, however, the simultaneous presence of informations coded by the different replicons resulted in delayed and reduced nodulation. To establish efficient nodulation on peas, the loss of information encoded by pRL1JI (one of the plasmids of *R. phaseoli*), was necessary. Nodulation plasmid of another *R. phaseoli* strain was incompatible with a *R. leguminosarum* plasmid, pIJ1001 [123]. In transconjugants, a high frequency of cointegrate formation between the two plasmids was observed, followed by the recombination of pIJ1001 and elimination of *R. phaseoli nod* plasmid.

As it is suggested by Brewin [127], bacteriocinogenic plasmids pRL1JI, pRL3JI, and pRL4JI, and some of the nodulation plasmids of different *R. leguminosarum* strains can be classed as follows: On the basis that they cannot coexist in the same cell as separate replicons, pRL1JI, pRL4JI, pRL3JI and pRL6JI belong to one incompatibility group. Since pRL10JI and pRL5JI (nodulation-conferring plasmids of *R. leguminosarum* strain 300 and TOM, respectively) could be maintained stably in the presence of the plasmids mentioned before, and could coexist also with each other, they belong to a second and third incompatibility group. Nodulation plasmids of *R. phaseoli* might belong to a fourth group.

It is interesting to note that when pPN1, a cointegrate of a *sym* plasmid of *R. trifolii* and R68.45, was introduced into different *Rhizobium* strains, the loss of resident *sym* plasmids was observed in certain cases [47]. The reason of this observation is not yet clear, since neither the *sym* plasmid of *R. trifolii* nor R68.45 did not express incompatibility towards resident *sym* plasmids. In *R. loti* or *R. meliloti* pPN1 did not cause the loss of resident plasmids, but its presence influenced the expression of their genetic information.

Opine utilization

It is well established that genes carried by the Ti plasmid of *A. tumefaciens* code for the production and utilization of opines [129-131]. Opine-like substances are present also in some nitrogen-fixing nodules [132]. There is some indication that genes coding for the catabolism of such compound in *R. meliloti* L5-30 is coded by the megaplasmid (J. Tempe, personal communication). Some *Rhizobium* strains are able to utilize a typical *Agrobacterium* opine, octopine and its related derivatives: in *R. meliloti* 41 this is coded by the *sym* megaplasmid [55]. It is not known whether opine metabolism is directly involved in symbiotic nitrogen fixation.

Other characteristics

Experiments suggesting that *Rhizobium* plasmids code for certain non-symbiotic functions such as multiple antibiotic resistance [133], or β-galacturonase-inducing ability [134] were described for different *Rhizobium* strains. Unfortunately, no physical evidence was shown for the presence or absence of plasmids.

Some correlation between the loss of infectiveness in *Rhizobium* and the ability to form mucous colonies was reported earlier [135]. Prakash et al. [12] later isolated "rough" mutants from *R. leguminosarum* LPR1705 by heat treatment. All "rough" derivatives were Nod$^-$ on peas. In addition, the absence of about a 110 Md plasmid was demonstrated in one of the mutants, though two other larger plasmids were still present. The loss of the smooth colony phenotype, however, does not necessarily correlate with the loss of infectiveness.

DNA HOMOLOGY OF PLASMIDS IN RHIZOBIA

Homologous DNA sequences on the large plasmids of fast-growing rhizobia

In earlier sections the localization, structure and function of *nif* and *nod* genes of rhizobia were discussed in detail. It was shown that according to DNA hybridization data and to physical mapping *nif* and *nod* regions were highly conserved. The extent of homology in regions outside the *nif* structural genes was studied after preparation of individual plasmids pRle1001a, pRtr5a and Rph3622b [16].

Southern blots of *Rhizobium* plasmids digested with restriction endonucleases were hybridized with ^{32}P-labelled plasmids of other *Rhizobium* species. High degree of homology was demonstrated between plasmids pRtr5a, pRle1001 and pRph3622b. pRmeV7, a plasmid of *R. meliloti* strain V7 which failed to hybridize to *Klebsiella nif* genes, does not share significant homology to these plasmids except to pRtr5a.

Construction of the physical map of the 150 Md pRle1001a [136] allowed the conclusion that restriction fragments homologous to pRtr5a including the one hybridizing with *K. pneumoniae nif* structural genes form one long contiguous region on the physical map of pRle1001a. A part of this region is also preserved in pRph3622b.

Transmissible plasmids of field isolates of *R. leguminosarum* 248, 306 and 309 share many common properties [27]. Nevertheless, there is a significant difference between them, since pRL1JI carries symbiotic gene functions (Nod, Fix). As it was demonstrated by Hombrecher [71] this plasmid carried *nif* structural genes, while pRL3JI and pRL4JI did not hybridize to *Klebsiella nif* genes. Based on transduction experiments some DNA sequence homology, however, was suggested between these plasmids in the region determining production of medium bacteriocin [46].

Since all of the *R. leguminosarum* strains examined so far harbour more than one plasmid, and no general method is available for the efficient purification of individual plasmids, other possible homologous regions between these plasmids are not known.

Jouanin et al. [137] analysed DNA sequence homology of six middle-size plasmids of *R. meliloti* strains L5-30, 41, 102F51, 12, 1322 and 54032. The molecular weights of these plasmids range from 89 to 143 M daltons. Plasmid DNA was purified by the alkaline denaturation method, which did not allow the copurification of megaplasmids of these strains. This was confirmed by the lack of hybridization using *K. pneumoniae nif* probe. DNA hybridization experiments between the plasmids listed above gave evidence that sequence homology is general in *R. meliloti* medium size plasmids studied, whatever the geographical origin of the strains was. Though the homology extended only to a few restriction fragments, in the absence of the physical map no conclusion could be drawn about the clustering of the hybridizing bands. The biological function of these common sequences is unknown.

Sequence homology of plasmids of slow-growing rhizobia

One ineffective and two effective strains of *Rhizobium japonicum* were studied by Haugland[23]. Plasmids from strain 61A76 and 61A24 were prepared, digested with restriction enzymes and cross-hybridized to the labelled probe of the other plasmid. Sequence homology, though limited, was demonstrated to a number of fragments. When total DNA was used as a probe a number of additional and more intense bands were found. Although plasmid preparation from strain *R. japonicum* 110 was unsuccessful, total DNA isolated from this strain also hybridized to the plasmid of strain 61A76 and particularly to 61A24 plasmid DNA. The presence of a megaplasmid in strain 110 which is sensitive to the preparation method cannot be excluded.

Homology between *Rhizobium* plasmids and Ti plasmids of *Agrobacterium tumefaciens*

Species of *Rhizobium* and *Agrobacterium* are closely related, they both belong to the family of Rhizobiaceae. It was of interest to study whether the taxonomical relatedness was reflected in common DNA sequences as well. Homology between *Agrobacterium* and *Rhizobium* plasmids was especially interesting, since tumour-inducing properties of *Agrobacterium* and many symbiotic functions of *Rhizobium* are located on indigenous plasmids.

Prakash and Schilperoort compared four *Rhizobium* plasmids (pRle1001a, pRtr5a, pRph3622b and the smaller plasmid of *R. meliloti* L5-30) to pTiAch5 (octopine) and pTiC58 (nopaline) plasmids of *Agrobacterium tumefaciens* [138]. Significant homology of *Rhizobium* plasmids to the common region B, C and D of pTiAch5 and pTiC58 was demonstrated (Fig. 2).

A large part of common region B is highly conserved in pRle1001a, pRtr5a and pRph3622b. These common sequences of Ti plasmids were shown to be involved in replicative functions and incompatibility [139, 140]. Moreover, B region probably carries part of Tra functions on nopaline Ti-plasmid [139].

C regions of Ti plasmids determine Tra functions [139]. Part of the common sequence C of both pTiAch5 and pTiC58 showed significant homology to pRtr5a and pRph3622b, but no homology to pRle1001a was found. It is worthwhile to mention that pRtr5a and pRph3622b are self-transmissible, while pRle1001a proved to be Tra^- [14, 43].

Region D, which is described to carry virulence functions [141, 142] both in octopine and nopaline plasmids, covers common and non-common sequences. pRtr5a carried fragments of considerable homology to the common region D of both Ti plasmids. pRle1001a hybridized only to D region of pTiC58; on the contrary, pRph3622b revealed hybridizing sequences only in the D region common to pTiAch5. These differences in hybridization within region D could be explained by the presence of non-common parts in this area of D.

The smaller (90 Md) plasmid of *R. meliloti* L5-30 does not share any homology to pTiC58. However, three regions of pTiAch5, which carry hybridizing fragments were identified. Several other regions of Ti plasmids were also shown to be conserved in the *nif* plasmids, but the function of these regions in Ti plasmids is not yet known.

On the basis of the detailed homology studies Prakash and Schilperoort [138] suggest, that *Rhizobium nif* plasmids and Ti plasmids of *Agrobacterium* may have derived from a common ancestor plasmid. According to them the highly conserved common region A of pTiAch5 and pTiC58 correspond to T-DNA although it did not show homology with any of the plasmids studied. In contrast, Hadley and Szalay[143] identified homologous regions between *Rhizobium* plasmids and T-DNA of octopine-type plasmid pTiB$_6$806.

T-DNA clones representing the right boundary region of T-DNA hybridized to rhizobial DNA fragments. Both the left side and the "core region" of T-DNA failed to show any homology, which seems to indicate that functions encoded by these regions are not conserved in *Rhizobium*.

DNA homology between plasmids of slow-growing *R. japonicum* strains and an octopine Ti plasmid was studied by Haugland [23]. Although some homology was detected between the plasmids, none of the fragments involved in the process of tumour induction showed any hybridization to *R. japonicum* plasmids. The role of the homologous sequences (especially of those coded by *Rhizobium* plasmids) remains to be determined.

CONCLUSIONS

The overwhelming majority of the *Rhizobium* species and strains harbour indigenous plasmids. Their number may be up to 8 or 9 in some strains. This amount of extrachromosomal DNA may represent at least 50 per cent of the total rhizobial DNA and may contain a large number of genes.

Until now vital genes, genes required for the general metabolic processes, have been localized only on one linkage map [144-147] which is considered as the bacterial chromosome.

Genes carried by the indigenous plasmids probably code for different and (under free-living conditions) non-vital functions. In many fast-growing species (*R. leguminosarum, R. trifolii, R. phaseoli, R. meliloti* and in some broad host range *Rhizobium* sp.) one of the plasmids (the *sym* plasmid) carries genes controlling nodulation and nitrogen-fixation processes and some of these genes are clustered. In slow-growers the situation is less clear, mainly due to techincal difficulties in demonstrating very large plasmids.

In addition to the essential symbiotic genes, the indigenous plasmids also carry a range of other traits. These might be involved in efficient nodulation or nitrogen fixation, or in the adaptation to soil conditions, in bacteriocin production, etc.

Although some chromosomal genes are expressed in the bacteroids, it seems that most of the genes directly involved in symbiosis are plasmid-borne. This kind of organization of symbiotic genes is especially interesting and there are several explanations for it. For instance, the presence and clustering of symbiotic genes on one *sym* plasmid can be a consequence of the late evolution and lateral distribution of these genes among rhizobia [90]. According to the "organelle genome" idea [11] the large symbiotic plasmids are equivalent of a chloroplast or mitochondrial genome and the bacteroids are the symbiotic organelles.

The majority of plant-associated bacteria contain large plasmids and in several cases genes involved in bacterium-plant interaction were shown to be plasmid-borne. The plant pathogen *Agrobacterium* and *Rhizobium* have especially many common features. This may reflect simply their evolutionary relatedness or their ability to interact with plants may also be related. For instance, the homology of T-DNA border fragment with DNA from several *Rhizobium* species, the presence of unusual compounds (opines) in the nodules or opine catabolism genes on *sym* plasmids are very intriguing, although the significance of these findings with respect to symbiosis has not been substantiated yet.

Identification of other plasmid-borne genes, studies on the function, control and organization of these genes may help us to answer the above questions.

ACKNOWLEDGEMENTS

We thank Miss Zsuzsanna Rácz for help in the preparation of the manuscript.

REFERENCES

1. Burns, R.C. and Hardy, R.W.F. *Nitrogen Fixation in Bacteria and Higher Plants,* Springer Verlag, New York (1975).
2. Phillips, D.A. Efficiency of symbiotic nitrogen fixation in legumes, *Ann. Rev. Plant. Physiol. 31,* 29-49 (1980).
3. Kondorosi, A. and Johnston, A.W.B. The genetics of *Rhizobium, Int. Rev. Cytol.* Suppl. *13,* 191-224 (1981).
4. Schwinghamer, E.A. In *A Treatise on Dinitrogen Fixation,* Section III (Editors, R.W.F. Hardy and W.S. Silver), John Wiley and Sons Inc., New York (1977).
5. Beringer, J.E., Brewin, N.J. and Johnston, A.W.B. The genetic analysis of *Rhizobium* in relation to symbiotic nitrogen fixation, *Heredity 45,* 161-186 (1980).
6. Nuti, M.P., Lepidi, A.A., Prakash, R.K., Schilperoort, R.A. and Cannon, F.C. Evidence for nitrogen fixation (*nif*) genes on indigenous *Rhizobium* plasmids, *Nature* (London) *282,* 533-535 (1979).
7. Ruvkun, G.B. and Ausubel, F.M. Interspecies homology of nitrogenase genes, *Proc. Natl. Acad. Sci. U.S.A. 77,* 191-195 (1980).
8. Mazur, B.J., Rice, D. and Haselkorn, B. Identification of blue-green algal nitrogen fixation genes by using heterologous DNA hybridization probes, *Proc. Natl. Acad. Sci. U.S.A. 77,* 186-190 (1980).
9. Denarie, J., Boistard, P., Casse-Delbart, F., Atherly, A.G., Berry, J.O. and Russell, P. Indigenous plasmids of *Rhizobium, Int. Rev. Cytol.* Suppl. *13,* 225-246 (1981).
10. Vincent, J.M. In *Nitrogen Fixation,* Vol. II (Editors, W.E. Newton and W.H. Orme-Johnson), Univ. Park Press, Baltimore, Maryland (1980).
11. Verma, D.P.S. and Long, S. The molecular biology of *Rhizobium*-legume symbiosis, *International Review of Cytology* Suppl. *14,* 211-245 (1983).
12. Prakash, R.K., Hooykaas, P.J.J., Ledeboer, A.M., Kijne, J.W., Schilperoort, R.A., Nuti, M.P., Lepidi, A.A., Casse, F., Boucher, C., Julliot, J.S. and Denarie, J. In *Nitrogen Fixation,* Vol. II (Editors, W.E. Newton and W.H. Orme-Johnson), University Park Press, Baltimore (1980).
13. Brewin, N.J., Beringer, J.E. and Johnston, A.W.B. Plasmid-mediated transfer of host-range specificity between two strains of *Rhizobium leguminosarum, J. Gen. Microbiol. 120,* 413-420 (1980).
14. Hooykaas, P.J.J., Snijdewint, F.G.M. and Schilperoort, R.A. Identification of the *Sym* plasmid of *Rhizobium leguminosarum* strain 1001 and its transfer to and expression in other Rhizobia and *Agrobacterium tumefaciens, Plasmid 8,* 73-82 (1982).
15. Djordjevic, M.A., Zurkowski, W. and Rolfe, B.G. Plasmids and stabil-

ity of symbiotic properties of *Rhizobium trifolii*, *J. Bact. 151*, 560–568 (1982).

16. Prakash, R.K., Schilperoort, R.A. and Nuti, M.P. Large plasmids of fast-growing rhizobia: homology studies and location of structural nitrogen fixation (*nif*) genes, *J. Bacteriol. 145*, 1129–1136 (1981).
17. Zurkowski, W. Conjugational transfer of the nodulation-conferring plasmid pWZ2 in *Rhizobium trifolii*, *Mol. Gen. Genet. 181*, 522–524 (1981).
18. Lamb, J.W., Hombrecher, G. and Johnston, A.W.B. Plasmid-determined nodulation and nitrogen-fixation abilities in *Rhizobium phaseoli*, *Mol. Gen. Genet. 186*, 449–452 (1982).
19. Banfalvi, Z., Sakanyan, V., Koncz, C., Kiss, A., Dusha, I. and Kondorosi, A. Location of nodulation and nitrogen fixation genes on a high molecular weight plasmid of *R. meliloti*, *Mol. Gen. Genet. 184*, 318–325 (1981).
20. Rosenberg, C., Casse-Delbart, F., Dusha, I., David, M. and Boucher, C. Megaplasmids in the plant associated bacteria *Rhizobium meliloti* and *Pseudomonas solanacearum*, *J. Bact. 150*, 402–406 (1982).
21. Nuti, M.P., Ledeboer, A.M., Lepidi, A.A. and Schilperoort, R.A. Large plasmids in different *Rhizobium* species, *J. Gen. Microbiol. 100*, 241–248 (1977).
22. Gross, D.C., Vidaver, A.K. and Klucas, R.V. Plasmids, biological properties and efficacy of nitrogen fixation in *Rhizobium japonicum* strains indigenous to alkaline soils, *J. Gen. Microbiol. 114*, 257–266 (1979).
23. Haugland, R. and Verma, D.P.S. Interspecific plasmid and genomic DNA sequence homologies and localization of *nif* genes in effective and ineffective strains of *Rhizobium japonicum*, *J. Mol. Appl. Genet. 1*, 205–217 (1981).
24. Cantrell, M.A., Hickok, R.E. and Evans, H.J. Identification and characterization of plasmids in hydrogen uptake positive and hydrogen uptake negative strains of *Rhizobium japonicum*, p. 371, in *Current Perspectives in Nitrogen Fixation* (Editors, A.H. Gibson and W.E. Newton), Aust. Acad. Sci., Canberra (1981).
25. Currier, T.C. and Nester, E.W. Isolation of covalently closed circular DNA of high molecular weight from bacteria, *Anal. Biochem. 76*, 431–441 (1976).
26. Ledeboer, A.M. Large plasmids in *Rhizobiaceae*. Ph.D. thesis, Leiden University, Leiden, The Netherlands (1978).
27. Hirsch, P.R., van Montagu, M., Johnston, A.W.B., Brewin, N.J. and Schell, J. Physical identification of bacteriocinogenic, nodulation and other plasmids in strains of *Rhizobium leguminosarum*, *J. Gen. Microbiol. 120*, 403–412 (1980).
28. Schwinghamer, E.A. A method for improved lysis of some Gram-

negative bacteria, *FEMS Microb. Lett.* 7, 157-162 (1980).

29. Casse, F., Boucher, C., Julliot, J.S., Michel, M. and Denarie, J. Identification and characterization of large plasmids in *Rhizobium meliloti* using agarose gel electrophoresis, *J. Gen. Microbiol. 113,* 229-242 (1979).
30. Eckhardt, T. A rapid method for the identification of plasmid deoxyribonucleic acid in bacteria, *Plasmid 1*, 584-588 (1978).
31. Beringer, J.E., Beynon, J.L., Buchanan-Wollaston, A.V. and Johnston, A.W.B. Transfer of the drug-resistance transposon Tn5 to *Rhizobium, Nature* (London) *276,* 633-634 (1978).
32. Buchanan-Wollaston, A.V., Beringer, J.E., Brewin, N.J., Hirsch, P.R. and Johnston, A.W.B. Isolation of symbiotically defective mutants in *Rhizobium leguminosarum* by insertion of the transposon Tn5 into a transmissible plasmid, *Mol. Gen. Genet. 178,* 185-190 (1980).
33. Meade, H.M., Long, S.R., Ruvkun, G.B., Brown, S.E. and Ausubel, F.M. Physical and genetic characterization of symbiotic and auxotrophic mutants of *Rhizobium meliloti* induced by transposon Tn5 mutagenesis, *J. Bacteriol.* 149, 114-122 (1982).
34. Rolfe, B.G., Gresshoff, P.M. and Shine, J. Rapid screening for symbiotic mutants of *Rhizobium* and white clover, *Plant. Sci. Lett. 19,* 277-284 (1980).
35. Forrai, T., Vincze, E., Banfalvi, Z. Kiss, G.B., Randhawa, G.S. and Kondorosi, A. Localization of symbiotic mutations in *Rhizobium meliloti, J. Bacteriol. 153,* 635-643 (1983).
36. Julliot, J.S., Dusha, I., Renalier, M.H., Terzaghi, B., Garnerone, A.M. and Biostard, P. *Mol. Gen. Genet.* (submitted) (1983).
37. Dart. P. pp. 425, in *A Treatise of Dinitrogen Fixation* (Editors, R.W.F. Hardy and W.S. Silver), John Wiley and Sons, New York (1977).
38. Sutton, W.D. Some features of the DNA of *Rhizobium* bacteroids and bacteria, *Biochem. Biophys. Acta. 366,* 1-10 (1974).
39. Tshitenge, G., Luyindula, N., Lurquin, P.F. and Ledouse, L. Plasmid deoxyribonucleic acid in *Rhizobium vigna* and *Rhizobium trifolii, Biochim. Biophys. Acta. 414,* 357-361 (1975).
40. Zurkowski, W. and Lorkiewicz, Z. Plasmid deoxyribonucleic acid in *Rhizobium trifolii, J. Bact. 128,* 481-484 (1976).
41. Higashi, S. Transfer of clover infectivity of *Rhizobium trifolii* to *Rhizobium phaseoli* as mediated by an episomic factor, *J. Gen. Appl. Microb. 13,* 391-403 (1967).
42. Dunican, L.K., O'Gara, R. and Tierney, A.B. pp. 77-90, in *Symbiotic Nitrogen Fixation in Plants* (Editor P.S. Nutman), Cambridge Univ. Press, London and New York (1976).
43. Johnston, A.W.B., Beynon, J.L., Buchanan-Wollaston, A.V., Setchell, S.M., Hirsch, P.R. and Beringer, J.E. High frequency transfer

of nodulating ability between strains and species of *Rhizobium, Nature* (London) *276,* 635-636 (1978).

44. Hooykaas, P.J.J., van Brussel, A.A.N., den Dulk-Ras, H., van Slogteren, G.M.S. and Schilperoort, R.A. Sym plasmid of *Rhizobium trifolii* expressed in different rhizobial species and *Agrobacterium tumefaciens, Nature 291,* 351-353 (1981).
45. Kondorosi, A., Kondorosi, E., Pankhurst, C.E., Broughton, W.J. and Banfalvi, Z. Mobilization of a *Rhizobium meliloti* megaplasmid carrying nodulation and nitrogen fixation genes into other *Rhizobium* and *Agrobacterium, Mol. Gen. Genet. 188,* 433-439 (1982).
46. Brewin, N.J., Beringer, J.E., Buchanan-Wollaston, A.V., Johnston, A.W.B. and Hirsch, P.R. Transfer of symbiotic genes with bacteriocinogenic plasmids in *Rhizobium leguminosarum, J. Gen. Microbiol. 116,* 261-270 (1980).
47. Scott, D.B. and Ronson, C.W. Identification and mobilization by cointegrate formation of a nodulation plasmid in *Rhizobium trifolii, J. Bact. 151,* 36-43 (1982).
48. Wong, C.H., Pankhurst, C.E., Kondorosi, A. and Broughton, W.J. *J. Cell Biol.* (submitted) (1983).
49. Banfalvi, Z., Randhawa, G.S., Kondorosi, E., Kiss, A. and Kondorosi, A. Construction and characterization of R-prime plasmids carrying symbiotic genes of *R. meliloti, Mol. Gen. Genet. 189,* 129-135 (1983).
50. Hirsch, A.M., Long, S.R., Bang, M., Haskins, N. and Ausubel, F.M. Structural studies of alfalfa roots infected with nodulation mutants of *Rhizobium meliloti, J. Bacteriol. 151,* 411-419 (1982).
51. Rolfe, B.G., Djordjevic, M., Scott, K.F., Hughes, J.E., Badenoch-Jones, J., Gresshoff, P.M., Cen, Y., Dudman, W.F., Zurkowski, W. and Shine, J. In *Current Perspectives in Nitrogen Fixation* (Editors, A.H. Gibson and W.E. Newton), Aust. Acad. Sci. Canberra (1981).
52. Long, S.R., Buikema, W.J. and Ausubel, F.M. Cloning of *Rhizobium meliloti* nodulation genes by direct complementation of Nod^- mutants, *Nature 298,* 485-488 (1982).
53. Kondorosi, E., Banfalvi, Z., Slaska-Kiss, C. and Kondorosi, A. in *UCLA Symposium on Molecular and Cellular Biology,* New Series, Vol. 12 (in press) (1983).
54. Friedman, A.M., Long, S.R., Brown, S.E., Buikema, W.J. and Ausubel, F.M., Construction of a broad host range cosmid cloning vector and its use in the genetic analysis of *Rhizobium* mutants, *Gene 18,* 289-296 (1982).
55. Kondorosi, A., Kondorosi, E., Banfalvi, Z., Broughton, W.J., Pankhurst, C.E., Randhawa, G.S., Wong, C.H. and Schell, J. in *Molecular Biology of Bacterium-Plant Interaction* (Editor, A. Pühler) (in press) (1983).
56. Pankhurst, C.E., Broughton, W.J., Bachem, C., Kondorosi, E. and

Kondorosi, A. in *Molecular Biology of Bacterium-Plant Interaction* (Editor, A. Pühler) (in press) (1983).
57. Fåhraeus, G. and Ljunggren, H. *The Ecology of Soil Bacteria* (Editors, T.R.G. Gray and D. Parkinson), Liverpool University Press (1968).
58. Badenoch-Jones, J., Summons, R.E., Djordjevic, M.A., Shine, J., Lethans, D.S. and Rolfe, B.G. Mass spectrometric quantification of indole-3-acetic acid in *Rhizobium* culture supernatants: Relation to root hair curling and nodule initiation, *Appl. Envir. Microbiol. 44,* 275-280 (1982).
59. Wang, T.L., Wood, E.A. and Brewin, N.J. Growth regulators, *Rhizobium* and nodulation in peas. Indole-3-acetic acid from the culture medium of nodulating and non-nodulating strains of *R. leguminosarum, Planta 155,* 345-349 (1982).
60. Wang, T.L., Wood, E.A. and Brewin, N.J. Growth regulators, Rhizobium and nodulation in peas. The cytokinin content of a wild-type and Ti-plasmid-containing strain of *R. leguminosarum, Planta 155,* 350-355 (1982).
61. Henson, I.E. and Wheeler, C.T. Hormones in plants bearing nitrogen-fixing root nodules: The distribution of cytokinins in *Vicia faba* L. *New Phytol. 76,* 433-439 (1976).
62. Syono, K. and Torrey, J.G. Identification of cytokinins of root nodules of the garden pea *Pisum sativum* L., *Plant Physiol. 57,* 602-606 (1976).
63. Buikema, W.B., Long, S.R., Brown, S.E., van den Bos, R.C., Earl, C. and Ausubel, F.M. *J. Molec. Appl. Genet.* (in press) (1983).
64. Paau, A.S. and Brill, W.J. Comparison of the genomic arrangement and the relative transcription of the nitrogenase genes in *Rhizobium meliloti* during symbiotic development in alfalfa root nodules, *Can. J. Microbiol. 28,* 1330-1339 (1982).
65. Rolfe, B.G. and Gresshoff, P.M. *Rhizobium trifolii* mutant interactions during the establishment of nodulation in white clover, *Aust. J. Biol. Sci. 33,* 491-504 (1980).
66. Napoli, C., Dazzo, F. and Hubbell, D. Production of cellulose microfibrils by *Rhizobium, Appl. Microbiol. 30,* 123-131 (1975).
67. Napoli, C. and Albersheim, P. *Rhizobium leguminosarum* mutants incapable of normal extracellular polysaccharide production, *J. Bacteriol. 141,* 1454-1456 (1980).
68. Cannon, F.C., Riedel, G.E. and Ausubel, F.M. Overlapping sequences of *Klebsiella pneumoniae nif* DNA cloned and characterized, *Mol. Gen. Genet. 174,* 59-66 (1979).
69. Rosenberg, C., Boistard, P., Denarie, J. and Casse-Delbart, F. Genes controlling early and late functions in symbiosis are located on a megaplasmid in *Rhizobium meliloti, Mol. Gen. Genet. 184,* 326-333 (1981).

70. Krol, A.J.M., Hontelez, J.G.J., Roozendaal, B. and van Kammen, A. On the operon structure of the nitrogenase genes of *Rhizobium leguminosarum* and *Azotobacter vinelandii, Nucl. Acids. Research 10,* 4147-4157 (1982).
71. Hombrecher, G., Brewin, N.J. and Johnston, A.W.B. Linkage of genes for nitrogenase and nodulation ability on plasmids in *Rhizobium leguminosarum* and *Rhizobium phaseoli, Mol. Gen. Genet. 182,* 133-136 (1981).
72. Ruvkun, G.B. and Ausubel, F.M. A general method for site-directed mutagenesis in prokaryotes, *Nature 289,* 85-88 (1981).
73. Johnston, A.W.B., Ma, Q-S., Hombrecher, G. and Downie, J.A. in *Molecular Genetics of the Bacteria-Plant Interaction* (Editor, A. Pühler), Springer Verlag (in press) (1983).
74. Quinto, C., de la Vega, H., Flores, M , Fernandez, L., Ballado, T., Soberon, G. and Palacios, R. Reiteration of nitrogen fixation gene sequences in *Rhizobium phaseoli, Nature 299,* 724-726 (1982).
75. Fuhrman, M. and Hennecke, H. Coding properties of cloned nitrogenase structural genes from *Rhizobium japonicum, Mol. Gen. Genet. 187,* 419-425 (1982).
76. Ruvkun, G.B., Sundaresan, V. and Ausubel, F.M. Directed transposon Tn5 mutagenesis and complementation analysis of *Rhizobium meliloti* symbiotic nitrogen fixation genes, *Cell 29,* 551-559 (1982).
77. Downie, J.A., Ma, Q-S., Knight, C.D., Hombrecher, G. and Johnston, A.W.B. Cloning of the symbiotic region of *Rhizobium leguminosarum:* the nodulation genes are between the nitrogenase genes and a *nifA*-like gene, *EMBO Journal 2,* 947-952 (1983).
78. Ditta, G., Stanfield, S., Corbin, D. and Helinski, D.R. Broad host range DNA cloning system for Gram-negative bacteria: Construction of a gene bank of *Rhizobium meliloti, Proc. Natl. Acad. Sci. U.S.A. 77,* 7347-7351 (1980).
79. Shine, J., Scott, K.F., Fellows, F., Djordjevic, M., Schofield, P., Watson, J.M. and Rolfe, B.G. In *Molecular Genetics of the Bacteria-Plant Interaction* (Editor, A. Pühler), Springer-Verlag (in press) (1983).
80. Hennecke, H. Recombinant plasmids carrying nitrogen fixation genes from *Rhizobium japonicum, Nature 291,* 354-355 (1981).
81. Török, I. and Kondorosi, A. Nucleotide sequence of the *R. meliloti* nitrogenase reductase (*nifH*) gene, *Nucleic Acids Res. 9,* 5711-5723 (1981).
82. Sundaresan, V. and Ausubel, F.M. Nucleotide sequence of the gene coding for the nitrogenase iron protein from *Klebsiella pneumoniae, J. Biol. Chem. 256,* 2808-2812 (1981).
83. Scott, K.F., Rolfe, B.G. and Shine, J. Biological nitrogen fixation: Primary structure of the *Klebsiella pneumoniae nifH* and *nifD* genes, *J. Mol. Appl. Genet. 1,* 71-81 (1981).

84. Mevarech, M., Rice, D. and Haselkorn, R. Nucleotide sequence of a cyanobacterial *nifH* gene coding for nitrogenase reductase, *Proc. Natl. Acad. Sci. U.S.A. 77*, 6476-6480 (1980).
85. Tanaka, M., Hanin, M., Yasunobu, K.T. and Mortenson, L.E. The amino acid sequence of *Clostridium pasteurianum* iron protein, a component of nitrogenase, *J. Biol. Chem. 252*, 7093-7100 (1977).
86. Hase, T., Nakano, T., Matsubara, H. and Zuruft, W.G. Correspondence of the larger subunit of the molybdenum-ion protein in clostridial nitrogenase to the *nifD* gene products of other nitrogen-fixing organisms, *J. Biochem. 90*, 295-298 (1981).
87. Lundell, D. and Howard, J.B. Isolation and partial characterization of two different subunits from the molybdenum-ion protein of *Azotobacter vinelandii* nitrogenase, *J. Biol. Chem. 253*, 3422-3426 (1978).
88. Kondorosi, A. In *Molecular Biology of Nitrogen Fixation* (Editor, W.J. Broughton), Oxford Univ. Press, Oxford (in press) (1983).
89. Weber, G. and Pühler, A. Expression of *Rhizobium meliloti* genes for nitrogen fixation in *Escherichia coli* minicells: Mapping of the subunit of the nitrogenase reductase, *Plant Molec. Biol. 1*, 305-320 (1982).
90. Postgate, J.R. In *Evolution in the Microbial World* (Editors, M.J., Carlile and J.J. Skehel), *Symp. Soc. Gen. Microbiol. 24*, 263-292 (1974).
91. Shine, J. and Dalgarno, L. Determinant of cistron specificity in bacterial ribosomes, *Nature 254*, 34-38 (1975).
92. Dixon, R., Kennedy, C., Kondorosi, A., Krishnapillai, V. and Merrick, M. Complementation analysis of *Klebsiella pneumoniae* mutants defective in nitrogen fixation, *Mol. Gen. Genet. 157*, 189-198 (1977).
93. MacNeil, T., MacNeil, D., Roberts, G.P., Supiano, M.A. and Brill, W.J. Fine-structure mapping and complementation analysis of *nif* (nitrogen fixation) genes in *Klebsiella pneumoniae*, *J. Bacteriol. 136*, 253-266 (1978).
94. Elmerich, C., Houmard, J., Sibold, L., Manheimer, I. and Charpin, N. Genetic and biochemical analysis of mutants induced by bacteriophage Mu DNA integration into *Klebsiella pneumoniae* nitrogen fixation genes, *Mol. Gen. Genet. 165*, 181-189 (1978).
95. Corbin, D., Barran, L. and Ditta, G. *Proc. Natl. Acad. Sci. U.S.A.* (in press) (1983).
96. Bisseling, T., Moen, A.A., van den Bos, R.C. and van Kammen, A. The sequence of appearance of leghaemoglobin and nitrogenase components I and II in root nodules of *Pisum sativum*, *J. Gen. Microbiol. 118*, 377-381 (1980).
97. Bisseling, T., van Staweren, W. and van Kammen, A. The effect of waterlogging on the synthesis of the nitrogenase components in bacteroids of *Rhizobium leguminosarum* in root nodules of *Pisum*

sativum, Biochem. Biophys. Res. Comm. 93, 687-693 (1980).

98. Hadley, R.G., Yun, A. and Szalay, A.A. in *Molecular Genetics of the Bacteria-Plant Interaction* (Editor, A. Pühler), Springer Verlag (in press) (1983).
99. Sundaresan, V., Jones, J.D.G., Ow, P.W. and Ausubel, F.M. *Klebsiella pneumoniae nifA* product activates the *Rhizobium meliloti* nitrogenase promoter, *Nature 301,* 728-732 (1983).
100. Rosenberg, M. and Court, D. Regulatory sequences involved in the promotion and termination of RNA transcription, *Annu. Rev. Genet. 13,* 319-353 (1979).
101. Riedel, G.E., Ausubel, F.M. and Cannon, F.C. Physical map of chromosomal nitrogen fixation (*nif*) genes of *Klebsiella pneumoniae, Proc. Natl. Acad. Sci. U.S.A. 76,* 2866-2870 (1979).
102. Merrick, M., Filser, M., Dixon, R., Elmerich, C., Sibold, L. and Houmard, J. The use of translocatable genetic elements to construct a fine-structure map of the *Klebsiella pneumoniae* nitrogen fixation (*nif*) gene cluster, *J. Gen. Microbiol. 117,* 509-520 (1980).
103. Corbin, D., Ditta, G. and Helinski, D.R. Clustering of nitrogen fixation (*nif*) genes in *Rhizobium meliloti, J. Bacteriol. 149,* 221-228 (1982).
104. Ma, Q-S., Johnston, A.W.B., Hombrecher, G. and Downie, J.A. Molecular genetics of mutants of *Rhizobium leguminosarum* which fail to *fix* nitrogen, *Mol. Gen. Genet. 187,* 166-171 (1982).
105. Dixon, R., Eady, R.R., Espin, G., Hill, S., Iaccarino, M. Kahn, D. and Merrick, M. Analysis of regulation of *Klebsiella pneumoniae* nitrogen fixation (*nif*) gene cluster with gene fusions, *Nature 286,* 128-132 (1980).
106. Krol, A.J.M., Hontelez, J.G.J., van den Bos, R.C. and van Kammen, A. Expression of large plasmids in the endosymbiotic form of *Rhizobium leguminosarum, Nucl. Acids Research 8,* 4337-4347 (1980).
107. Prakash, R.K., van Brussel, A.A.N., Quint, A., Mennes, A.M. and Schilperoort, R.A. The map position of sym-plasmid regions expressed in the bacterial and endosymbiotic form of *Rhizobium leguminosarum, Plasmid 7,* 281-286 (1982).
108. Kondorosi, A., Banfalvi, Z., Broughton, W.J., Forrai, T., Kiss, G.B., Kondorosi, E., Pankhurst, C.E., Randhawa, G.S., Svab, Z. and Vincze, E. In *Structure and Function of Plant Genomes* (Editors, O., Ciferri and L. Dure) (in press) (1983).
109. Riesenberg, D., Erdei, S., Kondorosi, E. and Kari, C. Positive involvement of ppGpp in derepression of the *nif* operon in *Klebsiella pneumoniae, Mol. Gen. Genet. 185,* 198-204 (1982).
110. Belitsky, B. and Kari, Cs. Absence of accumulation of ppGpp and RNA during amino acid starvation in *Rhizobium meliloti, J. Biol. Chem. 257,* 4677-4679 (1982).

111. Maier, R.J. and Brill, W.J. Ineffective and non-nodulating mutant strains of *Rhizobium japonicum, J. Bacteriol. 127,* 763-769 (1976).
112. Beringer, J.E., Johnston, A.W.B. and Wells, B. The isolation of conditional ineffective mutants of *Rhizobium leguminosarum, J. Gen. Microbiol. 98,* 339-343 (1977).
113. Denarie, J., Truchet, G. and Bergeron, B. Pp. 47-61 in *Symbiotic Nitrogen Fixation in Plants* (Editor, P.S. Nutman), Cambridge Univ. Press, London and New York (1976).
114. Kuykendall, L.D. Mutants of *Rhizobium* that are altered in legume interaction and nitrogen fixation, *Int. Rev. Cytol.* Suppl. 13, 299-309 (1981).
115. Scott, K.F., Hughes, J.E., Gresshoff, P.M., Beringer, J.E., Rolfe, B.G. and Shine, J. Molecular cloning of *Rhizobium trifolii* genes involved in symbiotic nitrogen fixation, *J. Mol. Appl. Gen. 1,* 315-326 (1982).
116. Drevon, J.J., Frazier, L., Russell, S.A. and Evans, H.J. Respiratory and nitrogenase activities of soybean nodules formed by hydrogen uptake negative (Hup^-) mutant and revertant strains of *Rhizobium japonicum* characterized by protein patterns, *Plant Physiol. 70,* 1341-1346 (1982).
117. Brewin, N.J., De Jong, T.M., Phillips, D.A. and Johnston, A.W.B. Co-transfer of determinants for hydrogenase activity and nodulation ability in *Rhizobium leguminosarum, Nature 288,* 77-79 (1980).
118. De Jong, T.M., Brewin, N.J., Johnston, A.W.B. and Phillips, D.A. Improvement of symbiotic properties in *Rhizobium leguminosarum* by plasmid transfer, *J. Gen. Microb. 128,* 1829-1838 (1982).
119. Ames, P., Schluederberg, S.A. and Bergman, K. Behavioral mutants of *Rhizobium meliloti, J. Bacteriol. 141,* 722-727 (1980).
120. Ames, P. and Bergman, K. Competitive advantage provided by bacterial motility in the formation of nodules by *Rhizobium meliloti , J. Bacteriol. 148,* 728-729 (1981).
121. Hirsch, P.R. Plasmid determined bacteriocin production by *Rhizobium leguminosarum, J. Gen. Microbiol. 113,* 219-228 (1979).
122. Wijffelman, C.A. Personal communication.
123. Johnston, A.W.B., Hombrecher, G., Brewin, N.J. and Cooper, M.C. Two transmissible plasmids in *Rhizobium leguminosarum* strain 300, *J. Gen. Microbiol. 128,* 85-93 (1982).
124. Pankhurst, C., Broughton, W. and Wienecke, U. *J. Gen. Microbiol.* (in press) (1983).
125. Vincze, E., Koncz, Cs. and Kondorosi, A. Construction *in vitro* of R-prime plasmids and their use for transfer of chromosomal genes and plasmids of *Rhizobium meliloti, Acta Biol. Acad. Sci. Hung. 32,* 195-204 (1981).
126. Dusha, I., Renalier, M.H., Terzaghi, B., Garnerone, A.M., Boistard, P. and Jolliot, J.S. In *Molecular Genetics of the Bacteria-Plant*

Interaction (Editor, A. Pühler), Springer Verlag (in press) (1983).

127. Brewin, N.J., Wood, E.A., Johnston, A.W.B., Dibb, N.J. and Hombrecher, G. Recombinant nodulation plasmids in *Rhizobium leguminosarum*, *J. Gen. Microbiol. 128*, 1817-1827 (1982).
128. Johnston, A.W.B. and Brewin, N.J. In *Current Perspectives in Nitrogen Fixation* (Editors, A.H. Gibson and W.E. Newton), Canberra (1982).
129. Genetello, C., Van Larebeke, N., Holsters, M., Depicker, A., Van Montagu, M. and Schell, J. Ti plasmids of *Agrobacterium* as conjugative plasmids, *Nature 265*, 561-563 (1977).
130. Kerr, A., Manigault, P. and Tempe, J. Transfer of virulence *in vivo* and *in vitro* in *Agrobacterium*, *Nature 265*, 560-561 (1977).
131. Montoya, A., Chiltin, M-D., Gordon, M.P., Sciaky, D., and Nester, E.W. Octopine and nopaline metabolism in *Agrobacterium tumefaciens* and crown gall tumor cells: Role of plasmid genes, *J. Bacteriol. 129* 101-107 (1977).
132. Tempe, J. in *Molecular Genetics of the Bacteria-Plant Interaction* (Editor, A. Pühler), Springer Verlag (in press) (1983).
133. Cole, M.A. and Elkan, G.H. Transmissible resistance to penicillin G, neomycin, and chloramphenicol in *Rhizobium japonicum*, *Antimicrobial Agents and Chemotherapy 4*, 248-253 (1973).
134. Olivares, J., Montoya, E. and Palomares, A. Some effects derived from the presence of extrachromosomal DNA in *Rhizobium meliloti*, pp. 375-385, in *Recent Developments in Nitrogen Fixation*, Proceedings of the Second International Symposium, Salamanca, 1976 (Editors, W. Newton, J.R. Postgate and C. Rodriguez-Barrueco), Academic Press (1977).
135. Sanders, R.E., Carlson, R.W. and Albersheim, P.A. *Rhizobium* mutant incapable of nodulation and normal polysaccharide excretion, *Nature 271*, 240-242 (1978).
136. Prakash, R.K., van Veen, R.J.M. and Schilperoort, R.A. Restriction endonuclease mapping of a *Rhizobium leguminosarum* Sym plasmid, *Plasmid 7*, 271-280 (1982).
137. Jouanin, L., De Lajudie, D., Bazetoux, S. and Huguet, T. DNA sequence homology in *Rhizobium meliloti* plasmids, *Mol. Gen. Genet. 182*, 189-195 (1981).
138. Prakash, R.K. and Schilperoort, R.A. Relationship between *nif* plasmids of fast-growing *Rhizobium* species and Ti plasmids of *Agrobacterium tumefaciens*, *J. Bact. 149*, 1129-1134 (1982).
139. Holsters, M., Silva, B., van Vliet, F., De Black, M., Dhaese, P., Depicher, A., Inze, D., Engler, G., Villarroel, R., van Montagu, M. and Schell, J. The functional organization of the nopaline *Agrobacterium tumefaciens* plasmid p-TiC58, *Plasmid 3*, 212-230 (1980).

140. Koekman, B.P., Hooykaas, P.J.J. and Schilperoort, R.A. Localization of the replication control region on the physical map of the octopine Ti-plasmid, *Plasmid 4,* 184-195 (1980).
141. Garfinkel, D.J. and Nester, E.W. *Agrobacterium tumefaciens* mutants affected in crown gall tumorigenesis and octopine catabolism, *J. Bact. 144,* 732-743 (1980).
142. Depicher, A., van Montagu, M. and Schell, J. Homologous DNA sequences in different Ti-plasmids are essential for oncogenecity, *Nature 275,* 150-153 (1978).
143. Hadely, R.G. and Szalay, A.A. DNA sequences homologous to the T DNA region of *Agrobacterium tumefaciens* are present in diverse *Rhizobium* species, *Mol. Gen. Genet. 188,* 361-369 (1982).
144. Kondorosi, A., Kiss, G.B., Forrai, T., Vincze, E. and Banfalvi, Z. Circular linkage map of *Rhizobium meliloti* chromosome, *Nature 268,* 525-527 (1977).
145. Beringer, J.B., Hoggan, S.A. and Johnston, A.W.B. Linkage mapping in *Rhizobium leguminosarum* by means of R plasmid-mediated recombination, *J. Gen. Microbiol. 104,* 201-207 (1978).
146. Meade, H.M. and Signer, E.R. Genetic mapping of *Rhizobium meliloti, Proc. Natl. Acad. Sci. U.S.A. 74,* 2076-2078 (1977).
147. Casadesus, J. and Olivares, J. Rough and fine linkage mapping of the *Rhizobium meliloti* chromosome, *Mol. Gen. Genet. 174,* 203-209 (1979).

7. *Frankia* and its Symbiosis in Non-legume (actinorhizal) Root Nodules

C.T. Wheeler

INTRODUCTION

Difficulties experienced in the development of reproducible methods for the isolation and culture of *Frankia* [1] gave rise in the early 1970's to suggestions that these organisms may be obligate symbionts of non-legume root nodules [2]. This misconception was dispelled in 1978 by the development of techniques for the culture *in vitro* of the endophyte of *Comptonia peregrina* (=*Myrica asplenifolia*) nodules [3], a major advance which has produced a considerable increase in the research effort devoted to the non-legume/*Frankia* symbiosis. It was proposed [4] that the terminology 'actinorhizal association' be used to distinguish this symbiosis from that between *Rhizobium* and non-legumes such as *Parasponia* and the term will be used in this sense in the current article.

Increased awareness of the importance of the actinorhizal association has resulted in the addition of six genera to the list of actinomycete nodulated plants prepared by Bond for the International Biological Programme [5]. These are, from N. America *Chamaebatia* [6] and *Cowania* [7]; from S. America *Trevoa, Kentrothamnus* [8] and *Talguena* [9]; from Pakistan *Datisca*, a rediscovery of nodulation [10], earlier reports originating from Europe [11,12] having been overlooked. A list of actinorhizal genera known to the author in early 1983 is shown in Table 1; information on nodulation of individual species is given in other reviews [5,13,14].

While some plant genera such as *Alnus* almost always show nodulation wherever examined, nodulation in other genera can be highly variable. For example, in Australia no nodules were found on *Casuarina* growing more than 70 km from the coast, nor did coastal *Casuarina* always bear nodules [15]. Greenhouse tests confirmed that soil from such localities did not support nodulation, presumably due to the absence of endophyte from the soil or to inhibitory properties of certain soil types. Nodulation of actinorhizal species may vary not only with locality but also with plant age. For

Table 1. Plant genera bearing actinorhizal nodules

Family	Genus
Casuarinaceae	*Casuarina*
Myricaceae	*Myrica* (including *Comptonia*)
Betulaceae	*Alnus*
Rosaceae (tribe *Dryadeae*)	*Dryas, Cercocarpus, Purshia, Chamaebatia, Cowania.*
(tribe *Rubeae*)	*Rubus*
Coriariaceae	*Coriaria*
Rhamnaceae (tribe *Colletieae*)	*Colletia, Discaria, Trevoa, Talguenea, Kentrothamnus*
(tribe *Rhamneae*)	*Ceanothus*
Eleagnaceae	*Eleagnus*
Datiscaceae	*Datisca*

example, searches by the author and by Dr. J.C. Gordon in an old stand of *Alnus rubra*, growing in the Oregon coastal range, found few nodules on trees believed to bear nodules some ten years earlier. In old, natural stands of *Hippophae rhamnoides*, low nodulation may result from degeneration of the root system following nematode infection [16]. Excavation of young plants is usually the easiest and the most certain method of determining nodulation habit in the field.

CULTURE OF *FRANKIA IN VITRO*

The first isolate of *Frankia*, from nodules of *Comptonia peregrina*, was obtained using a fairly complicated technique involving enzymic digestion of nodules to release the endophyte from host plant cells [3]. Some simpler techniques which have been developed subsequently and which are more suited to the development of routine procedures for isolation, are as follows:

1) Centrifugation of nodule homogenates on sucrose density gradients to separate vesicle clusters from contaminating micro-organisms, followed by planting out of separated fractions suspended in nutrient agar [17].

2) Microdissection of vesicle clusters from homogenates of surface-sterilised nodules, passage through several changes of sterile water and transfer to liquid media [18].

3) Planting out of crushed nodule fragments from axenically grown plants onto nutrient agar [19, 20].

4) Sterilisation of the outermost nodule layers with 3 per cent osmium tetroxide, followed by fragmentation of nodules into liquid media and sub-culture of fragments showing *Frankia*-like outgrowths.

Over 200 isolates of *Frankia* from stands of *Alnus crispa* and *Alnus rugosa* in Canada have been obtained employing this last technique [21]. While the method clearly has considerable potential for the routine isolation of Frankiae from actinorhizal nodules, great caution is required in the use of the highly toxic osmium tetroxide.

A number of culture media have been developed which have proved

Table 2. Some media for culture of *Frankia*

'*Frankia*' broth [17]		QMOD medium [28]		Bennett's medium [23, 25]		Burgraaf's medium [26]	
			g.l^{-1}		g.l^{-1}		g.l^{-1}
Yeast extract	0.5% (w/v)	Yeast extract	0.5	Yeast extract	1.0	Casamino acids	1.0
Dextrose	1.0% (w/v)	Glucose	10.0	Beef extract	1.0	Na propionate	1.0
Casamino acids	0.5% (w/v)	Bactopeptone	5.0	Glucose	10.0	Biotin	0.002
Vitamin B_{12}	1.6 mgl^{-1}	K_2HPO_4	0.3	Casamino acids	2.0	Fe-EDTA	0.01
H_3BO_3	1.5 mgl^{-1}	$MgSO_4 \cdot 7H_2O$	0.2	Agar	15 g	K_2HPO_4	1.0
$ZnSO_4 . 7H_2O$	1.5 mgl^{-1}	KCl	0.2	pH 7.3		$NaH_2PO_4 \cdot 2H_2O$	0.67
$MnSO_4 . H_2O$	4.5 mgl^{-1}	Ferric citrate	0.01g			$CaCl_2 \cdot 2H_2O$	0.1
$NaMoO_4 \cdot 2H_2O$	0.25 mgl^{-1}		mg l^{-1}			$MgSO_4 \cdot 7H_2O$	0.2
$CuSO_4 . 5H_2O$	0.04 mgl^{-1}	H_3BO_3	1.5			Trace elements as	
Agar if required	0.8% (w/v)	$MnSO_4$ $7H_2O$	0.8			QMOD pH6.8	
pH 6.4		$ZnSO_4 . 7H_2O$	0.6				
		$CuSo_4 . 7H_2O$	0.1				
		$(NH_4)_6MO_7O_{24} \cdot 4H_2O$	0.2				
		$CoSO_4 . 7H_2O$	0.01				
		L--lecithin	0.5–50				
		Agar if required	15 g				
		Adjust to pH 6.8–7.0 and add					
		0.1 g l^{-1} $CaCO_3$					

generally useful for isolation of Frankiae. Although strains of *Frankia* so far isolated appear to be relatively undemanding in their nutritional requirements, differences in substrate utilisation [23] suggest some variation in growth requirements between strains. The compositions of four media which may be used for isolation and culture of *Frankia* are shown in Table 2, while some others are given by Lechevalier et al. [23, 24].

MORPHOLOGICAL VARIATION IN *FRANKIA IN VITRO* AND *IN VIVO*

Frankiae *in vitro* are septate, filamentous organisms which produce sporangia in submerged culture. These organisms show considerable variation in form. Vegetative hyphae range from white to dark brown and some Frankiae produce soluble pigments, for example *Myrica pensylvanica* isolate MpI1, green; *Purshia tridentata* isolate PtI1, yellow brown [23, 24, 27]; *Alnus nitida* WgAn2 (isolate of A.D.L. Akkermans), orange-pink. Some strains may produce coloured spores, for example *Alnus incana* ssp. *rugosa* isolate AirI2 black [24].

The early classification systems of Becking [2] proposed separation of *Frankia* into ten species, based on cross inoculation studies with crushed nodule inoculum and on the morphology of the endophyte *in vivo*. Further studies with strains grown *in vitro* have shown this system to be untenable since the host plant may influence the morphology of *Frankia* and the specificity of isolates for host plant genera may be different in some instances from that suggested by studies with crushed nodule inocula. For example, the *Comptonia peregrina* isolate CpI1 will nodulate *Alnus glutinosa* [28] whereas Becking [2] suggested specificity for infection of *Myrica* and inability to nodulate *Alnus* as a characteristic of his species *Frankia brunchorstii*. *Frankia* CpI1 forms club-shaped vesicles in *Comptonia* but spherical vesicles in *Alnus* and sporangia are not formed in either *Alnus* or *Comptonia* nodules, although these are evident in *in vitro* cultures of the endophyte [29]. Baker [27] has produced a cumulative listing of many of the Frankiae currently in culture, which compares several of their properties.

SEROLOGICAL AND METABOLIC RELATIONSHIPS BETWEEN FRANKIAE

Preliminary investigation of interrelationships between a number of *Frankia* isolates suggests that these may be divided into at least two groups [27]. Group I Frankiae cross reacted homologously with antiserum to the *Comptonia peregrina* isolate CpI1 and Group II strains cross reacted heterologously with CpI1 antiserum and homologously or heterologously with antiserum to the *Eleagnus umbellata* isolate EuI1 (Table 3). Group I Frankiae nodulated *Alnus* and *Myrica* but Group II was much more heterogenous. It contains in addition to strains which show homology with EuI1 and nodulate *Eleagnus, Hippophae* and *Shepherdia*, the *Alnus*

Table 3. Serological classification of some *Frankia* isolates from data of Baker and co-workers [27, 30]

Serogroup I		Serogroup II	
Homologous cross reaction with CpI1 antiserum		Heterologous cross reaction with CpI1 antiserum Homologous cross reaction with EuI1 antiserum	
Alnus incana ssp. *rugosa*	AirI1	*Alnus incana* ssp. *rugosa*	AirI2 [h](+nod. − N_2ase)
Alnus rubra	ArI3 and others	*Casuarina equisetifolia*	G2 [h](− nod. + N_2ase)
			DI1 (− nod. + N_2ase)
Alnus viridis ssp. *crispa*	AvcI1	*Ceanothus americanus*	CaI1 (N.D.)
Alnus viridis ssp. *sinuata*	AvsI2 (+nod. − N_2ase)	*Eleagnus umbellata*	EuI1 (+nod. − N_2ase)
	AvsI3		EuI5 (+nod. − N_2ase)
Comptonia peregrina	CpI and others	*Myrica gale*	MgI5
		Purshia tridentata	PtI1
Myrical cerifera	McI1		
Myrica pensylvanica	MpI1		

h = heterologous reaction to both CpI1 and EuI1 antisera. nod. = ability to nodulate the original host plant species. N_2ase = effectivity of strain in nitrogen fixation. Strains not designated are either both infective and effective in nitrogen fixation or data is not published (N.D.).

incana ssp. *rugosa* strain AirI2, the *Casuarina equisetifolia* strain G2 (both of which give heterologous reactions to CpI1 and EuI1 antisera) and the *Myrica gale* isolate MgI5. Strains in both groups give rise to nodules which are ineffective in nitrogen fixation (AvsI2, AirI2, EuI1, EuI5). The *Casuarina* strains G2 and DI1 did not reinfect *Casuarina.* They showed nitrogenase activity, *in vitro,* gave rise to nitrogenase positive nodules on *Hippophae* and conceivably may be 'contaminants' of the *Casuarina* nodules rather than the 'true' endophyte.

With respect to their cell chemistry, a number of Frankiae of both serogroups have been shown to have the same cell wall type and phospholipid type [23]. However, the whole cell sugar pattern of Group II Frankiae was again more variable than Group I in which xylose predominates [23]. Frankiae of Group I show poor utilisation of carbohydrates for growth whereas Group II strains can mostly utilise glucose, maltose xylose and arabinose. Nearly all strains hydrolysed starch to some extent but only EuI1 hydrolysed casein and about half the strains tested in each group showed nitrate reductase activity [23].

CARBON METABOLISM IN RELATION TO NITROGEN FIXATION

The carbon requirements of the Group I strain AvcI1 have been investigated further by Akkermans, Blom and co-workers. This strain can utilise NH_4, NO_3 and several amino acids for growth, with ammonia assimilation catalyzed by glutamine synthetase [31, 32]. Amino acids cannot serve as the sole C source, nor can sugars or alcohols [33]. Cells show activity of the glyoxylate cycle enzymes isocitrate lyase and malate synthase when grown on acetate or fatty acids, however, and these enzymes presumably are involved in the production of dicarboxylic acids for further metabolism via the citric acid cycle [34]. Glyoxylate cycle enzymes were not detected in cells grown on propionate [32] which presumably may be metabolized to succinate by carboxylation to methyl malonyl CoA followed by isomerization of this compound by the deoxyadenosyl cobalamin (vitamin B_{12} derivative) — requiring enzyme, methyl malonyl mutase.

Studies of vesicle clusters isolated from *Alnus* root nodules containing AvcI1 as the microsymbiont suggest that this *Frankia* strain does not utilize glucose, either *in vitro* as already discussed, or *in vivo* since activity of the enzymes hexokinase and pyruvate kinase could not be detected [35]. However, activity of citric acid cycle enzymes was detected in extracts [35, 36] and the demonstration of succinate enhanced oxygen uptake for isolated vesicle clusters from *Alnus, Hippophäe* and *Datisca* suggests that succinate, or another organic dicarboxylic acid, may be the plant carbon compound utilized to support endophyte metablism *in vivo*, at least for some *Frankia* strains [35]. It is of interest, therefore, that cytoimmunochemical techniques have shown activity of phosphoenolpyruvate carboxylase in *Alnus glutinosa* nodules to be located predominantly in the cytosol of cortical cells containing the endophyte vesicles [37]. The oxaloacetate generated by the action of this enzyme could be used to support metabolism of the endophyte, either by itself or after conversion into malate by the action of malate dehydrogenase. Accumulation of radioactivity in malic acid, separable from extracts of *Alnus glutinosa* and *Myrica gale* nodules by paper chromatography, can be observed readily if nitrogen-fixing nodules are incubated in $NaH^{14}CO_3$ [38].

The importance of organic acids for the metabolism of *Frankia* has been shown further in studies of *in vitro* acetylene reduction by isolated strains [19, 39, 40]. Malate or fumarate could replace succinate as carbon sources which would support nitrogenase activity but some other organic acids such as acetate, citrate or 2-oxoglutarate were not utilized nor were sugars or lipids such as Tweens. A requirement for EDTA in the growth medium for nitrogenase activity by strain AvcI1 was demonstrated, but the reasons for this requirement were not established [40]. These observations support earlier suggestions of the localization of nitrogenase activity in endophyte vesicles, based on the association of vesicle production with the emergence of nitrogen fixation in developing *Alnus* nodules [41], on the association of nitrogenase activity with a vesicle fraction isolated from *Alnus* nodule homogenates [42] and on the absence of vesicle production in the *Eleagnus umbellata* EuI1 strain, which does not show nitrogenase activity [43].

The thick-walled vesicle structure thus provides a specialized environment in which nitrogenase can function. Analogies have been drawn between the vesicle and the heterocyst of blue-green algae in providing protection from oxygen. Studies of cell-free extracts of nitrogenase from actinorhizal endophyte preparations have shown this enzyme to have oxygen sensitivity similar to that of other nitrogen-fixing systems [42, 44]. However, while nitrogenase activity in free-living *Rhizobium* is only observed under low-oxygen partial pressures [45], free living *Frankia* shows optimum nitrogenase activity under oxygen partial pressures close to atmospheric [40]. Variation of acetylene-reducing activity with oxygen partial pressure in *Frankia in vitro* is similar to that reported for intact actinorhizal nodules, e.g., from *Alnus rubra* [46]. This supports suggestions that there are no structural modifications [47] or special proteins [46] of the host plant which aid in protection from oxygen. Tjepkema [48] however, has recently presented spectrophotometric evidence which adds weight to earlier observations of the occurrence of bound haemoglobins in actinorhizal root nodules [49, 50]. If these proteins are indeed present then their role in the oxygen relations of the endophyte within the nodules remains to be determined.

ASSIMILATION OF AMMONIA

Nutritional studies with *Frankia* AvcI1 have shown that this organism can utilize N_2 NH_4, NO_3^- and various amino acids as nitrogen sources for growth *in vitro* [33]. Detection of glutamine synthetase (GS) activity in cell-free extracts, as well as activity of glutamateoxoloacetate transaminase, and the absence of glutamate dehydrogenase (GDH) activity suggested that ammonia may be assimilated via the GS-GOGAT pathway [51]. However, confirmation of this has not been obtained since glutamate synthase (GOGAT) was not detected, although this enzyme may be readily inactivated during extraction [52].

Enzymes of ammonia assimilation are suppressed in, or absent from the microsymbiont *in vivo* since there is little GS or GDH activity in vesicle cluster fractions isolated from nodule homogenates [53] and cytoimmunochemical tests suggested that GS is not present in the endophyte in nodule sections [54]. Both GS and GDH activity in the nodules are probably of host plant origin so that, as in the *Rhizobium*-legume association, ammonia produced by nitrogen fixation in the endophyte must be secreted for assimilation into organic combination in the host plant cytosol.

In alder nodules the association of high levels of GDH activity with host plant organelles such as mitochondria led Akkermans and co-workers [33] to suggest a route for assimilation of ammonia into citrulline (the prominent amino acid in nodules of this species) in which mitochondrial GDH would catalyze the oxidation of glutamate, thus producing ammonia for carbamoyl phosphate formation. However, Schubert and Coker [55] suggest that alder nodule GDH catalyzes the reductive amination of 2-oxoglutarate to glutamate, rather than the oxidative deamination of glutamate, since radioactivity from $^{13}NH_4^+$ fed to nodules was incorporated apparently independently into glutamate and into the amide group of glutamine. Low concentrations of NH_4^+ could be assimilated via GS and higher concentrations via GDH since azaserine (an inhibitor of GOGAT) did not prevent incorporation of ^{13}N into glutamate while methionine sulphoximine (an inhibitor of GS) increased incorporation into glutamate. The amide group of glutamine conceivably may be utilized in the synthesis of citrulline via the formation of carbamoyl phosphate [56]. Further enzymic and radiotracer work is required to resolve these different viewpoints and also to establish the routes of ammonia assimilation in the nodules of actinorhizal plants other than *Alnus,* which synthesize and export large amounts of amides [57, 58]. The relative amounts of the nitrogenous constituents exported from the nodules into xylem sap collected from detopped *Alnus glutinosa* and from *Myrica gale* (an "amide" plant) are compared in Fig. 1.

Tentative flow charts for the incorporation of ammonia into citrulline via the formation of glutamine and into asparagine are presented in Figures 2a and 2b. The ATP requirements for citrulline biosynthesis suggest that this process is much more energy demanding than asparagine biosynthesis, which utilizes one-third less energy (as ATP equivalents) per ammo-

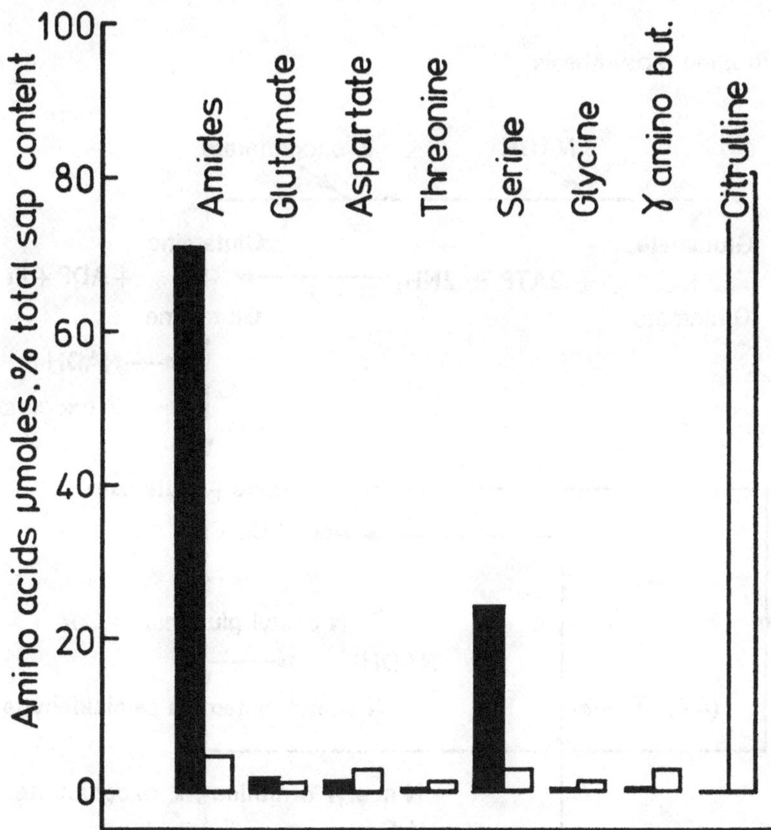

Fig. 1. Free amino acids and amides of the pooled bleeding sap, collected over a 2h period in mid-afternoon from 20 plants each of 6-months-old *Alnus glutinosa* (□) and *myrica gale* (■). The average weight of bleeding sap per plant collected from *Alnus* was 0.046g and from *Myrica* 0.020g. Results presented are for compounds with concentrations greater than 10 nmole g^{-1} bleeding sap. The amide fraction of *Myrica gale* contained both asparagine and glutamine but *Alnus glutinosa* contained detectable amounts of glutamine only. Analyses were essentially as described previously [57].

nia assimilated. The ratio of C:N (2:1) is the same in both molecules. However, the difference of 1.8 ATP utilized per NH_3 assimilated is small compared with the ATP required for the reduction of N_2 to NH_3 by nitrogenase (8 ATP per N reduced). Only the most careful comparisons of carbon respired per N_2 fixed would distinguish small differences of this nature. It is not surprising therefore, that in a comparison of the energy requirements for nitrogen fixation of actinorhizal nodules (including *Alnus* and *Myrica*) and nodules of legumes, by measurement of the ratio of respiratory to nitrogenase activity, Tjepkema and Winship [59] concluded that the energy costs were broadly similar. Within the actinorhizal plant,

(a) Citrulline biosynthesis

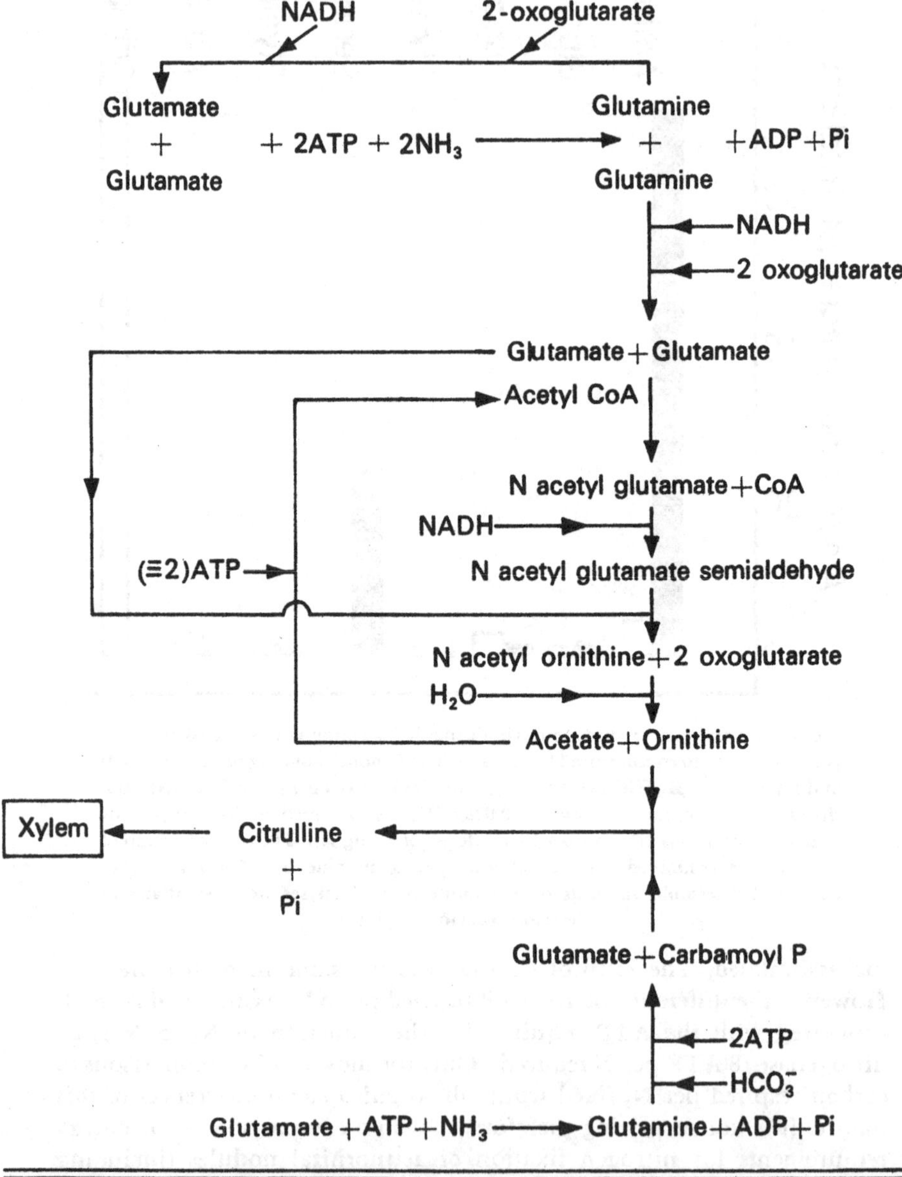

$3NH_3 \longrightarrow$ Citrulline $= 16$ ATP equivalents

$= 5.3$ ATP/NH_3

b) Asparagine biosynthesis

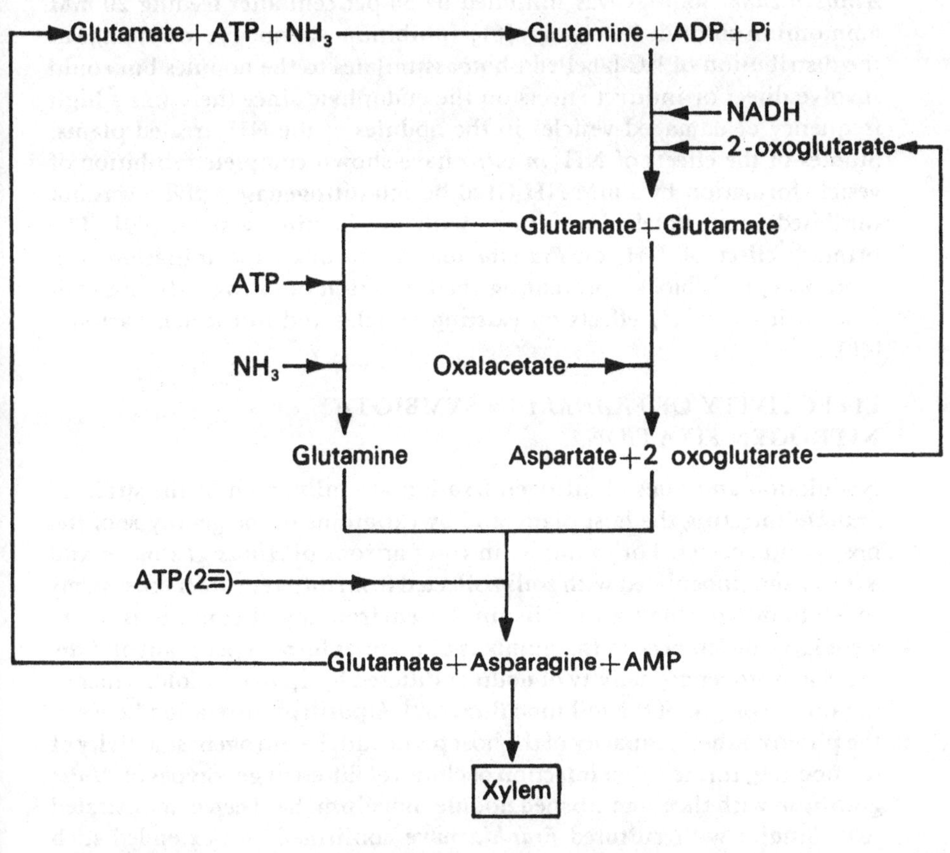

$$2NH_3 \longrightarrow \text{Asparagine} = 7 \text{ ATP equivalents}$$
$$= 3.5 \text{ ATP}/NH_3$$

Fig. 2. Energy utilisation, in ATP equivalents, during the assimilation of ammonia into organic combination.

the transport of the relatively N-rich citrulline (3N per molecule) may be less energy consuming than the transport of asparagine (2N per molecule) and help to offset any differences in expenditure of energy in their biosynthesis in the nodules.

Nitrogen fixation in actinorhizal root nodules, as in nodules of legumes, is inhibited by combined nitrogen [60]. Nitrogenase activity of *Alnus incana* nodules was inhibited by 50 per cent after feeding 20 mM ammonium chloride for one day [61]. Inhibition was not due to a change in the distribution of ^{14}C-labelled photoassimilates to the nodules but could involve direct or indirect effects on the endophyte since there was a high frequency of damaged vesicles in the nodules of the NH_4 treated plants. Studies of the effects of NH_4 *in vitro* have shown complete inhibition of vesicle formation by 1 mM NH_4Cl although nitrogenase activity was not inhibited immediately in cultures with pre-existing vesicles [40]. The primary effect of NH_4 on *Frankia* may be through the initiation of a morphological 'block', preventing the formation of nitrogen-fixing vesicles, with secondary effects on existing vesicles and nitrogenase activity [40].

EFFECTIVITY OF *FRANKIA* IN SYMBIOTIC NITROGEN FIXATION

Nodulation and rates of nitrogen fixation are influenced by the strain of *Frankia* infecting the host plant, and by variations in the genotype of the host plant species. For example, in comparisons of *Alnus glutinosa* and *Alnus rubra* inoculated with soils collected from different alder ecosystems in Scotland and then grown in similar environmental conditions, there were large differences in the numbers of nodules formed per plant and the specific nitrogenase activity of nodules differed by up to two-fold, depending on the origin of the soil inoculum [62]. A positive correlation between the photosynthetic capacity of the host plant and the nitrogenase activity of the nodules, formed after infection of clones of different genotypes of *Alnus glutinosa* with the same crushed nodule inoculum, has been demonstrated [63]. Studies with cultured *Frankia* have confirmed and extended such observations to show that different strains of *Frankia*, in symbiosis with both different host plant species and different genotypes of the same host plant species, may influence plant growth rates and nodule nitrogenase activity [64, 65].

Recent findings of a connection between effectivity in symbiotic nitrogen fixation and *in vivo* spore production in *Alnus* are of particular interest to researchers wishing to improve nitrogen fixation in this genus. Although production of sporangia in submerged culture is a characteristic of *Frankia*, sporangia are not always evident in nodules so that these may be differentiated into 'spore positive' or 'spore negative' types [66]. Spore positive nodules when crushed give inoculum which is up to a thousand

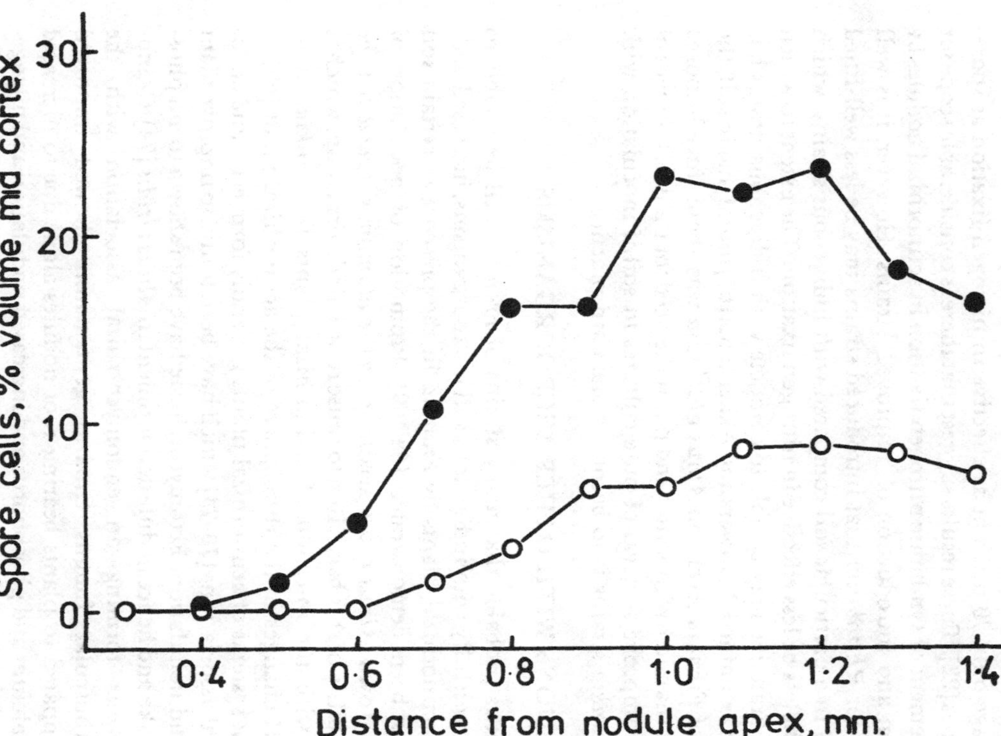

Fig. 3. Host plant cells containing the spore form of the endophyte as a percentage volume of the mid-cortex of nodules of *Alnus glutinosa* (○) and *Alnus rubra* (●). Both species were inoculated with a suspension of *Alnus glutinosa* crushed nodules. Volume assessments were by serological techniques as described previously [62].

times more infective, weight for weight, than spore negative nodules [67]. However, high spore frequency in nodules of *Alnus rubra,* formed in response to infection with inoculum from *Alnus glutinosa* spore-positive nodules, is associated with lower nitrogen fixation in the nodules of the former species (Fig. 3) [62].

Recently, evaluation of *Frankia* isolates from *Alnus rugosa* and *Alnus crispa* in Canada has shown that isolates from spore positive nodules were on an average only 70 per cent as effective in nitrogen fixation as spore-negative types [22]. These results suggest that there is considerable scope for the improvement of symbiotic nitrogen fixation in actinorhizal nodules by the selection and introduction of 'improved' strains. However, it is well established for *Rhizobium* that introduced strains may be less well-fitted for survival in a particular soil compared with indigenous strains, which may nevertheless be less effective in nitrogen fixation. The extent to which introduced Frankiae may be able to compete with indigenous strains has not yet been examined. Houwers and Akkermans [67] found that in soil, the infectivity of *Frankia* AvcI1 on *Alnus glutinosa* was about three hundred times lower than in hydroponics and point the need for the development of methods to improve survival of pure cultures in soil if inoculation with cultured *Frankia* strains is to become a standard practice.

INTERACTIONS WITH OTHER MICROORGANISMS

Although it is probable that many strains of *Frankia* will be unable to compete successfully with indigenous soil microorganisms, increased nodulation of actinorhizal plants by *Frankia* in the presence of certain soil bacteria has been demonstrated [68, 69]. Promotion of nodulation of aseptically-grown *Alnus rubra* seedlings by *Pseudomonas cepacia* may be due to the ability of this bacterium to cause root-hair deformation, which is associated with the infection of actinorhizal plants by *Frankia,* or to "buffer" pH changes in the rhizosphere of the actinorhizal plant [69].

The root systems of actinorhizal plants generally form ecto- and endomycorrrhizal associations [70–73] which have been found to improve plant nutrition and nodulation. Ectomycorrhizae have been shown to aid phosphorus uptake and hence nodule development in *Alnus viridis* [74]. *Ceanothus velutinus* forming an endomycorrhizal association with the vesicular-arbuscular fungus *Glomus gerdemanni* showed improved growth compared to plants bearing root nodules alone due to increased uptake of calcium and phosphorus and increased nodulation and nitrogenase activity [75]. Clearly both actinorhizal and mycorrhizal organisms are important for host plant nutrition and growth and further research is required to determine the conditions which will optimize the benefits to the host plant of these symbiotic associations.

NITROGEN FIXATION AND THE LIFE CYCLE OF ACTINORHIZAL PLANTS

Comprehensive studies of changes in the patterns of nodulation and nitrogen fixation during the life of actinorhizal plants are largely lacking. This is due mainly to the difficulties engendered by the perennial growth habit and long life of many actinorhizal species, which dictate that such studies be conducted on similar sites, bearing plants of a range of ages which cover the life span of a particular species. Variability in nodule frequency per plant, even within a relatively uniform stand, creates difficulties for obtaining reliable estimates of changes in nitrogen fixation with plant age [76, 77]. An added problem for sampling of nodules in the field is the wide variation in the age distribution of perennial nodules (survival of *Alnus glutinosa* nodules up to eight-year-olds has been noted [81]), some of which will be borne on young roots while others may be borne on roots several years old within the same plant [78]. These latter difficulties prompted Sprent et al. [79] to express nitrogenase activity of field nodules of *Myrica gale* in terms of acetylene reduced per nodule lobe, rather than nodule weight.

While comparing stands of 0 to 3- and 13 to 16-year-old *Hippophäe rhamnoides*, nitrogen gains of 27 and 139 kg hectare^{-1} annum^{-1} have been reported [80]. However, nitrogenase activity decreases rapidly as nodules age within a stand so that 120-day-old *Hippophäe* nodules reduce acetylene at only a quarter of the rate of 60-day-old nodules [78]. Increased area rates of nitrogen accretion with age must come from greater nodulation of plants, therefore, which as noted above is notoriously difficult to assess accurately. In a homogeneous, 6 to 7-year-old *Hippophäe* stand [77] numbers of nodules per square metre varied from 30 to 281. It is clear that even fairly reliable estimates of nodulation and field fixation will probably be obtained only by the analysis of a large number of large samples from a particular population of plants.

Nitrogenase activity of actinorhizal plants also changes considerably during a single growing season so that estimates of N accretion in the field, obtained from ^{15}N fixation or C_2H_2 reduction measurements require the integration of the area under nitrogenase activity curves, obtained from assays conducted at frequent intervals during the growing season.

Ambient temperature is a major factor affecting seasonal activity, together with photoperiod [60, 81–83]. In both *Myrica gale* and in *Alnus* species, there is a decline in nitrogenase activity in the summer, even though conditions are still favourable for photosynthesis. This decline coincides with the cessation of elongation growth of the shoot [62, 83] and may be due to diversion of photosynthates away from the nodules to establish overwintering carbohydrate reserves in other plant parts. In the

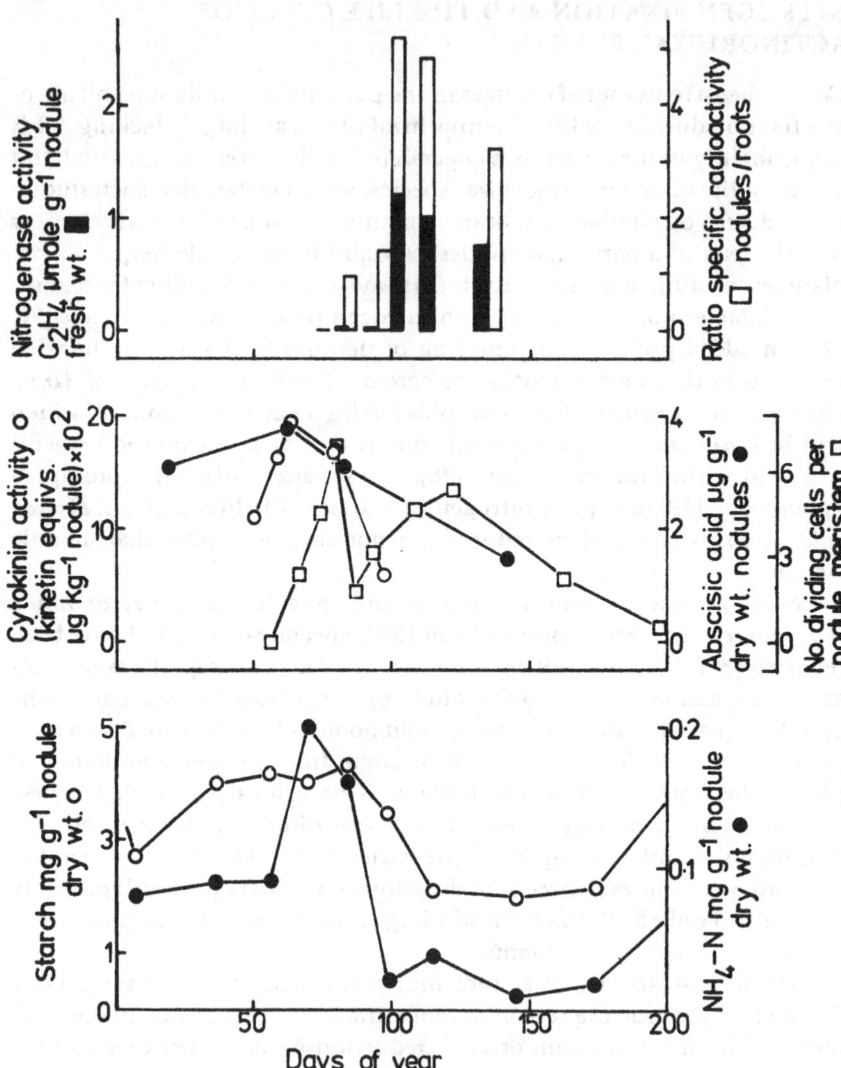

Fig. 4. Chronology of changes in meristematic activity, nitrogenase activity, starch, NH_4-N and levels of cytokinins and abscisic acid in nodules of *Alnus glutinosa* on emergence from winter dormancy.

nodules themselves, starchy reserves accumulate from mid-summer until leaf-fall [62, 79, 83].

The factors which govern seasonal changes in the direction of translocation and in the utilization of photosynthates in the nodulated plant are largely unknown. However, detailed study of the physiology and metabolism of *Alnus glutinosa* nodules during the recommencement of growth and nitrogenase activity in the spring has shown the complexity of the events which accompany renewed nodule activity, following the stimulation of the host plant by favourable environment. Some of the changes which take place in the nodules during emergence from dormancy are summarized in Fig. 4. Following the onset of low nitrogenase activity at bud burst, high levels of NH_4 in the nodules develop which may suppress further increase in nitrogenase activity until leaf expansion is well advanced [84, 85]. The renewal of tree growth probably depends initially on the utilization of nitrogenous reserves within the plant so that the limited supplies of photosynthates synthesized by the expanding leaves can be metabolized to support new plant growth rather than the energy demanding nitrogen-fixing processes.

Recommencement of nodule growth in the spring is associated with high levels of cytokinins in the nodules although it is not clear whether this is a cause or consequence of renewed meristematic activity in the nodules. Overwintering nodules also contain high levels of abscisic acid [86] and it is conceivable that part of the role of elevated cytokinin levels may be to overcome inhibitory effects of this compound [87]. Such basic studies of the control of seasonal nodule activity may assist in the development of an understanding of the mechanisms which govern the organization and development of nodule tissues as well as suggesting ways in which the annual period of nitrogen fixation might be prolonged by manipulation of the physiology of the host plant.

REFERENCES

1. Bond, G. Fixation of nitrogen by higher plants other than legumes, *Annual Review Plant Physiology 18*, 107-126 (1967).
2. Becking, J.H. Frankiaceae Fam. Nov. (*Actinomycetales*) with one new combination and six new species of the genus *Frankia* Brunchorst 1886, 174, *International Journal of Systematic Bacteriology 20*, 201-220 (1970).
3. Callaham, D., Del Tredici, P. and Torrey, J.G. Isolation and cultivation *in virto* of the actinomycete causing root nodulation in *Comptonia, Science 199*, 899-902 (1978).
4. Torrey, J.G. and Tjepkema, J.D. Preface, *Botanical Gazette 140*, Si-Sv (1979).
5. Bond, G. The results of the IBP survey of root-nodule formation in

non-leguminous angiosperms, in *Symbiotic Nitrogen Fixation in Plants* (Editor, P.S. Nutman), Cambridge University Press, London (1976).

6. Heisey, R.M., Delwiche, C.C., Virginia, R.A., Wrona, A.F. and Bryan, B.A. A new nitrogen fixing non-legume: *Chamaebatia foliolosa* (Rosaceae), *American Journal Botany 67*, 429-431 (1980).
7. Rhighetti, T.L. and Munns, D.N. Nodulation and nitrogen fixation in cliffrose [*Cowania mexicana* var. *stansburiana* (Torr.) Jeps.], *Plant Physiology 65*, 411-412 (1980).
8. Medan, D. and Tortosa, R.D. Nodulos actinomicorricicos en especies argentinas de los generos *Kentrothamnus, Trevoa* (Rhamnaceae) y *Coriaria* (Coriariaceae), *Boletin Societas Argentina Botanica 20*, 71-81 (1981).
9. Kummerow, J. In Bond, G. and Becking, J.H. Root nodules in the genus *Colletia, New Phytologist 90*, 57-65 (1982).
10. Chaudhary, A.H. Nitrogen fixing root nodules in *Datisca cannabina, Plant Soil 51*, 163 (1979).
11. Trotter, A. Intorno a tubercoli radicali di *Datisca cannabina, Bulletia Societe Botanico Italia* 50-52 (1902).
12. Severini, G. Sui tubercoli radicali di *Datisca cannabina, Annali Botanico (Roma) 15*, 29-51 (1922).
13. Akkermans, A.D.L. and van Dijk, C. In *Nitrogen Fixation. I. Ecology* (Editor, W.J. Broughton), Clarendon Press, Oxford, 1981.
14. Bond, G. Taxonomy and distribution of non-legume nitrogen-fixing systems, in *Biological Nitrogen Fixation in Forest Ecosystems; Foundations and Applications* (Editors, J.C. Gordon and C.T. Wheeler), Martinus Nijhoff, The Hague.
15. Lawrie, A.C. Field nodulation in nine species of *Casuarina* in Victoria, *Australian Journal Botany 30*, 447-460 (1982).
16. Oremus, P.A.I. and Otten, H. Factors affecting growth and nodulation of *Hippophae rhamnoides* L. ssp. *rhamnoides* in soils from two successional stages of dune formation, *Plant Soil 63*, 317-331 (1981).
17. Baker, D. and Torrey, J.G. The isolation and cultivation of actinomycetous root nodule endophytes, in *Symbiotic Nitrogen Fixation in the Management of Temperate Forests* (Editors, J.C. Gordon, C.T. Wheeler and D.A. Perry), Forest Research Laboratory, Oregon State University, Corvallis (1979).
18. Berry, A. and Torrey, J.G. Isolation and characterisation 'in vivo' and 'in vitro' of an actinomycetous endophyte from *Alnus rubra* Bong, in *Symbiotic Nitrogen Fixation in the Management of Temperate Forests* (Editors, J.C. Gordon, C.T. Wheeler and D.A. Perry), Forest Research Laboratory, Oregon State University, Corvallis (1979).
19. Gauthier, D., Diem, H.G. and Dommergues, Y. 'In vitro' nitrogen fixation by two actinomycete strains isolated from *Casuarina*

nodules, *Applied Environmental Microbiology 41*, 306-308 (1981).

20. Diem, H.G., Gauthier, D. and Dommergues, Y.R. Isolation of *Frankia* from nodules of *Casuarina equisetifolia, Canadian Journal of Microbiology 28*, 526-530 (1982).
21. Lalonde, M., Calvert, H.E. and Pine, S. Isolation and use of *Frankia* strains in actinorhizae formation, in *Current Perspectives in Nitrogen Fixation* (Editors, A.H. Gibson and W.E. Newton), Australian Academy of Science, Canberra (1981).
22. Normand, P. and Lalonde, M. Evaluation of *Frankia* strains isolated from provenances of two *Alnus* species, *Canadian Journal of Microbiology 28*, 1133-1142 (1982).
23. Lechevalier, M.P., Horriere, F. and Lechevalier, H. The biology of *Frankia* and related organisms, in *Developments in Industrial Microbiology 23*, The Society for Industrial Microbiology (1982).
24. Lechevalier, M.P., Horriere, F. and Lechevalier, H. The biology of *Frankia* and related organisms, in *The Prokaryotes*, (Editors, M.P. Star, H. Stolp, H.G. Truper, A. Balows and H.G. Schlegel), Springer Verlag, Berlin (1981).
25. Jones, K.L. Fresh isolates of actinomycetes in which the presence of sporogenous aerial mycelia is a fluctuating characteristic, *Journal Bacteriology 57*, 141-145 (1949).
26. Shipton, W.A. and Burgraaf, A.J.P. A comparison of the requirements for various C and N sources and vitamins in some *Frankia* isolates, *Plant Soil 69*, 149-161 (1982).
27. Baker, D. A cumulative listing of isolated Frankiae, the symbiotic, nitrogen-fixing actinomycetes, *The Actinomycetes 17*, 35-42.
28. Lalonde, M. and Calvert, H.E. Production of *Frankia* hyphae and spores as infective inoculant for *Alnus* species, in *Symbiotic Nitrogen Fixation in the Management of Temperate Forests* (Editors, J.C. Gordon, C.T. Wheeler and D.A. Perry), Forest Research Laboratory, Oregon State University, Corvallis (1979).
29. Newcomb, W., Callaham, D., Torrey, J.G. and Peterson, R.L. Morphogenesis and fine structure of actinomycetous endophyte of nitrogen fixing root nodules of *Comptonia peregrina, Botanical Gazette 140*, 522-534.
30. Baker, D. and McCain, R. and Seling, E. Serological and host capability relationships among the isolated Frankiae, *Canadian Journal of Botany* (in press).
31. Blom, J. Utilisation of fatty acids and NH_4^+ by *Frankia* AvcI 1, *FEMS Microbiology Letters 10*, 143-145 (1981).
32. Blom, J. Carbon and nitrogen source requirements of *Frankia* strains, *FEMS Microbiology Letters 13*, 51-55 (1981).
33. Akkermans, A.D.L., Blom, J., Huss-Danell, K. and Roelofsen, W. The carbon and nitrogen metabolism of *Frankia* in pure culture and

in root nodules, in *The Second National Symposium on Biological Nitrogen Fixation, Proceedings*, The Finnish National Fund for Research and Development, Helsinki (1982).

34. Blom, J. and Harkink, R. Metabolic pathways for gluconeogenesis and energy generation in *Frankia* AvcI 1, *FEMS Microbiology Letters 11*, 221-224.
35. Huss-Danell, K., Roelofsen, W., Akkermans, A.D.L. and Meyer, P. Carbon metabolism of *Frankia* ssp. in root nodules of *Alnus glutinosa* and *Hippophae rhamnoides, Physiologia Plantarum 54*, 461-466 (1982).
36. Akkermans, A.D.L., Huss-Danell, K. and Roelofsen, W. Enzymes of the tricarboxylic acid and malate-aspartate shuttle in the N_2-fixing endophyte of *Alnus glutinosa, Physiologia Plantanum 53*, 289-294 (1981).
37. Perrot-Rechenmann, C., Vidal, J., Maudinas, B. and Gadal, P. Immunocytochemical study of phosphoenolpyruvate carboxylase in nodulated *Alnus glutinosa, Planta 153*, 14-17 (1981).
38. Wheeler, C.T. Unpublished observations.
39. Tjepkema, J.D., Ormerod, W. and Torrey, J.G. On vesicle formation and 'in vitro' acetylene reduction by *Frankia, Nature 287*, 633-635 (1980).
40. Tjepkema, J.D., Ormerod, W. and Torrey, J.G. Factors affecting vesicle formation and acetylene reduction (nitrogenase activity) in *Frankia* sp. CpI1, *Canadian Journal Microbiology 27*, 815-823 (1981).
41. Mian, S. and Bond, G. The onset of nitrogen fixation in young alder plants and its relation to differentiation in the nodular endophyte, *New Phytologist 80*, 187-192 (1978).
42. Akkermans, A.D.L., van Straten, J. and Roelofsen, W. Nitrogenase activity of nodule homogenates of *Alnus glutinosa* in a comparison with the *Rhizobium* pea system, in *Recent Developments in Nitrogen Fixation* (Editors, W. Newton, J.R. Postgate and C. Rodriguez-Barrueco), Academic Press, London (1977).
43. Baker, D., Newcomb, W. and Torrey, J.G. Characterisation of an ineffective actinorhizal microsymbiont, *Frankia* sp. EuI1 (Actinomycetales), *Canadian Journal Microbiology 26*, 1072-1109 (1980).
44. Benson, D.R., Arp, D.J. and Burris, R.H. Cell-free nitrogenase and hydrogenase from actinorhizal root nodules, *Science 205*, 688-689 (1979).
45. Bergersen, F.J., Turner, G.L., Gibson, A.H. and Dudman, W.F. Nitrogenase activity and respiration of cultures of *Rhizobium* with special reference to the concentration of dissolved oxygen, *Biochemica Biophysica Acta 444*, 164-174 (1976).
46. Wheeler, C.T., Gordon, J.C. and Ching, T-M. Oxygen relations of

the root nodules of *Alnus rubra* Bong, *New Phytologist 82*, 449-457 (1979).

47. Tjepkema, J.D. Oxygen relations in leguminous and actinorhizal nodules, in *Symbiotic Nitrogen Fixation in the Management of Temperate Forests* (Editors J.C. Gordon, C.T. Wheeler and D.A. Perry), Forest Research Laboratory, Oregon State University, Corvallis (1979).
48. Tjepkema, J.D. Hemoglobins in actinorhizal nodules, in *The Biology of Frankia and its Associations with Higher Plants*, Abstracts of Meeting, University of Wisconsin, Madison (1982).
49. Egle, K. and Munding, H. Uber den gehalt an haminkorpern in den wurzelknollchen von nicht-leguminosen, *Naturwissenschaften 38*, 548-549 (1951).
50. Davenport, H.E. Haemoglobin in the root nodules of *Casuarina cunninghamiana*, *Nature 186*, 653-654 (1960).
51. Miflin, B.J. and Lea, P.L. The pathway of nitrogen assimilation in plants, *Phytochemistry 15*, 855-873 (1976).
52. Blom, J. Carbon and nitrogen metabolism of free-living *Frankia* ssp. and of *Frankia-Alnus* symbioses, Ph.D. thesis, The Agricultural University, Wageningen (1982).
53. Blom, J., Roelofsen, W. and Akkermans, A.D.L. Assimilation of nitrogen in root nodules of alder (*Alnus glutinosa*), *New Phytologist 89*, 321-326 (1981).
54. Hirel, B., Perrot-Rechenmann, C., Maudinas, B. and Gadal, P. Glutamine synthetase in alder (*Alnus glutinosa*) root nodules. Purification, properties and cytoimmunochemical localisation, *Physiologia Plantarum 55*, 197-203 (1982).
55. Schubert, K.R. and Coker, G.T. Ammonia assimilation in *Alnus glutinosa* and *Glycine max*. Short-term studies using [^{13}N] ammonium, *Plant Physiology 67*, 662-665 (1981).
56. Kolloffel, C. and Verkerk, B.C. Carbamoyl phosphate synthetase activity from cotyledons of developing and germinating pea seeds, *Plant Physiology 69*, 143-145 (1982).
57. Wheeler, C.T. and Bond, G. The amino acids of non-legume root nodules, *Phytochemistry 9*, 705-708 (1970).
58. Dixon, R.O.D. and Wheeler, C.T. Biochemical, physiological and environmental aspects of symbiotic nitrogen fixation, in *Biological Nitrogen Fixation in Forest Ecosystems: Foundations and Applications* (Editors, J.C. Gordon and C.T. Wheeler), Martinus Nijhoff, The Hague (in press).
59. Tjepkema, J.D. and Winship, L.J. Energy requirements for nitrogen fixation in actinorhizal and legume root nodules, *Science 209*, 279-281 (1980).
60. Wheeler. C.T. and McLaughlin, M.E. Environmental nodulation of

nitrogen fixation in actinomycete nodulated plants, in *Symbiotic Nitrogen Fixation in the Management of Temperate Forests* (Editors, J.C. Gordon, C.T. Wheeler and D.A. Perry), Forest Research Laboratory, Oregon State University, Crovallis (1979).

61. Huss-Danell, K., Sellstedt, A., Flower-Ellis, A. and Sjostrom, M. Ammonium effects on the function and structure of nitrogen-fixing root nodules of *Alnus incana* (L.) Moench, *Planta 156*, 332-340 (1982).
62. Wheeler, C.T., McLaughlin, M.E. and Steele, P. A comparison of symbiotic nitrogen fixation in Scotland in *Alnus glutinosa* and *Alnus rubra, Plant Soil 61*, 169-188 (1981).
63. Gordon, J.C. and Wheeler, C.T. Whole plant studies on photosynthesis and acetylene reduction in *Alnus glutinosa, New Phytologist 80*, 179-186 (1978).
64. Dawson, J.O. and Sun, S-H. The effect of *Frankia* isolates from *Comptonia peregrina* and *Alnus crispa* on the growth of *Alnus glutinosa, A. cordata* and *A. incana* clones, *Canadian Journal of Forest Research 11*, 758-762 (1981).
65. Dillon, J.T. and Baker, D. Variations in nitrogenase activity among pre-cultured *Frankia* strains tested on actinorhizal plants as an indication of symbiotic compatibility, *New Phytologist 92*, 215-220 (1982).
66. van Dijk, C. Spore formation and endophyte diversity in root nodules of *Alnus glutinosa* (L.) Vill., *New Phytologist 81*, 601-615 (1978).
67. Houwers, A. and Akkermans, A.D.L. Influence of inoculation on yield of *Alnus glutinosa* in the Netherlands, *Plant Soil 61*, 189-202 (1981).
68. Knowlton, S., Berry, A. and Torrey, J.G. Evidence that associated soil bacteria may influence root hair infection of actinorhizal plants by *Frankia, Canadian Journal of Microbiology 26*, 971-977 (1980).
69. Knowlton, S. and Dawson, J.O. Effects of *Pseudomonas cepacia* and cultural factors on the nodulation of *Alnus rubra* roots by *Frankia, Canadian Journal of Botany* (in press).
70. Rose, S.L. Mycorrhizal associations of some actinomycete-nodulated nitrogen-fixing plants, *Canadian Journal of Botany 58*, 1449-1454 (1980).
71. Molina, R. Ectomycorrhizal specificity in the genus *Alnus, Canadian Journal of Botany 59*, 325-334 (1981).
72. Riffle, J.W. First report of vesicular-arbuscular mycorrhizae on *Eleagnus angustifolia, Mycologia 69*, 1200-1203.
73. Diem, H.G. and Gauthier, D. Effect of endomycorrhizal infection (*Glomus mossae*) on the nodulation and growth of *Casuarina equisetifolia, Comptes Rendu de l'Academie Sciences Series III 294*, 215-218 (1982).

74. Mejstrik, V. and Benecke, U. The ectotrophic mycorrhizae of *Alnus viridis* (Chaix.) D.C. and their significance in respect to phosphorus uptake, *New Phytologist 68*, 141-149.
75. Rose, S.L. and Youngberg, C.T. Tripartite associations in snowbrush (*Ceanothus velutinus*): Effect of vesicular-arbuscular mycorrhizae on growth, nodulation and nitrogen fixation, *Canadian Journal of Botany 59*, 34-39 (1981).
76. Fessenden, R.J., Knowles, R. and Brouzes, R. Acetylene-ethylene assay studies on excised root nodules of *Myrica asplenifolia* L., *Soil Science Society America Proceedings 37*, 893-898 (1973).
77. Oremus, P.A.I. A quantitative study of nodulation in *Hippophae rhamnoides* ssp. rhamnoides in coastal dune area, *Plant Soil 52*, 59-68 (1979).
78. Oremus, P.A.I. Growth and nodulation of *Hippophae rhamnoides* L. in the coastal sand dunes of the Netherlands, Ph.D. thesis, University of Utrecht (1982).
79. Sprent, J.I., Scott, R. and Perry, K.M. The nitrogen economy of *Myrica gale* in the field, *Journal of Ecology 66*, 657-668 (1978).
80. Stewart, W.D.P. and Pearson, M.C. Nodulation and nitrogen fixation by *Hippophae rhamnoides* in the field, *Plant Soil 26*, 348-360 (1967).
81. Akkermans, A.D.L. Nitrogen fixation and nodulation of *Alnus* and *Hippophae* under natural conditions, Ph.D. thesis, University of Leiden (1971).
82. Wheeler, C.T., Perry, D.A., Helgerson, O. and Gordon, J.C. Winter fixation of nitrogen in Scotch broom (*Cytisus scoparius* L.), *New Phytologist 82*, 697-701 (1979).
83. Schwintzer, C.R., Berry, A.M. and Disney, L.M.S. Seasonal patterns of root nodule growth, endophyte morphology, nitrogenase activity and shoot development in *Myrica gale, Canadian Journal of Botany 60*, 746-757 (1982).
84. Wheeler, C.T., Watts, S.H. and Hillman, J.R. Changes in carbohydrates and nitrogenous compounds in the root nodules of *Alnus glutinosa, New Phytologist* (submitted for publication).
85. Skeffington, R. Studies on nitrogen fixation in non-legume plants, Ph.D. thesis, University of Dundee (1975).
86. Watts, S.H., Wheeler, C.T., Hillman, J.R., Berrie, A.M.M., Crozier, A. and Math, V.B. Abscisic acid in the nodulated root system of *Alnus glutinosa, New Phytologist*, (submitted for publication).
87. Henson, I.E. and Wheeler, C.T. Hormones in plants bearing nitrogen-fixing root nodules: distribution and seasonal changes in levels of cytokinins in *Alnus glutinosa* (L.) Gaertn, *Journal of Experimental Botany 28*, 1076-1086 (1977).

8. Nitrogen Fixation by Lichens

J.W. Millbank

INTRODUCTION

Since the publication of the comprehensive study "The Lichens" [1] in 1973, knowledge of the nitrogen fixation process has made great strides, but reviews of investigations concerning the process in lichens have been relatively few. Chapters dealing with lichens, mostly as one of a family of blue-green algal symbioses, are to be found in "A Treatise on Dinitrogen Fixation" [30] and "The Biology of Nitrogen Fixation" [73], and lichens are alluded to briefly in "The Biology of Nitrogen Fixing Organisms" [83]. Most other books confine themselves to a few words, and although a number of symposium reports [9, 25, 26, 79, 85] include studies of nitrogen fixation by lichens, the subject tends to be referred to only briefly. Other aspects of lichen physiology and ecology have been dealt with more thoroughly; so there is perhaps some justification for the balance to be redressed by the present chapter. One feature is that the great bulk of the investigations have been carried out by relatively few individuals and groups. It is to be hoped that this state of affairs will improve.

Nitrogen fixation is confined to prokaryotes and therefore in lichens the process is only found in those with a cyanophyte phycobiont, some 7 per cent of the whole group. For this reason also, details of metabolism of the fungal partner do not greatly concern us.

PHYSIOLOGY AND BIOCHEMISTRY

(a) Diazotrophic lichens and techniques for study

Since the list of nitrogen-fixing lichens was published in 1977 [61], more genera and species have been reported, mostly using the acetylene reduction technique. An updated list is given in Table 1. Besides additional species of genera already listed, a notably comprehensive report [27] on the Stictaceae of New Zealand, is the principal addition. Recent studies using $15N_2$ are of course fewer in number [38, 62, 63, 64, 76, 86] and tend to use established species of diazotrophs. Respecting techniques, the widely known acetylene reduction test [88] is naturally pre-eminent, but its serious limitations for quantitative studies are becoming more generally recognized

Table 1. The names of established nitrogen-fixing lichens

Lichen	Reference	Lichen	Reference
Coccocarpia cronia	27	*Pseudocyphellaria amphisticta*	
Collema auriculatum	36	*P. aurata*	
C. coccophorus	78	*P. billardierii*	
C. crispum	37	*P. cf. carpoloma*	
C. fluviatile	36	*P. colensoi*	
C. furfuraceum	36	*P. coriacea*	
C. granosum	8	*P. coronata*	
C. pulposum	22	*P. crocata*	
C. subconveniens	27	*P. degelii*	
C. subfervum	36	*P. delisea*	
C. tenax	81	*P. dissimilis*	
C. tuniforme	34	*P. faveolata*	
Ephebe lanata	36	*P. flavicans*	
Lecidea sp.	81	*P. granulata*	27
Leptogium burgessii	36	*P. hamata*	
L. cyanescens	48	*P. hirsutula*	
L. lichenoides	8	*P. homeophylla*	
L. sinuatum	36	*P. hookeri*	
L. teretiusculum	36	*P. intricata*	
Lichina confinis	84	*P. multifida*	
Li. pygmaea	84	*P. polyschista*	
Lobaria laetevirens	36	*P. rubella*	
L. oregana	17	*P. thouarsii*	36
L. pulmonaria	66	*Solarina crocea*	45
L. scrobiculata	36	*Stereocaulon paschale*	42
Massalongia carnosa	36	*Sticta caperata*	27
Nephroma arcticum	45	*S. filix*	27
N. australe	27	*S. fuliginosa*	36
N. laevigatum	36	*S. latifrons*	27
Pannaria pezizoides	36	*S. limbata*	36
P. rubiginosa	36	*S. variabilis*	27
Parmeliella atlantica	36	*S. weigelii*	48
P. plumbea	36		
Peltigera aphthosa var. *variolosa*	65	**Heterocyst frequency determined, but acetylene reduction not tested [36]**	
P. canina	66	*Lobaria amplissima*	
P. dolichoriza	27	*Dendriscocaulon umhausense*	
P. evansiana	49	*Psoroma hypnorum*	
P. membranacea	62	*Solorina saccata*	
P. polydactyla	90	*Solorina spongiosa*	
P. praetextata	80	*Stereocaulon vesuvianum*	
P. pruinosa	90	*Sticta dufourii*	
P. rufescens	34	*Sticta canariensis*	
P. scabrosa	55		
Placopsis gelida	37		
Placynthium nigrum	36		
P. pannariellum	36		
Polychidium muscicola	36		

[31, 38, 62, 77]. The use of $15N_2$ gas in the field, however, is hardly ever attempted, as it presents daunting problems, and this aspect will be dealt with later when fixation under field conditions is discussed.

One novel procedure has been reported, that of Crittenden and Kershaw [13] who describe a technique for the simultaneous assay of carbon dioxide exchange and nitrogenase activity (acetylene reduction), using gas chromatography. Essentially, the technique exploits the high sensitivity to hydrocarbons of the flame ionization detector by converting the carbon dioxide to methane by means of a nickel catalyst. A programmed electromagnetic valve system ensures that substrate acetylene in the mixture to be analyzed is not passed over the catalyst. This technique of simultaneous measurement gives a very worthwhile gain in speed of analysis and comparability of data for the two activities of a given lichen, at the expense of technical complexity.

(b) Rates of nitrogen fixation under laboratory conditions

In the earlier surveys [1, 61] the rapid rate of nitrogen fixation by the cyanophyte was emphasized, and more recent studies have confirmed this characteristic. Rates are at least equal to those of the free-living form [86]; total cell protein is commonly used as the basis for expressing rates, although it is not certain that free living and symbiotic cyanobacteria have similar contents. Environmental factors affecting the rate in lichens include pH, temperature, light intensity and thallus moisture content; the season at which the lichen material is collected for investigation is also of importance. What emerges most strikingly when reports of rates are studied is how important the environmental conditions can be, not only during the actual experiment but also during the previous history of the material and its period of "pretreatment" and/or storage. For this reason it is not profitable to provide a table of "rates of nitrogenase activity" from which comparisons will inevitably be drawn. I will confine myself to the statement that most samples of *Peltigera canina* can reduce acetylene at rates of the order of 10 nanomoles · $hour^{-1}$mg dry wt^{-1} during spring and autumn, at optimal moisture content, temperature, light and pH. Expressed in terms of chlorophyll *a*, such rates are about 8 nanomoles C_2H_4 μg $chla^{-1}$· hr^{-1}. This represents approximately 50 μg nitrogen · hr^{-1}· gm dry $thallus^{-1}$. For the *Nostoc* in the cephalodia of *P. aphthosa* the rates are of the order of 20 nanomoles C_2H_4 · μg $chla^{-1}$ · hr^{-1}. Most of this difference can be accounted for by the greatly increased proportion of heterocysts.

Returning to the specific effects of light, moisture, temperature and pH, most of the investigations into the effect of these factors have been concerned with establishing how the field environment influences nitrogen-fixing activity. They will be discussed in detail later, but it is appropriate here to recount the general conclusions.

The effect of pH on nitrogen fixation as opposed to more general observations of substrate preference by lichen species, was noted by Fogg and Stewart [22] who observed that the cephalodia of *Stereocaulon* did not develop on thalli from acid areas; thus in the absence of the cyanophyte nitrogen fixation did not take place. Cephalodia-free *Stereocaulon* has also been observed in SO_2-polluted areas of inner London by the author. Englund [18] reported that excised cephalodia of *Peltigera aphthosa* had an optimum pH range of 5–7. Stewart [87] reports that the *Nostoc* of this lichen has a pH optimum of 7–8.

Respecting temperature, lichens fix nitrogen over a very wide range from approximately 0°C [6, 20, 22, 58] to 35 or even 46°C for short periods [37, 54]. The temperature optimum for temperate lichens is mostly 15–20°C, and 28–30°C for tropical species.

Light is not consistent in its effect. Some lichens (*Lichina, Nephroma, Solorina* and *Stereocaulon*) have been reported [37, 42, 45] to show increasing rates of nitrogenase activity with increasing light intensity up to 20,000 lux, while Peltigera are reported to saturate at 3000 lux [19, 37] and Mcfarlene and Kershaw [54] quoting absolute units of photosynthetically-active energy showed that there is no marked increase of nitrogenase activity in *Peltigera canina* and *P. rufescens* at light energy levels greater than about 100 $\mu E \cdot cm^{-2} \cdot sec^{-1}$. It is much more important to realize that the optimum probably varies with the natural habitat of the lichen [48] and the season of the year [54]. Furthermore, high light intensities in the field are often (but not always) associated with high temperatures and low relative humidity; lichen thalli lose water rapidly and cease to metabolize under these conditions. Results obtained for the response to high light intensity at high moisture contents should be treated with caution as it is difficult to achieve high light intensities in the laboratory concomitantly with high moisture contents but at moderate or low temperatures. At the other extreme, nitrogen fixation rates in darkness have been reported to vary very markedly [37, 45, 57] and this undoubtedly depends on the availability of endogenous carbon reserves to provide the necessary energy. Under such circumstances polyglucose reserves in the cyanobacteria are mobilized and degraded via the oxidative pentose phosphate pathway, thus providing ATP and reductant for, *inter alia*, nitrogen fixation. However, dark fixation of carbon dioxide also plays a part [75]. In *Peltigera aphthosa*, at least, phosphoenolpyruvate carboxylase (PEP) in the fungus plays an important part in providing carbon skeletons for ammonia assimilation. Ribulose diphosphate carboxylase (RUBP), abundant in the *Nostoc*, as carboxysomes, seems to play no part in dark CO_2-fixation. The continuing ability to assimilate the ammonia formed prevents the inhibition of nitrogenase activity which would otherwise take place, and thus active fixation can be prolonged.

Finally, *moisture content* is of course an extremely important factor governing the rate of nitrogen fixation by lichens. After prolonged periods of desiccation nitrogenase activity can recover to the "normal" value on remoistening; the comprehensive studies by Kershaw and collaborators [49, 54, 55] have established that a moisture content of 150–200 per cent oven-dry weight is critically necessary in *Peltigera* spp; Denison [17] and Huss-Danell [40] have shown that similar moisture contents are optimal for *Lobaria* and *Stereocaulon* respectively.

(c) Acetylene reduction and 15 N_2 incorporation; H_2 evolution

Rates of nitrogen fixation derived from the acetylene test for nitrogenase have been widely quoted, but it was soon recognized that quantitative estimates of nitrogen fixation by a given organism or system necessitate the establishment of a specific conversion factor for acetylene reduction and "true" nitrogen incorporation using 15N [31]. This conversion factor can be greatly influenced by the experimental conditions applied to the nitrogen-fixing system concerned. Very few estimates indeed have been published for lichens; the results for these are set out in Table 2.

Table 2. Relation between acetylene reduction and nitrogen fixation

Lichen	mols ethylene formed / mols nitrogen fixed	Temperature	Reference
Lobaria oregana	4.78	20°C	38
L. pulmonaria	4.27	20°C	38
L. pulmonaria	3.9	15°C	63
Peltigera membranacea	6.1	15°C	63
P. polydactyla	5.9	15°C	63

The variation from the "theoretical" value of three comes as no surprise, as the multisubstrate nature of nitrogenase is well known and the reduction of hydrogen ions to hydrogen gas concurrent with the reduction of N_2 to NH_3 is now recognised as a normal part of the fixation process. The proportion of the total electron throughput devoted to H_2 production is always substantial, and, under certain conditions of pH, temperature and enzyme component ratio can become very large [10]. Fortunately, however, many organisms have a mechanism for "recycling" the hydrogen released and thus reducing or perhaps even eliminating the serious energy loss. This mechanism involves a "undirectional" or "uptake" hydrogenase, coupling the oxidation of hydrogen with phosphorylation and the production of ATP.

Such a system has been demonstrated to occur in *Peltigera membranacea, P. polydactyla* and *Lobaria pulmonaria* [63]. At the temperatures normal for the lichens' habitats the recycling mechanism was virtually completely effective, with no net evolution of hydrogen. At higher temperatures a significant evolution of hydrogen occurred in *Peltigera.* Thus, there is evidently an effective mechanism for countering energy loss *in vivo*, but it is emphasized that the absence of net hydrogen evolution is no indication that the nitrogenase is fixing nitrogen efficiently and the only definitive way of relating acetylene reduction to nitrogen fixation is by simultaneous or immediate successive measurement by the two techniques under identical conditions.

(d) Enzyme changes in the symbiotic phycobiont

It is accepted that all cyanophilic lichens fix nitrogen, many of them rapidly, and by far the greater part of the fixed nitrogen is lost from the phycobiont, to the benefit of the mycobiont. The relative growth rates of the symbionts are closely coordinated, although in most cases very slow compared to free-living algae and fungi. Nitrogen reserves, normally present in cyano-bacteria in the form of cyanophycin granules, are notably absent in phycobionts. Stewart and Rowell [86] studied the activities of ammonia-assimilating enzymes in the *Nostoc* phycobionts of *Peltigera aphthosa* and *P. canina* and compared them with the activities of the free-living organism. They investigated glutamine synthetase (GS), the key ammonia-assimilating enzyme in heterocystous cyanobacteria; glutamic dehydrogenase (GDH), the corresponding enzyme in many eukaryotes, including fungi; alanine dehydrogenase (ADH) and glutamicaspartic aminotransferase (GOT) which occur in eukaryotes and prokaryotes. GS was very active in free-living *Nostoc*, and GDH of negligible activity. GOT and ADH activities were appreciable. In cyanobacteria isolated from the lichen, however, GS activity was reduced by 94 per cent, GOT and ADH activities were decreased and GDH activity was very low. In the intact thallus of *P. canina* GS activity was negligible and GDH activity very high. The cephalodia (see later) of *P. aphthosa*, which contain 50–60 per cent *Nostoc* cells showed very low GS activity, but high GDH activity, compared with that of the main thallus. Thus, in the symbiotic state, the blue-green phycobionts suffer a major loss of ammonia assimilating capability, which could explain their slow growth rate and the reduced effect of NH_4^+ ions on nitrogenase activity. The state of "nitrogen starvation" of the symbiotic algae may be important in sustaining particularly active nitrogenase.

By the use of digitonin, a detergent which largely disorganizes the membrane permeability characteristic of fungi, but has little effect on cyanobacterial cells, the form of nitrogen released by the symbiotic *Nostoc* was studied.

Ammonia was liberated by intact lichen discs only in the presence of digitonin, and with N_2 as nitrogen source. The digitonin had no effect on the nitrogenase activity of the alga. By the use of $^{15}N_2$ labelling together with digitonin, Stewart and Rowell also showed with *P. canina* that after a period of nitrogen fixation, 50 per cent of the total amount of nitrogen fixed remained in the discs, and of the 50 per cent which was released half was organic nitrogen and half was ammonia. The study was extended to *P. aphthosa* [74, 76] and, following Millbank and Kershaw [65], exploited the thallus morphology as a means of obtaining intact *Nostoc* with "symbiotic state" metabolism. The results complemented their findings with *P. canina* and further showed that ammonia released by the phycobiont *Nostoc* as a result of low GS activity was rapidly assimilated by the cephalodial fungus via GDH. Various transaminases then gave rise to substantial 'pools' of aspartate and alanine. These latter were then transferred to the main thallus, via the fungal hyphae. It is possible that Millbank's findings [59], that polypeptides of two groups, having molecular weights of the order of 1100 and 200, were apparently released by excised cephalodia, can be reconciled to these. The substances lost from Millbank's material contained much alanine and aspartic acid, and could have arisen from ruptured fungal hyphae; the form of translocated nitrogen within lichen thalli is yet to be established, although Stewart, Rowell and Rai [87] consider alanine probable. Stewart's group also revealed that Methionine sulphoximine (MSX), an inhibitor of GS, prevented the inhibition of nitrogenase by added ammonium ions in discs of *Peltigera aphthosa* and also in free-living *Nostoc* isolated from *P. aphthosa.* Symbiotic *Nostoc* (excised cephalodia) were not affected by added ammonium ions. These *Nostoc* cells lack active GS enzyme. Exogenous glutamine also inhibited nitrogenase activity, but alanine and glutamate did not. Evidently an *active* GS system was needed for ammonia to inhibit nitrogenase, and they concluded that glutamine produced in the main thallus could be involved in regulating nitrogenase in the *Nostoc* in the cephalodia. Its mode of action is unknown. Thus, it remains to be shown whether the regulations of GS or other ammonia assimilating enzymes in the *Nostoc* phycobiont is by specific effectors from the eukaryote and whether nitrogenase is itself regulated by the GS enzyme or a product. Postgate [72] reports that the control of nitrogenase activity, *via* the *nif* gene cluster is a delicate process believed to be regulated by combinations of products of the GS regulator gene (*ntr*) cluster. Thus, regulation of nitrogenase is evidently brought about by the GS regulator gene cluster and not by the enzyme itself or one of its products. In other words, nitrogenase and glutamine synthetase (GS) are regulated together, rather than one regulating the other.

(e) Heterocyst frequency in the phycobiont

The nitrogenase enzyme system in filamentous nitrogen-fixing cyanophytes is known to be located in the heterocysts [21] with very few exceptions, e.g. *Plectonema.* In all symbiotic associations with cyanophytes, a species of *Nostoc* is the diazotroph concerned with the single exception of *Azolla* (which has *Anabaena*). The heterocysts, thick-walled cells with enhanced respiratory capability but lacking those components of the photosynthetic system concerned with the photolysis of water and assimilation of CO_2, normally form about 3-5 per cent of the total cell population of the filaments in free-living cyanophytes. In symbioses, however, the proportion is much greater, up to 50 per cent being reported in the axillary glands of the angiosperm *Gunnera* [82]. In such cases the overall photosynthetic ability of the *Nostoc* is drastically reduced, but carbon sources for energy and synthetic requirements are derived from the photosynthetic products of the "host". Lichens are exceptions to this general rule in that the "host" mycobiont is nonphotosynthetic; the proportion of heterocysts is of the order of 3-4 per cent. However, those relatively few lichens containing *two* phycobionts have one cyanophyte and the other a member of the chlorophyceae; hence products of photosynthetic carbon assimilation are available from the other phycobiont. In these circumstances the proportion of heterocysts in the cyanophyte is greatly increased and may reach 30 per cent or more [36]. The diazotroph partner is found in aggregates within structures known as cephalodia, which may be external (as in *Peltigera aphthosa, Stereocaulon paschale* and *Placopsis gelida*) or internal (as in *Lobaria pulmonaria*). When external the *Nostoc* can form about 50 per cent of the cell mass of the cephalodia, and a very convenient system for study is often available as they can be dissected off in an intact state and thus provide a source of active symbiotic *Nostoc*. As has been described above, the enzyme complement and metabolism of symbiotic *Nostoc* differs greatly from that in the free-living form, but attempts to separate active nitrogen-fixing suspensions of cells from lichen thalli for study have failed. This is due to the fact that the filaments are convoluted and intimately associated with fungal hyphae. On homogenization they are disrupted and the junctions between the heterocysts and vegetative cells broken. These junctions are vital for the supply and removal of metabolites from the heterocysts and, further, once ruptured permit oxygen in the suspending medium to gain access to the heterocyst contents and destroy nitrogenase.

The mechanism controlling the differentiation of heterocysts in the filaments of cyanobacteria is not yet established, but it is presumed that chemical species are involved derived normally from the cyanophyte, and also from the other phycobiont and perhaps the mycobiont in the case of lichens. Variation in the proximity of the cephalodia to the green alga phycobiont has been exploited in a few lichens to demonstrate the likelihood of a chemical stimulus controlling heterocyst differentiation [36]. In *Lobaria amplissima*, when closely adjacent to the green phycobiont the *Nostoc* cells in the cephalodia have a heterocyst population of 20 per cent; *Nostoc* filaments in cephalodia located remote from the green phycobiont only differentiate heterocysts to the extent of 3-5 per cent.

(f) Oxygen tension in lichen thalli

The rate of nitrogenase activity in lichens is notably high, and the possibility of vegetative cell fixation in lichens with low heterocyst populations has been considered. Such an attribute would demand a low environmental oxygen tension and a study was made [60] of the internal pO_2 of thalli and cephalodia, using micro-oxygen electrodes, in darkness and when illuminated at various intensities. The results indicated that the environment of the phycobiont zone in *Peltigera polydactyla, P. canina* and *P. aphthosa* was slightly hypoaerobic rising to aerobic when strongly illuminated. In darkness it became substantially hypoaerobic ($1-1.5\times10^4$ Pa; aerobic $=2\times10^4$ Pa). However, in cephalodia, the aggregation of *Nostoc* cells, even though with 30 per cent heterocysts, gave rise to considerably hyperaerobic conditions (3×10^4 Pa; $pO_2=0.3$) under strong illumination. It was concluded that conditions in lichens never became sufficiently microaerobic to permit significant vegetative cell fixation, and the very active nitrogenase was a result of cellular nitrogen deficiency, as already stated.

(g) The requirement for molybdenum

The Mo requirement of diazotrophs has of course been well established for over 50 years, and its part as a constituent of the high-molecular weight fraction of nitrogenase is well known. Horstmann et al. [38] investigated the effect of added Mo in the irrigating medium of some corticolous lichens prior to testing for nitrogenase activity and demonstrated very striking increases in the rates of activity. Addition of 1 ppm Mo enhanced acetylene reduction by 180 per cent in *Lobaria pulmonaria* growing on white oak trees near Corvallis, Oregon, U.S.A. and by 50 per cent in *Lobaria oregana* on Douglas fir. Mo at 10 ppm was evidently inhibitory. It is clear that the nitrogenase system in the lichens was severely limited by Mo deficiency; other Mo-requiring enzymes such as nitrate reductase may well also be limited. Only 0.01 ppm Mo was detected in water falling through the canopy of Douglas fir trees.

(h) Recognition mechanisms for the mycobiont/phycobiont symbiosis

Although the mechanisms involved in the development of lichen symbioses are not directly concerned with nitrogen fixation it is perhaps worth mentioning that four species of the nitrogen-fixing lichen *Peltigera—P. canina, P. horizontalis, P. polydactyla* and *P. praetextata*—have been shown to contain protein fractions containing phytolectins [53, 71]. These evidently originate from the mycobiont and do not bind any of the freshly isolated phycobionts. They bind, however, to the phycobionts cultured after isolation from the lichens, thus indicating that there is a distinct modification to the cell wall of the lichenized alga *in situ*. It would appear that the phytolectin mechanism is very likely to be involved in the recognition or initial interaction between compatible symbionts.

THE EFFECT OF NATURAL ENVIRONMENTAL FACTORS

So far, the effects of moisture, light and temperature on nitrogen fixation by lichens have been outlined, as factors concerned in the general physiology of the organisms. The factors are now discussed in greater detail as their study has been extensively concerned in attempts to establish the rates of nitrogen fixation by lichens in the field. These studies have re-emphasised the fundamental handicap of investigators of lichen physiology: laboratory growth and long-term maintenance of material is impractical, and field material has to be used. Thus, the problems of inherent variability of lichens are compounded by ignorance of the previous environmental history of any given batch of material.

Moisture

When air-dry thalli of *Peltigera* spp. are remoistened nitrogenase activity recommences, sharply at first and then more slowly, reaching maximum values after three to five hours. This effect is manifest even though saturated moisture content is achieved within one hour. Henriksson and Simu [34] showed *Peltigera rufescens* and *Collema tuniforme* to possess ability to recover from prolonged dry storage, but did not report the time course of recovery. Hitch [35] reported on *Lichina confinis*, *Peltigera canina* and *Collema crispum*; however, he did not examine material for longer than two hours. The re-establishment of rates probably varies widely both within and between species, and certainly with the previous moisture history of a given specimen. Once established and stabilized, the relationship between rate and moisture content is very constant for many species, however. At moisture contents less than 100 per cent of the thallus dry weight, activity is usually zero or very low. At moisture contents above 100 per cent, the activity rises rapidly, becoming constant at and above 150–200 per cent in many species of *Peltigera* and *Stereocaulon*, and at 300 per cent in *Collema furfuraceum* [49, 51, 54]. This relationship is not consistent for all species, however, and it has been suggested by Kershaw [49] that much variation could be interpreted as a crude adaptation to a specific environment. However, results from a study of *Collema* [51] do not indicate such an adaptation, and Kershaw is now disinclined to support this general conclusion.

When lichens are subject to long-term drought, their response to remoistening can still be very rapid. Kershaw and Dzikowski [50] studied *Peltigera polydactyla* and found the recovery time to be short (four hours) after a three-day dry period, becoming longer (12 hours) after extended (66-day) periods. It is important to illuminate the material during the rehydration process, and the recovery could very well be due to the illumination as much as to the restoration of moisture. The study also showed that the effect on recovery of an insufficient period of rehydration was to produce thalli in an unstable and varying state of enzyme activity, critically dependent on thallus moisture content over a much wider range than usual. Thus, the effect of natural varying periods of wetness and dryness is likely to be extremely complex, and obviously different investigators, who may have been unaware of the effect of light and the previous history of the material, produced results not comparable with one another.

Huss Danell [40], working with *Stereocaulon*, subjected it to very long periods (up to 75 weeks) of dry storage at 4°C in darkness and at very low moisture contents indeed (3–10 per cent of the oven dry weight). After 36 hours of soaking, although the rates of nitrogenase activity were low (0.15 μmole ethylene gm^{-1} hr^{-1}), they were still equal to those before storage and consistent with the view that *Stereocaulon* is one of the most drought-resistant lichens.

Table 3. Temperature optima for lichen nitrogenase activity

Lichen	Temperature, °C	Reference
Collema furfuraceum	25	51
Lichina confinis	20-25	37
Leptogium cyanescens	30	48
Lobaria pulmonaria	30	48
Nephroma arcticum	15	45
Peltigera aphthosa	20-30	19,43,44
P. canina	16-21	58
P. praetextata	25-30	54
P. rufescens	30	54
Solorina crocea	15	45
Stereocaulon paschale	25	15, 42
Sticta weigelii	30	48

Temperature

Temperature optima for lichens have been reported as shown in Table 3. Many of these results are suspect because insufficient attention has been paid to the maintenance of thallus moisture content at high temperatures, as the activity can be critically dependent on moisture content. The period of pretreatment, to obtain physiological stability before estimating the enzyme activity is also occasionally suspect (Kershaw, personal communication). Further, the temperature within the thallus of a lichen can be significantly higher than that of its surroundings, and so optima below 20°C are unlikely, and this value is probably a fairly universal one in lichens.

Light

As has been stated already, lichens are able to fix nitrogen effectively at low light intensities, and it would appear that they are generally adapted to shade conditions. When conditions are extremely bright, they are generally also warm, thallus moisture rapidly becomes low and metabolic activity minimal. Conflicting results on the ability of nitrogen-fixing lichens to fix in darkness [37, 45] can undoubtedly be explained by the pretreatment of the material used. Hitch and Stewart reported *in situ* field measurements whereas Kallio et al. used a long dark pretreatment. Thus, endogenous energy reserves were low, and the energy-demanding enzyme system would soon come to a halt in darkness. The findings of Kershaw et al. [52] that nitrogenase activity continues in darkness for much longer at temperatures below 15°C than at 25°C or higher, is explicable when it is recognized that the rate of nitrogenase activity will be less at low temperatures. Thus, the requirement for energy reserves is relatively small, they last longer, and nitrogen fixation can consequently continue throughout the night in winter under conditions of good illumination by day. Dark CO_2-fixation can also enable nitrogen fixation to continue for prolonged periods in darkness under appropriate circumstances. When lichens are completely screened from light by snow, however, they ultimately exhaust their reserves, and activity recommences only after snowmelt when a sufficient period has elapsed for the regeneration of energy reserves and the establishment of enzyme balance [39, 56].

Seasonal effects

Kershaw and his collaborators, in their intensive study of the effects of the environment on lichen physiology, have been noteworthy for their production of 'matrices' of graphical results in which the effect of, for example, thallus moisture content upon nitrogenase activity are repeated for a range of temperatures and light intensities. Complete sets of data of this kind have been entirely replicated using material gathered during the various seasons of the year characteristic of the area. *Peltigera praetextata* and *P. rufescens* have been examined in this way [54] as have *Stereocaulon paschale* [15, 41] and *Collema furfuraceum* [51]. Earlier studies reporting seasonal effects have been carried out by Hitch and Stewart [37] and Kallio and Kallio [43].

Kershaw was at first firmly of the view that a pronounced seasonal effect was a general feature of lichens characterized by a low level of nitrogenase activity during midsummer, and a complete cessation of activity in winter as a result of snow cover. His more recent studies [51], however, have prompted a considerable revision of this view. *Collema furfuraceum* was examined from a corticolous habitat in northern Canada where it remained moist throughout the summer and never became snow covered in winter, although exposed to extreme cold (down to about -40°C). No apparent seasonal variation in potential nitrogenase activity was observed. Kershaw now considers that the drops in the potential rate of nitrogenase activity in summer that he had previously noted in *Peltigera* were due to prolonged drought incapacitating the organism's metabolic systems to such an extent that the standard pretreatment was insufficient to achieve a stable state for study. This is an instance of the inherent difficulty of obtaining absolute physiological data for lichens, because of their total functional dependence on environment. Thus, their metabolic condition at any one time and place is conditioned by climatic factors, over which the investigator has no control and often an inadequate knowledge.

NITROGEN FIXATION UNDER FIELD CONDITIONS: THE INPUT OF NITROGEN TO THE ECOSYSTEM

A number of studies have been made of the contribution of lichens to the nitrogen economy of ecosystems where they are significant. Arctic systems have been studied by Alexander and her collaborators [3, 4, 6] in Alaska, Kallio in Finland and Kershaw's group in Canada. Carroll and collaborators in Oregon, U.S.A. [11, 16, 17] have reported on the *Lobaria* of the Douglas fir forests of the Pacific Northwest. Huss-Danell [39] has estimated the contribution by *Stereocaulon paschale* in Northern Sweden, using data derived from numerous estimates of acetylene reduction, taken in the field. She estimated the contribution as being of the order of 1 KgN ha^{-1}. yr^{-1}; if winter fixation was significant this figure might be increased somewhat. Denison [17], further to his informative popular article [16] has suggested that *Lobaria oregana* can contribute about 3.5 KgN $ha^{-1}.yr^{-1}$ to the Oregon forest ecosystems. Alexander, in a brief summary [2] of the results of her extensive studies in the Alaska tundra suggested that the contribution of *Sterocaulon*, *Peltigera aphthosa* and *P. canina* was of the order of 0.5 to 1.5 kg ha^{-1} yr^{-1}, using the acetylene technique.

The benign environment of the Oregon forests clearly permits greatly increased growth of the epiphytic lichens, and consequently their significant nitrogen contribution, which can be estimated by taking into account the "drop-off" of the thallus and hence decay. Similar studies on the nitrogen contribution by lichens to the ecosystems in North Carolina [5], Columbia [23] and New Mexico [24] have also been made.

Crittenden and Kershaw [14] discuss the role of *Stereocaulon paschale* in the nitrogen economy of subarctic woodlands in some detail. In a refreshingly critical and perceptive survey, they point out that the activity of the nitrogen-fixing system is critically dependent on the influential environmental factors and further, the history (or "pretreatment") of the lichen before and after collection is also of great importance. Simple predictive models, based on laboratory estimates of nitrogenase activity (acetylene reductions) under defined conditions, after collection and transfer of material for study, are unable to describe levels of nitrogen fixation in nature. Estimates of nitrogen input per annum based on such data are "precarious" to say the least. Information on the quantities of nitrogenous material leached from the thalli, and death, decay and decomposition of the thalli, are listed as areas requiring attention. Crittenden [12] has provided some data on the losses of nitrogen from *Stereocaulon paschale* brought on by rainfall episodes. The losses were compared with those from the non-nitrogen-fixing lichen *Cladonia stellaris* under identical conditions. His results were obtained by careful analysis of rainfall before and after passage through known weights of lichen material supported by stainless steel mesh above collecting vessels in the field. During rainfall, both species absorbed ammonia nitrogen from rain and released organic nitrogen. Losses of organic nitrogen from *S. paschale* were 2.1 times those from *C. stellaris*, on a dry weight basis (6.5 times on an area basis). Both species effectively scavenged ammonium ions from rain, but *Stereocaulon* suffered an overall net loss of nitrogen in the organic form, to the extent of 9.4 mg · m^{-2} of pure lichen mat, during the seven-week period studied. Losses of nitrogen were highest following the resumption of rainfall on thalli that had remained moist between showers compared with those thalli which had become air-dry. The nature of the organic nitrogen leached is largely unknown; rather surprisingly, only about 15 per cent is composed of polypeptides, and the manner of its loss is also not known. The contribution of the lichen mat to the nitrogen economy of the area amounts to about the same as that of rainfall and Crittenden estimates that the leaching losses amount to about 100 mg · m^{-2} of mat (equivalent to 1kg ha^{-1} yr^{-1}) per growing season. This represents up to 12 per cent of the nitrogen fixed by the lichen. Estimates of losses of nitrogenous compounds during artificial rainfall "episodes" have also been made by the author [64], using a controlled environment chamber and $^{15}N_2$. Recently fixed nitrogen was released in such 'episodes' and much of it was inorganic. The losses from one event could be large, more than the total nitrogen fixed in 24 hours.

It is clear that the contribution of lichens to the nitrogen status of their environment is difficult to estimate and liable to error. One major problem is the use of the acetylene technique. Though convenient, rapid and deli-cate, the technique provides data that do not bear a constant relationship with "true" nitrogen fixation, and the ratio of ethylene formed: nitrogen fixed is influenced by all important environmental factors relevant to lichens. On the other hand, the use of $^{15}N_2$ gas is expensive and also beset with technical difficulties. The necessity to use enclosed vessels precludes its use in field studies except for very short experimental periods but long-period incubations with $^{15}N_2$ gas are essential to provide an overall integration of constantly varying rates of fixation which is the most vital need for a realistic estimate of the amount of nitrogen fixed. An attempt has been made [67] to overcome this difficulty by the construction in the author's laboratory of a controlled environment chamber to simulate the lichen's natural environment as far as possible whilst providing an atmosphere enriched with $^{15}N_2$. A duplicate chamber enables periodic measurements of acetylene reduction to be carried out. In this way a long term incubation under $^{15}N_2$ for up to four weeks can be achieved and compared with frequent "spot" analyses of acetylene reduction to obtain an entirely arbitrary but hopefully relevant conversion factor. The apparatus seeks to provide programmed daylength, light intensity, temperature, CO_2 concentration and thallus moisture content, through rainfall and air movement induced by fans. The last factor is by far the most difficult to control satisfactorily as the thallus is inaccessible and visual observation is the only resort to obtain data. Within these constraints, conversion factors have been obtained for *Peltigera membranacea, P. polydactyla* and *Lobaria pulmonaria* collected from sites in the U.K. [62]. Conversion factors of the order of 10 seem appropriate for seasons where the thalli are likely to be moist for much of the time; when long periods of low moisture content are normal, factors of up to 22 may need to be applied; less extreme conditions indicate a factor of 12. Estimates of nitrogen fixed, in kg/ha/yr have been reported assuming a 10 per cent thallus coverage. These estimates varied and were 2.4 in a predominantly dry habitat and 5.8 in a very moist habitat [62].

Subsequent improvements of the apparatus have enabled the collection of rainfall run-off and estimation of the ^{15}N labelling of any combined nitrogen. Preliminary results show that recently-fixed nitrogen is present in the run-off water, but the moss fronds intimately associated with the lichen thalli have undoubtedly absorbed and presumably assimilated nitrogen compounds that have been eluted from the lichen thalli by 'rain'.

THE EFFECT OF ATMOSPHERIC POLLUTION AND AGROCHEMICALS

(a) Sulphur dioxide pollution

Although many studies have been made on the general effects of atmospheric pollution, especially sulphur dioxide, on lichen growth and metabolism most of them have been concerned with respiration and photosynthesis and very few reports relate to the effects of pollutants upon nitrogen fixation. In the majority of metabolic studies, solutions of sulphur dioxide in water, rather than sulphur dioxide gas, have been used. Although much more convenient, and although a correlation between aqueous and gaseous exposure has been described [69] it is very improbable that experiments involving total immersion of lichen thalli in aqueous solutions will give results that are generally applicable to the organism in its natural state where total immersion rarely occurs.

Hallgren and Huss [29] reported that nitrogenase activity was markedly decreased after exposure to aqueous solutions of bisulphite (HSO_3^-). Sheridan [81] used bisulphite in aqueous solution equivalent to gaseous SO_2 at concentrations between 0.1 and 100 ppm, on *Collema tenax*. Both authors found adverse effects on nitrogenase. The effect of exposure to *gaseous* SO_2 was reported by Kallio and Varheenma [46], who transported specimens of *Stereocaulon paschale* and *Nephroma arcticum* from their natural (unpolluted) habitat to the University of Turku, Finland. Experiments involving transportation of lichens are notoriously liable to give variable results, and further the pollution levels cannot be controlled by the investigators; however, there was no question that the nitrogenase activity had been severely affected during the three-four weeks exposure. Laboratory studies of the effect of gaseous SO_2 on nitrogenase activity have been reported by Henriksson and Pearson [33] who exposed *Peltigera canina* thalli to various concentrations of SO_2 from 0.1 to 500 ppm, in closed flasks.

Unfortunately, incubations of thalli in closed vessels for long periods, and the absorption of SO_2 by the vessel surfaces, prevents accurate interpretation of results. Moreover, the actual gaseous concentration in the vessels is unknown and besides the very high concentrations of SO_2 deliberately used by investigators to accelerate the effects also contributes to the difficulty in the interpretation of results. In spite of these limitations the effect of SO_2 was wholly deleterious to the nitrogen-fixation abilities of lichens.

To expose lichen thalli to SO_2 gas at realistic concentrations (25–200 $\mu g/m^3$) embracing unpolluted rural areas and industrial city centres presents great technical problems, and until recently the only reports were those of Turk, Wurth and Lange [89] and O'Hare and Williams [32]. Neither studies reported on nitrogen-fixation. Others (e.g. Black and Unsworth [7]) describe the apparatus for exposure of plant material to SO_2, but without the need for high humidity.

Mimmack [68] has carried out a study on the effects of gaseous SO_2 at concentrations between 50 and 200 $\mu g/m^3$ on *Peltigera membranacea* from a range of localities and simulating all seasons. The effect on nitrogenase activity is manifest after about 20 days under winter conditions; much more quickly under summer conditions, and strain or habitat variation seems to be important. It was not possible to establish with certainty whether the effect on nitrogenase precedes that on the photosynthetic system; both are profoundly and irreversibly damaged.

(b) Effect of pesticides and agricultural chemicals

Hallbom and Bergman [28] reported on the effect of ammonium nitrate and various herbicides (MCPA; 2, 4D; "Garlone 3A" and "Krenite") upon acetylene reduction by *Peltigera praetextata*. The herbicides had little or no effect, but the application of nitrogenous fertilizer gave rise to a reduction in nitrogenase activity and ultimately proved destructive to the lichen. This was considered to be due to bacterial invasion and disruption of the mycobiont. Normally there is no deleterious effect when combined nitrogen is applied to lichen at low concentrations, under laboratory conditions, but the application rates common in forestry practice (150 kg ha^{-1}), representing 15 gm of dry NH_4NO_3 powder per sq metre are very likely to cause serious damage. The author has experience of collecting *Peltigera membranacea* from Scottish forests which had recently been air sprayed with urea or NH_4NO_3; the lichens were devoid of nitrogenase activity. Kauppi [47] confirmed the generally deleterious effects of fertilizer dressings on lichens, but he did not study nitrogen-fixing species.

REFERENCES

1. Ahmadjian, V. and Hale, M.E. (Editors). *The Lichens*, Academic Press, New York (1973).
2. Alexander, V. Nitrogen fixing lichens in tundra and teiga ecosystems, p. 256, in *Current Perspectives in Nitrogen Fixation* (Editors, A.H. Gibson and W.E. Newton), Canberra, Australian Academy of Science (1981).

3. Alexander, V., Billington, M. and Schell, D.M. Nitrogen fixation in arctic and alpine tundra, in *Vegetation and Production Ecology of an Alaskan Arctic Tundra, Ecological Studies 29*, 539-558 (Editor, L.L. Tieszen), Stuttgart, Germany, Springer Verlag (1978).
4. Alexander, V. and Schell, D.M. Seasonal and spatial variation of nitrogen fixation in the Barrow (Alaska) tundra, *Arctic and Alpine Research 5*, 77-88 (1973).
5. Becker, V.E. Nitrogen-fixing lichens in the forests of the Southern Appalachian Mountains of North Carolina, *The Bryologist 83*, 29-39 (1980).
6. Billington, M. and Alexander, V. Nitrogen fixation in a black spruce (*Picea mariana*) forest in Alaska, *Ecological Bulletins* (Stockholm) 26, 209-215 (1978).
7. Black, V.J. and Unsworth, M.H. A system for measuring effects of sulphur dioxide on gas exchange of plants, *Journal of Experimental Botany 30*, 81-88 (1979).
8. Bond, G. and Scott, G.D. An examination of some symbiotic systems for fixation of nitrogen, *Annals of Botany 19*, 67-77 (1955).
9. Brown, D.H., Hawksworth, D.L. and Bailey, R.H. (Editors). *Lichenology Progress and Problems*, Academic Press, London, U.K. (1976).
10. Burris, R.H., Arp, D.J., Benson, D.R., Emerich, D.W., Hageman, R.V., Jones, T., Ludden, P.W. and Sweet, L.J. The biochemistry of nitrogenase, pp. 37-54, in *Nitrogen Fixation* (Editors, W.D.P. Stewart and J.R. Gallon), Academic Press, London, U.K. (1980).
11. Carroll, G.C. Forest canopies; complex and independent subsystems, pp. 87-107 in *Forests: Fresh Perspectives from Ecosystems Analysis* (Editor, R.H. Waring), Oregon State University Press, Corvallis, Ore., U.S.A. (1980).
12. Crittenden, P.D. The role of lichens in the nitrogen economy of subarctic woodlands: Nitrogen loss from the nitrogen-fixing lichen *Stereocaulon paschale* during rainfall, in *Nitrogen as an Ecological Factor* (Editors, I.H. Rorison, J.A. Lee and S. McNeill), Blackwell, Oxford, U.K. (1983).
13. Crittenden, P.D. and Kershaw, K.A. A procedure for simultaneous measurement of carbon dioxide exchange and nitrogenase activity in lichens, *New Phytologist 80*, 393-401 (1978).
14. Crittenden, P.D. and Kershaw, K.A. Discovering the role of lichens in the nitrogen cycle in boreal-arctic ecosystems, *The Bryologist 81*, 258-267 (1978).
15. Crittenden, P.D. and Kershaw, K.A. Studies on lichen-dominated systems, 22. The environmental control of nitrogenase activity in *Stereocaulon paschale* in spruce lichen woodland, *Canadian Journal of Botany 57*, 236-254 (1979).

16. Denison, W.C. Life in tall trees, *Scientific American 228*, 75–80 (1973).
17. Denison, W.C. *Lobaria oregana*, a nitrogen fixing lichen in old growth Douglas fir forests, pp. 266–275 in *Symbiotic Nitrogen Fixation in the Management of Temperate Forests* (Editors, J.C. Gordon, C.T. Wheeler and D.A. Perry), Oregon State University Press, Corvallis, Ore., U.S.A. (1979).
18. Englund, B. The physiology of the lichen *Peltigera aphthosa* with special reference to the blue-green phycobiont, *Nostoc* sp., *Physiologia Plantarum 41*, 298–304 (1977).
19. Englund, B. Effects of environmental factors on acetylene reduction by intact thallus and excised cephalodia of *Peltigera aphthosa, Ecological Bulletins* (Stockholm) *26*, 234–246 (1978).
20. Englund, B. and Meyerson, H. In situ measurement of nitrogen fixation at low temperatures, *Oikos 25*, 283–287 (1974).
21. Fay, P. Nitrogen fixation in heterocysts, pp. 121–165, in *Recent Advances in Biological Nitrogen Fixation* (Editor, N.S. Subba Rao), Oxford & I.B.H. Publishers, New Delhi (1979).
22. Fogg, G.E. and Stewart, W.D.P. In situ determination of biological nitrogen fixation in Antarctica, *British Antarctic Survey Bulletin 15*, 39–46 (1968).
23. Forman, R.T.T. Canopy lichens with blue-green algae; a nitrogen source in a Colombian rain forest, *Ecology 56*, 1176–1184 (1975).
24. Forman, R.T.T. and Dowden, D.L. Nitrogen fixing lichen roles, from desert to alpine, in the Sangre de Cristo mountains, New Mexico, *The Bryologist 80*, 561–570 (1977).
25. Gibson, A.H. and Newton, W.E. (Editors). *Current Perspectives in Nitrogen Fixation*, Australian Academy of Science, Canberra (1981).
26. Granhall, U. (Editor). *Environmental Role of Nitrogen Fixing Blue-green Algae and Asymbiotic Bacteria, Ecological Bulletins* (Stockholm) *26*, (1978).
27. Green, T.G.A., Horstmann, J., Bonnett, H., Wilkins, A. and Silvester, W.B. Nitrogen fixation by members of the Stictaceae (lichens) of New Zealand, *New Phytologist 84*, 339–348 (1980).
28. Hallbom, L. and Bergman, B. Influence of certain herbicides and a forest fertilizer on the nitrogen fixation by the lichen *Peltigera praetextata, Oecologia 40*, 19–27 (1979).
29. Hallgren, J.E. and Huss, K. Effects of sulphur dioxide on photosynthesis and nitrogen fixation, *Physiologia Plantarum 34*, 171–176 (1975).
30. Hardy, R.W.F. and Silver, W.S. (Editors). *A Treatise on Dinitrogen Fixation, Section III, Biology*, John Wiley and Sons, Inc., New York (1977).
31. Hardy, R.W.F., Burns, R.C. and Holsten, R.D. Applications of the

acetylene-ethylene assay for measurement of nitrogen fixation, *Soil Biology and Biochemistry 5*, 47-81 (1973).

32. O'Hare, G.P. and Williams, P. Some effects of sulphur dioxide flow on lichens, *Lichenologist* 7, 116-120 (1975).
33. Henriksson, E. and Pearson, L.C. Nitrogen fixation rate and chlorophyll content of the lichen *Peltigera canina* exposed to sulphur dioxide, *American Journal of Botany 68*, 680-684 (1981).
34. Henriksson, E. and Simu, B. Nitrogen fixation by lichens, *Oikos 22*, 119-121 (1971).
35. Hitch, C.J.B. A study of some environmental factors affecting nitrogenase activity in lichens, M.Sc. thesis, University of Dundee (1971).
36. Hitch, C.J.B. and Millbank, J.W. Nitrogen metabolism in lichens, VII, Nitrogenase activity and heterocyst frequency in lichens with blue green phycobionts, *New Phytologist 75*, 239-244 (1975).
37. Hitch, C.J.B. and Stewart, W.D.P. Nitrogen fixation by lichens in Scotland, *New Phytologist 72*, 509-524 (1973).
38. Horstmann, J.L., Denison, W.C. and Sylvester, W.B. $^{15}N_2$ fixation and molybdenum enhancement of acetylene reduction by *Lobaria* spp., *New Phytologist 92*, 235-241 (1982).
39. Huss-Danell, K. Nitrogen fixation by *Stereocaulon paschale* under field conditions, *Canadian Journal of Botany 55*, 585-592 (1977).
40. Huss-Danell, K. Nitrogenase activity in the lichen *Stereocaulon paschale*; recovery after dry storage, *Physiologia Plantarum 41*, 158-161 (1977).
41. Huss-Danell, K. Seasonal variation in the capacity for nitrogenase activity in the lichen *Stereocaulon paschale, New Phytologist 81*, 89-98 (1978).
42. Kallio, S. The ecology of nitrogen fixation in *Stereocaulon paschale, Reports of the Kevo Subarctic Research Station 10*, 34-42 (1973).
43. Kallio, S. and Kallio, P. Nitrogen fixation in lichens at Kevo, North Finland, pp. 292-304, in *Fennoscandian Tundra Ecosystems, Part I. Plants and Microorganisms* (Editor, F.E. Wielgolaski), Springer, New York (1978).
44. Kallio, S. and Kallio, P. Adaptation of nitrogen fixation to temperature in the *Peltigera aphthosa* group, *Ecological Bulletins* (Stockholm) *26*, 225-233 (1978).
45. Kallio, P., Suhonen, S. and Kallio, H. The ecology of nitrogen fixation in *Nephroma arcticum* and *Solorina crocea, Reports of the Kevo Subarctic Research Station 9*, 7-14 (1972).
46. Kallio, S. and Varheenma, T. On the effect of air pollution on nitrogen fixation in lichens, *Reports of the Kevo Subarctic Research Station 11*, 42-46 (1974).
47. Kauppi, M. The influence of nitrogen rich pollution components on

lichens, *Acta Universitatis Ouluensis* Series A, No. 101 (1980).

48. Kelly, B.B. and Becker, V.E. Effects of light intensity and temperature on nitrogen fixation by *Lobaria pulmonaria, Sticta weigelii, Leptogium cyanescens* and *Collema subfurvum, The Bryologist 78*, 350-355 (1975).
49. Kershaw, K.A. Dependence of the level of nitrogenase capacity on the water content of the thallus in *Peltigera canina, P. evansina, P. polydactyla* and *P. praetextata, Canadian Journal of Botany 52*, 1423-1427 (1974).
50. Kershaw, K.A. and Dzikowski, P.A. Physiological environmental interactions in lichens, 6. Nitrogenase activity in *Peltigera polydactyla* after a period of desiccation, *New Phytologist 79*, 417-421 (1977).
51. Kershaw, K.A. and MacFarlane, J.D. Physiological environmental interactions in lichens, 13. Seasonal constancy of nitrogenase activity, net photosynthesis and respiration in *Collema furfuraceum, New Phytologist 90*, 723, 734 (1982).
52. Kershaw, K.A., MacFarlane, J.D. and Tysiaczny, M.J. Physiological environmental interactions in lichens, 5. The interaction of temperature with nitrogenase activity in the dark, *New Phytologist 79*, 409-416 (1977).
53. Lockhart, C.M., Rowell, P. and Stewart, W.D.P. Phytohaemaglutinins from the nitrogen fixing lichens *Peltigera canina* and *P. polydactyla, F.E.M.S. Microbiology Letters 3*, 127-130 (1978).
54. MacFarlane, J.D. and Kershaw, K.A. Physiological environmental interactions in lichens, 4. Seasonal changes in the nitrogenase activity of *Peltigera praetextata* and *P. rufescens, New Phytologist, 79*, 403-408 (1977).
55. MacFarlane, J.D. and Kershaw, K.A. Physiological environmental interactions in lichens, 9. Thermal stress and lichen ecology, *New Phytologist 84*, 669-685 (1980).
56. MacFarlane, J.D. and Kershaw, K.A. Physiological environmental interactions in lichens, 11. Snowcover and nitrogenase activity, *New Phytologist 84*, 703-710 (1980).
57. MacFarlane, J.D., Maikawa, E., Kershaw, K.A. and Oaks, A. Environmental physiological interactions in lichens, 1. The interaction of light/dark periods and nitrogenase activity in *Peltigera polydactyla, New Phytologist 77*, 705-711 (1976).
58. Maikawa, E. and Kershaw, K.A. The temperature dependence of thallus nitrogenase activity in *Peltigera canina, Canadian Journal of Botany 53*, 527-529 (1975).
59. Millbank, J.W. Nitrogen metabolism in lichens, V. The forms of nitrogen released by the blue green phycobiont of *Peltigera* spp., *New Phytologist 73*, 1171-1181 (1974).
60. Millbank, J.W. The oxygen tension within lichen thalli, *New Phytologist 79*, 649-657 (1977).

61. Millbank, J.W. Lower plant associations, pp. 126-151, in *A Treatise on Dinitrogen Fixation, Section III, Biology* (Editors, R.W.F. Hardy and W.S. Silver), John Wiley and Sons, New York (1977).
62. Millbank, J.W. The assessment of nitrogen fixation and throughput by lichens, 1. The use of a controlled environment chamber to relate acetylene reduction estimates to nitrogen fixation, *New Phytologist 89*, 647-655 (1981).
63. Millbank, J.W. Nitrogenase and hydrogenase in cyanophilic lichens, *New Phytologist 92*, 221-228 (1982).
64. Millbank, J.W. The assessment of nitrogen fixation and throughput by lichens, 3. Losses of nitrogenous compounds by *Peltigera membranacea, P. polydactyla* and *Lobaria pulmonaria* in simulated rainfall episodes, *New Phytologist 92*, 229-234 (1982).
65. Millbank, J.W. and Kershaw, K.A. Nitrogen metabolism in lichens, 1. Nitrogen fixation in the cephalodia of *Peltigera aphthosa, New Phytologist 68*, 721-729 (1969).
66. Millbank, J.W. and Kershaw, K.A. Nitrogen metabolism in lichens, III. Nitrogen fixation by internal cephalodia of *Lobaria pulmonaria, New Phytologist 69*, 595-597 (1970).
67. Millbank, J.W. and Olsen, J.D. The assessment of nitrogen fixation and throughput by lichens, 2. Construction of an enclosed growth chamber for the use of $^{15}N_2$, *New Phytologist 89*, 657-665 (1981).
68. Mimmack, A. Nitrogenase activity and carbon metabolism in the lichen *Peltigera* under gaseous sulphur dioxide, Ph.D. thesis, University of London (1983).
69. Nieboer, E., Tomassini, F.D., Puckett, K.J. and Richardson, D.H.S. A model for the relationship between gaseous and aqueous sulphur dioxide concentration in lichen exposure studies, *New Phytologist 79*, 157-162 (1977).
70. Pearson, L.C. and Henriksson, E. Air pollution damage to cell membranes in lichens, 2. Laboratory experiments, *The Bryologist 84*, 515-520 (1981).
71. Petit, P. Phytolectins from the nitrogen fixing lichen *Peltigera horizontalis*—The binding pattern of primary protein extract, *New Phytologist 91*, 705-710 (1982).
72. Postgate, J.R. *The Fundamentals of Nitrogen Fixation*, Cambridge University Press, Cambridge, U.K. (1982).
73. Quispel, A. (Editor). *The Biology of Nitrogen Fixation*, North Holland, Amsterdam (1974).
74. Rai, A.N., Rowell, P. and Stewart, W.D.P. Ammonia assimilation and nitrogenase activity in the lichen *Peltigera aphthosa, New Phytologist 85*, 545-556 (1980).
75. Rai, A.N., Rowell, P. and Stewart, W.D.P. Nitrogenase activity and dark carbon dioxide fixation in the lichen *Peltigera aphthosa, Planta 151*, 256-264 (1981).

76. Rai, A.N., Rowell, P. and Stewart, W.D.P. $^{15}N_2$ incorporation and metabolism in the lichen *Peltigera aphthosa*, *Planta 152*, 544–552 (1981).
77. Rennie, R.J., Rennie, D.A. and Fried, M. Concepts of ^{15}N usage in dinitrogen fixation studies, pp. 107–133, in *Isotopes in Biological Dinitrogen Fixation*, I.A.E.A., Vienna (1978).
78. Rogers, R.W., Lange, R.T. and Nicholas, D.J.D. Nitrogen fixation by lichens of arid soil crusts, *Nature* (London) *209*, 96–97 (1966).
79. Rorison, I.H., Lee, J.A. and McNeill, S. (Editors). *Nitrogen as an Ecological Factor*, Blackwell, Oxford, U.K. (1983).
80. Scott, G.D. Further investigations of some lichens for fixation of nitrogen, *New Phytologist 55*, 111–116 (1956).
81. Sheridan, R.P. Impact of emissions from coal fired electricity generating facilities on N-fixing lichens, *The Bryologist 82*, 54–58 (1979).
82. Silvester, W.B. Endophyte adaptation in *Gunnera-Nostoc* symbiosis, pp. 521–538, in *Symbiotic Nitrogen Fixation in Plants* (Editor, P.S. Nutman), Cambridge University Press, Cambridge, U.K. (1975).
83. Sprent, J.I. *The Biology of Nitrogen Fixing Organisms*, McGraw Hill, Maidenhead, U.K. (1979).
84. Stewart, W.D.P. Algal fixation of atmospheric nitrogen, *Plant and Soil 32*, 555–588 (1970).
85. Stewart, W.D.P. and Gallon, J.R. (Editors). *Nitrogen Fixation*, Academic Press, London (1980).
86. Stewart, W.D.P. and Rowell, P. Modifications of nitrogen fixing algae in lichen symbioses, *Nature* (London) *265*, 371–372 (1977).
87. Stewart, W.D.P., Rowell, P. and Rai, A.N. *Symbiotic Nitrogen Fixing Cyanobacteria*, pp. 239–277, in *Nitrogen Fixation* (Editors, W.D.P. Stewart and J.R. Gallon), Academic Press, London (1980).
88. Subba-Rao, N.S. (Editor). *Recent Advances in Biological Nitrogen Fixation*, Oxford & IBH Publishing Co., New Delhi (1979).
89. Turk, R., Wirth, V. and Lange, O.L. Carbon dioxide exchange measurements for determination of sulphur dioxide resistance of lichens, *Oecologia 15*, 33–64 (1974).
90. Watanabe, A. and Kiyohara, T. Symbiotic blue-green algae of lichens, liverworts and cycads, pp. 189–196, in *Plant and Cell Physiology (Tokyo)*, Special Volume, *Studies on Microalgae and Photosynthetic Bacteria* (1963).

9. Biological Nitrogen Fixation in Sugar Cane

A.P. Ruschel and P.B. Vose

Practical observations lead to a growing conviction that sugar cane must have some type of nitrogen fixation associated with it. There is, for example, the fact that in Brazil a crop of 100 t ha^{-1} typically removes about 132 kg of nitrogen and a further 35-50 kg N ha^{-1} of nitrogen is lost when the leaves are burned off prior to harvest [1]. On the other hand, crops frequently do not receive more than 50 kg nitrogen fertilizer, so each year there is a theoretical net loss of nitrogen. There are also instances known where sugar cane crops have been grown on soils with naturally low nutrient levels for 15-30 years without the application of nitrogen fertilizer [2]. Finally, there has been the frequent experience that the crop, especially in the planting year, responds rather poorly to nitrogen fertilizer [2-5].

EVIDENCE FOR NITROGEN FIXATION ASSOCIATED WITH SUGAR CANE

Döbereiner [6] reported that sugar cane roots produced ethylene from acetylene up to 5 mol/g/h, showing nitrogenase activity. Acetylene reduction activity is now quite normally demonstrated in soil + root cores taken from the field, and in young plants grown in pots. It should be noted here that sugar cane is normally propagated vegetatively. In commercial practice whole stalks are often placed in the planting furrow, but in experimental work, usually a short length of stem, a 'sett' or 'seed piece' comprising a piece of stem including a node is used. If such a sett is surface sterilized and checked for nitrogenase activity, none will be found. If the sett is allowed to 'germinate' i.e., to develop a young root and shoot from the nodal bud the sett will then demonstrate intense nitrogenase activity, indicating the bacterial activity in the stem.

Ruschel [7] maintained seedlings of the cultivar L 61-41, grown in compost/soil mixture, in a $^{15}N_2$ atmosphere for 30 hours and obtained the first direct evidence of fixation. The relatively high amount of ^{15}N which accumulated in the leaves suggested fairly rapid translocation of the fixed nitrogen from the roots, although this has not been the usual experience.

A later $^{15}N_2$ gas experiment [8] compared the effect of intact versus disturbed soil systems, and it was found that there was little difference in $^{15}N_2$-fixation between plants of intact systems and plants merely with soil adhering to the roots. Addition of sucrose increased nitrogen fixation by disturbed plants. In these experiments, except in one case of a disturbed system, fixed nitrogen was not found in the leaves of the plants after 24 hours in $^{15}N_2$ atmosphere. Intact systems which were kept under simulated normal atmosphere for six days did show considerable amounts of fixed $^{15}N_2$ presumably due to the transfer of fixed nitrogen from the roots.

Surface sterilized setts when grown in nutrient solution culture, without soil, also showed the capacity to fix nitrogen. Vose et al. [9] used data from a nitrogen balance study, ^{15}N isotope dilution, and determination of natural ^{15}N abundance to demonstrate that plants grown in nutrient solution had about 20 per cent of nitrogen due to fixation. Acetylene reduction studies showed that even very low levels of NH_4^+ inhibited nitrogen fixation. A parallel experiment [10] measured nitrogen fixation by sugar cane in culture solution by direct measurement in $^{15}N_2$ enriched atmosphere over a period of 72 hours. Nitrogen fixation associated with roots was observed with and without nitrate in the solution. When plants had the roots exposed to $^{15}N_2$ gas for only two days there was no $^{15}N_2$ incorporation in the leaves, but the label appeared in the leaves when plants were exposed in a chamber for three days. Presumably the fixed $^{15}N_2$ was retained initially by the bacteria for later translocation. In one cultivar inoculation with a mixed inoculum. obtained by incubating sugar cane roots, increased nitrogen fixation.

Definite proof of nitrogen fixation associated with the sugar cane rhizosphere in the field was obtained by Matsui et al. [11]. They enclosed the root system of a nine-month-old sugar cane plant, growing in a normal sugar cane field, with a steel cylinder open at the bottom end. The stems of the plant were sealed with mastic at the top end of the cylinder and $^{15}N_2$ gas was infiltrated into the cylinder. The gas in the soil system was subsequently circulated by a pump and the excess CO_2 removed for a period of five days, which coincided with the very rapid elongation stage of the cane (Fig. 1). Significant ^{15}N enrichment occurred in soil taken from close to big as well as small roots. ^{15}N enrichment was not found in the leaves or stems in this experiment, possibly due to the great dilution by native plant nitrogen or due to the rapid development of the plant during the experimental period. More likely dilution of nitrogen from the atmosphere may have caused the lack of ^{15}N enrichment. Later work however, showed very rapid (about 30–40 minutes) transfer of gas from the shoots to the soil and as a result of this the $^{15}N_2$ soil atmosphere must have been considerably diluted during the course of the experiment. Gas transfer is greatly reduced in the dark, presumably due to closure of the stomata.

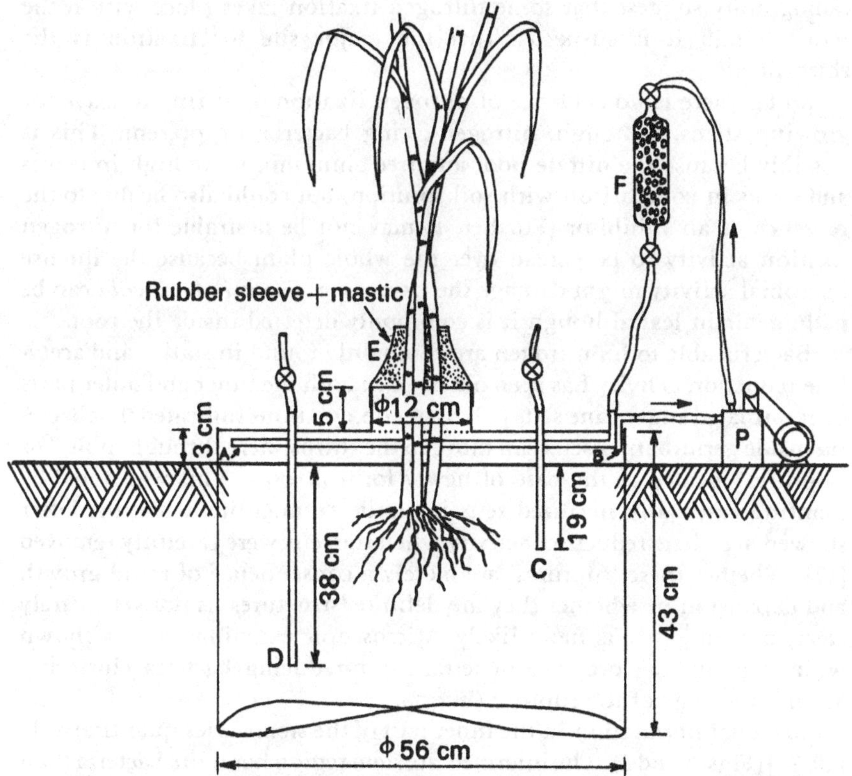

Fig. 1. Root system of a sugarcane plant growing in the field enclosed in a steel cylinder with open ends. Circulation of $^{15}N_2$-enriched atmosphere permitted this demonstration of N_2-fixation in the rhizosphere. From Matsui *et al.* (1981).

LOCATIONS OF NITROGEN-FIXING BACTERIA IN SUGAR CANE

Nitrogen-fixing bacteria in sugar cane occur in the rhizosphere, stalks and phyllosphere. The first report by Döbereiner [12] indicated the effect of cane roots on nitrogen-fixing populations by comparing inter-row soil and rhizosphere soil samples. Freitas et al. [13] used radiorespirometry to show that microbiological activity was much greater in sugar cane rhizosphere than in soil between rows. Microbial activity in rattoon cane was very intense in the soil areas where most of the active roots were present. Döbereiner [14] observed that 95 per cent of soil samples had *Beijerinckia* spp. in contrast to only 62 per cent in inter-row soil, although *Beijerinckia* is not now regarded in our laboratory as of major importance to the system.

Bacteria have been reported inside roots [15–17]. The fact that nitrogen fixation can occur in plants growing in culture and the results of a study [18] using tritiated acetylene reduction technique and electron microautoradiography suggest that some nitrogen fixation takes place within the root, though it is suspected that the major site for fixation is the rhizosphere.

So far there is no evidence of nitrogen fixation occurring in leaves or growing stems, although, nitrogen-fixing bacteria are present. This is possibly because the nitrate pool and free ammonium are high in leaves and stems in comparison with soil solution, but could also be due to the presence of an inhibitor. Further, it may not be desirable for nitrogen fixation activity to be spread over the whole plant because the intense microbial activity might damage the tissues e.g. *Erwinia herbicola* can be pathogenic in leaf although it is commonly detected inside the root.

Bacteria able to fix nitrogen are commonly found in stalks, and acetylene reduction activity has been observed in both the inner and outer parts of germinated sugar cane setts [19]. Later observations indicated that bacteria inside germinating setts can move to the rhizosphere through 'holes' or 'ruptures' located at the base of newly formed roots. This was noted by planting externally sterilized setts in sterile vermiculite, when the latter showed acetylene reduction activity when the setts were carefully removed [17]. Whether these 'ruptures' are merely a consequence of rapid growth and expansion or whether they are definite 'structures' is not yet entirely clear, but the latter is more likely. Microscopic examination has shown with certainty the presence of tetrazolium-reducing bacteria clustering around the edge of the rupture (Fig. 2).

Bacterial population in the inner part of the stem varies quantitatively [20, 21] (Figs. 3 and 4). The intermediate stem region has more bacteria than

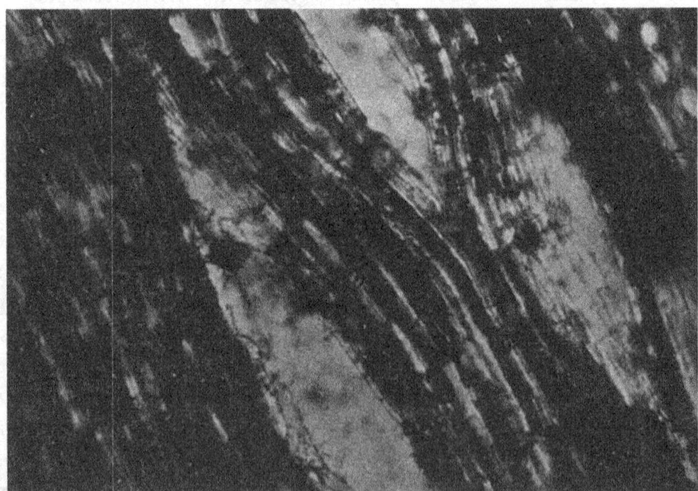

Fig. 2. 'Pores' or 'ruptures' at the base of a root developing from a sugarcane sett (piece of stalk with node.) From Patriquin *et al.* (1980).

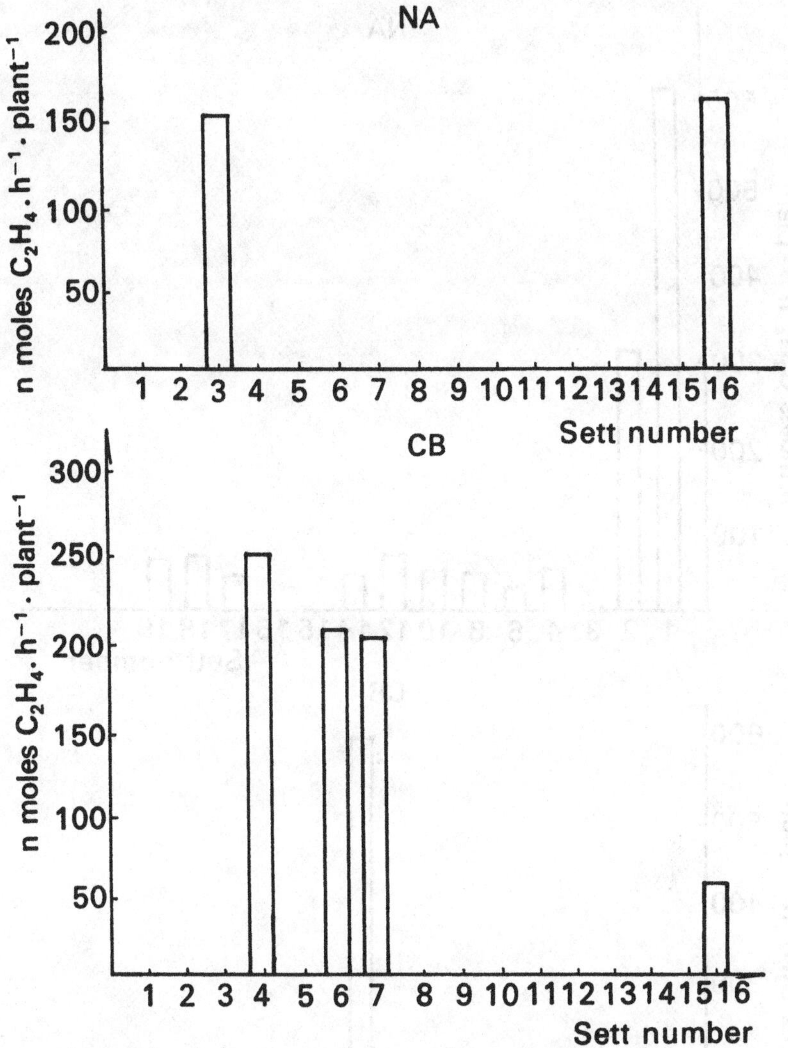

Fig. 3. Nitrogenase activity (nmol C_2H_4 hr) of setts of cultivars NA and CB taken from the greenhouse, surface sterilized and aseptically planted in sterile vermiculite. Numbered from the ground upwards. 3.5 weeks after planting. From Graciolli and Ruschel (1981).

the basal and/or apical regions. The basal part of stalks appears to have either least number of nitrogen fixing bacteria or none at all (Table 1).

After 'germination', plants obtained from different nodal pieces of the same sugar cane stalk showed different levels of acetylene reduction activity, again indicating a differential distribution of bacteria in stalks (Fig. 4). This suggests that the most efficient propagation of bacteria in newly established cane fields could be done by selecting the most suitable pieces of

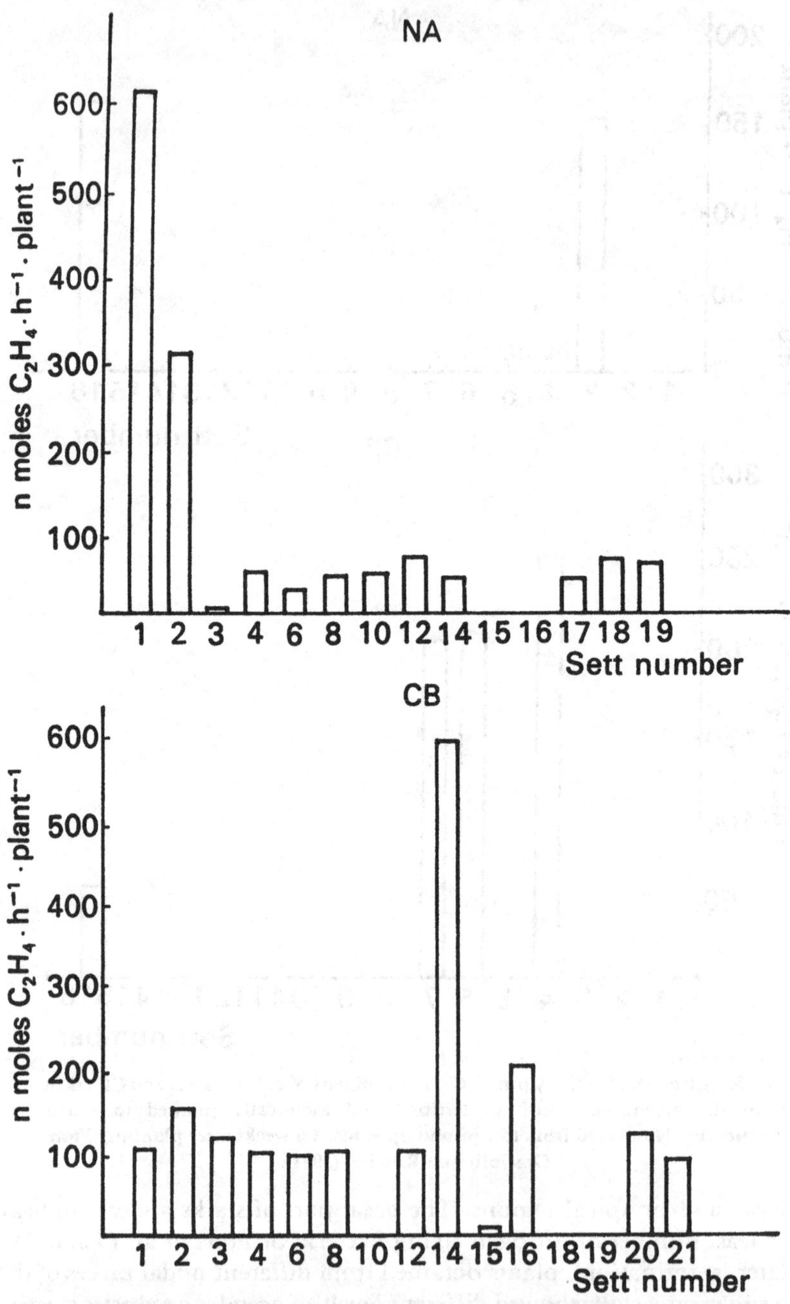

Fig. 4. Nitrogenase activity (nmol C_2H_4/hr) of setts of cultivars NA and CB taken from the field, surface sterilized and aseptically planted in sterile vermiculite. Numbered from the ground upwards. 3.5 weeks after planting. From Graciolli and Ruschel (1981).

Table 1. Number and per cent of N_2-fixing bacteria in stalks divided into parts, in relation to distance from soil

	Basal	Intermediate	Apical
Number bacteria/g (fresh mat.)×10^5	0.4	10.0	0.5
Per cent isolates Acetylene Reduction Activity positive	0	100	60

stalk for planting. As the nitrogen-fixing bacteria in the stalk are influenced by host variety and also subject to seasonal variation in bacterial activity, it has been concluded that some studies on the potential effect of inoculation need to be done [22].

Table 2. Acetylene Reduction Activity (nmol C_2H_4/h) after 48h incubation of bacteria from dilutions (w:v) of leaves of cultivars CO 515, IAC 48/65 and CB 50/41 after 72h growth period

	Media for	CO-513*	IAC 48/65*	CB 50/41*
Leaves (seedlings)	Aerobes	4 (2)	0 (1)	0 (0)
	Facultative anaerobes	332 (3)	163 (3)	5 (3)
	Microaerophylls	152 (2)	109 (2)	0 (0)
Leaves (mature)	Aerobes	113 (3)	14 (3)	106 (2)
	Facultative anaerobes	104 (3)	4 (2)	0 (1)
	Microaerophylls	0 (1)	26 (2)	0 (0)
Dead leaves	Aerobes	622 (6)	187 (5)	133 (2)
	Facultative anaerobes	129 (5)	216 (6)	288 (1)
	Microaerophylls	123 (6)	0 (1)	20 (0)

*log of highest positive dilution with Acetylene Reduction Activity.

There is a large number of nitrogen-fixing bacteria in the phyllosphere of sugar cane (Table 2), being higher in dead leaves than in living leaves. Here again the effect of host varieties is noted, CB 50/41 always showing acetylene reduction activity at lower dilutions than CO 513 and IAC 48/65. It seems that bacterial populations occur in living leaves independently of their oxygen needs. In dead leaves the numbers of all sorts of bacteria (aerobes, facultative anaerobes, microaerophilic) appear to increase, which is again dependent on host varieties.

NITROGEN-FIXING BACTERIA IN SUGAR CANE

Sugar cane has an integral system of bacteria which are perpetuated through normal vegetative propagation just as virgin Amazon jungle soils which show strong nitrogenase activity without any artificial inoculation. Populations of nitrogen-fixing bacteria, including non-spore forming and spore-forming aerobic and facultative anaerobic bacteria were found in sugar cane rhizosphere at 40, 80 and 120 cm depths [8]. Inoculation of 1 cm root sections on surface media gave colonies of *Azotobacter, Beijerinckia, Caulobacter* and *Vibrio*, with some evidence of clostridia. Subsequent work noted the occurrence of *Azospirillum*-like organisms and *Klebsiella* [23].

Isolation of bacteria actually responsible for nitrogen fixation in sugar cane from natural conditions has been a continuing problem [24]. Using different media it is possible to separate many nitrogen-fixing and a large number of non-nitrogen-fixing bacteria from any types of plant sample, whether roots, stalks, or leaves. The non-nitrogen fixers may be important for cross-feeding and to provide growth factors. For example, it has been noted that there is a tendency for isolates which at first showed nitrogenase activity, cease to do so later when they were purified. However, it was noted that nitrogenase activity could be restored in 80 per cent of those isolates which ceased to fix nitrogen if yeast extract was added to the medium. Isolates on nitrogen-free medium showed growth and nitrogenase activity if given 2 ppb cobalt [24].

In Egypt, Hegazi et al. [16] found *Azotobacter vinelandii, Klebsiella* spp., *Bacillus* and *Azospirillum* abundant in the rhizosphere of sugar cane but not *Beijerinckia* spp. In South Africa Purchase [25], on the basis of morphological evidence, reported *Azospirillum*-like bacteria from the roots of rattoon cane. Singh et al. [26] noted that at least eight genera of nitrogen-fixing bacteria were associated with sugar cane.

In a later work from our laboratory, using more critical computer-assisted API biochemical tests, bacterial populations in soil and in the root have been carefully distinguished and compared with those on the root surface. Rennie [27] examined sugar cane rhizosphere soil samples from Brazil and found them to contain equal populations of *Derxia gummosa, Enterobacter cloacae, Bacillus polymyxa* and *Azotobacter vinelandii*.

Rennie [28] reported that acetylene-reducing bacteria isolated from sugar cane were facultative anaerobes of the families Entero-bacteriaceae and Bacillaceae, with *Klebsiella pneumoniae, Enterobacter cloasae, Erwinia herbicola*, and *Bacillus polymyxa* present in the sett as well as roots, while *E. herbicola* was the dominant bacteria on the root surface. Neither *Beijerinckia* spp. nor *Azotobacter* were found associated with either sett or roots. Even so, 25 per cent of the acetylene-reducing isolates could not be identified, although only two of the 17 unidentified bacteria were not presumptive members of the Enterobacteriaceae.

Also using API identification, Graciolli [21] identified *Enterobacter cloacae, Bacillus polymyxa, Erwinia herbicola, Azotobacter vinelandii, Klebsiella pneumoniae*, and *Derxia gummosa* which were associated with the root surface. *Azospirillum brasiliense* was isolated from surface sterilized young leaves; *B. polymyxa, K. pneumoniae, E. herbicola, A. vinelandii*, and *D. gummosa* were isolated from old dry leaves. Bacterial species found in the inner stems varied in accordance with the distance of cuttings

Table 3. Per cent of bacteria isolated from sugarcane in different parts of the plant

		Cultivar		
		RB 705146	CB 41.76	NA 56-79
Root (internal + external)	*E. cloacae*	100	—	37
	E. herbicola	0	67	13
	B. polymyxa	—	17	—
	K. pneumoniae	—	17	—
	A. vinelandii	—	—	37
	Unknown	—	—	13
Internal	*E. cloacae*	100	—	—
	E. herbicola	0	0	—
	B. polymyxa	—	80	—
	K. pneumoniae	—	20	—
	A. vinelandii	—	—	—
	Unknown	—	—	—
Setts (internal + external)	*E. cloacae*	75	—	—
	E. herbicola	25	50	29
	B. polymyxa	—	33	—
	K. pneumoniae	—	17	—
	A. vinelandii	—	—	16
	Unknown	—	—	55
Internal	*E. cloacae*	100	50	—
	E. herbicola	0	33	—
	B. polymyxa	—	17	—
	K. pneumoniae	—	—	—
Leaves	*A. brasilense*	—	—	none
	E. cloacae	—	—	none
	E. herbicola	—	—	33
	A. vinelandii	—	—	17
	D. gummosa	—	—	17
	Unknown	—	—	33 (100)*

*bacteria in green leaves.

from soil. *E. herbicola* was found in intermediate and apex positions of stems whereas *A. vinelandii* was found only in the apex of stem.

As summarized in Table 3, sugar cane supports a heterogenous nitrogen-fixing population in different varieties studied but further research will have to be carried out to determine their relationship. The high population of unknown bacteria observed in NA 56-79 has to receive special attention in future work. Dart and Wani [29], working with sorghum, have noted an apparent continuum of types overlapping in properties between named genera and species, and this is a phenomenon which may have also some relevance to sugar cane populations.

FACTORS AFFECTING NITROGEN FIXATION

Nitrogen fixation in sugar cane is influenced by both physiological and external factors. Varietal effect and stage of plant growth seem to be specially more important than external factors which include fertilizer, temperature and moisture.

Ruschel and Ruschel [30] demonstrated varietal effect on nitrogenase activity. Commercially planted varieties NA 56-79 and CB 46-47 were found to have significantly greater nitrogenase activity than varieties CB 47-355, CP 51-22 and IAC 51-205 when tested with intact system at normal levels of oxygen. In contrast, at low oxygen levels CB 47-355, CB 46-47 and NA 56-76 showed high nitrogenase activity. The variety CB 41-76 had the lowest activity at all levels of oxygen.

Later, Ruschel [23] reported the detection and isolation of a potential genotypic effect on nitrogen fixation in sugar cane by identifying cultivars with high and low capacity for supporting nitrogenase activity. Germinated setts from F_1 and F_2 and clones from F_1 seedlings were analyzed for

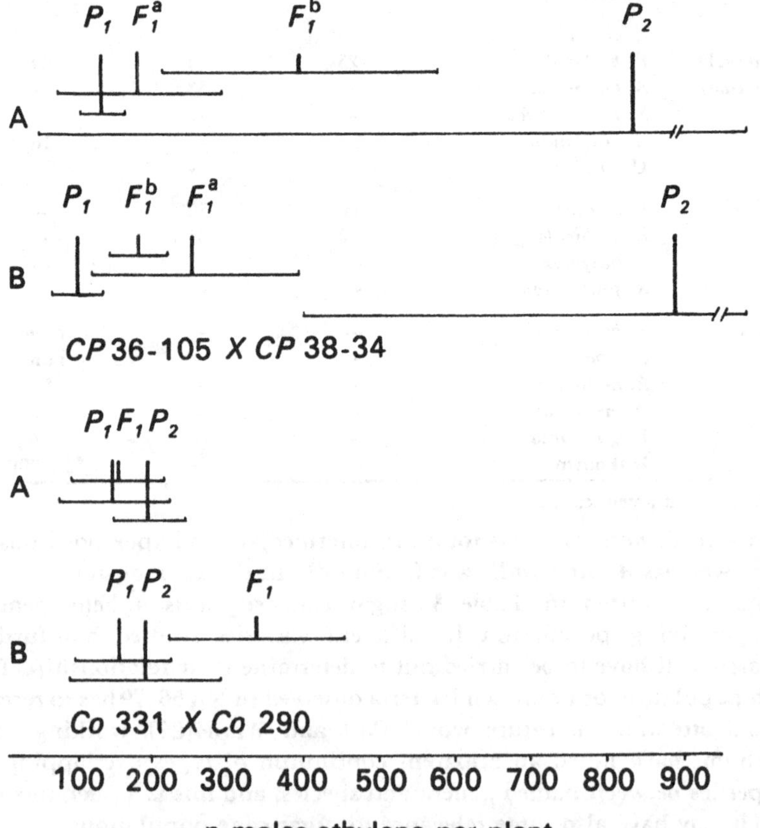

Fig. 5. Nitrogenase activity (nmol of ethylene per plant) of crosses CP 36-105×CP38-36 and Co 331×Co 290 in two experiments, A and B. From Ruschel and Ruschel (1981).

nitrogenase in an intact plant-soil system at different sampling dates. It will be seen from Fig. 5 that in the case of the cross Co 331×Co 290 the F_1 plants showed ethylene production similar to or a little greater than the parents. The F_1 progeny of the cross CP 36-105×CP 38-34 involving parents with widely differing capacity for nitrogenase activity showed much lower nitrogenase activity than the parent. These results suggest the partial dominance of nitrogen-fixing ability (Fig. 5).

The facts that the distribution of nitrogen-fixing bacteria in sugar cane stalks varies [20] and the existence of a seasonal effect on this distribution [22] indicate that physiological factors may influence nitrogen fixation in the field, especially in Brazil where there are two planting times, between fall and winter and during spring. Ruschel et al. [31] observed that nitrogenase activity of irrigated plants in the field differed from non-irrigated ones during mid-winter and mid-summer at different levels of nitrogen, phosphorus and potassium containing fertilizers. Nitrogenase activity of sugar cane roots was low during the initial plant development, probably due to low temperatures, but after 12 and 15 months a 100-fold increase was observed.

The nitrogenase activity of roots sampled after six and nine months was higher in irrigated than non-irrigated plants at all fertilizer levels. van Dillewijn [32] cites that the amount of root exudate decreases during drought but resumes vigorously after copious watering, which may explain the differences observed in irrigated plants. Twelve months after planting the highest nitrogenase activity was found in the roots of non-irrigated plants. After 15 months there was no significant difference between irrigated and non-irrigated plants, possibly due to the effect of abundant rain, characteristic of the month of January in the sugar cane area of Sao Paulo State, Brazil.

Nitrogenase activity of irrigated plants decreased with increasing ammonium sulphate fertilization, but nitrogen fertilization did not seem to affect nitrogenase activity in non-irrigated plots. Nitrogenase activity was not affected by variations in phosphate and potassium fertilization at any level investigated [31]. More work needs to be done in this area, because we do not have good knowledge of the effect of crop growth stage on nitrogen-fixation, or the number of days during the plant growth period required for significant fixation. Nitrogen fixation is known to be reduced by drought. Similarly, it could well be reduced during periods of extremely rapid growth, when it is possible that all the plant's carbohydrate may be required to sustain new development.

DETERMINING THE AMOUNT OF NITROGEN FIXED BY SUGAR CANE

Estimating the amounts of nitrogen fixed under field conditions has proven to be a very difficult proposition. Relative nitrogenase activity at any given time is easily obtained by acetylene reduction assay on soil cores, but obtaining an integrated value for a whole season is difficult. $^{15}N_2$-labelled gas has been used to demonstrate positive nitrogen-fixation associated with seedlings [7, 8] and to show ^{15}N enrichment of rhizosphere soil in field grown cane, but these approaches cannot give us an integrated value. The nitrogen 'difference' techniques adopted in legume studies are not appropriate for sugar cane, as there is no possibility of not leaving any one cultivar uninoculated, or of massive soil applications of ineffective bacteria to suppress the effective ones. The fact that sugar cane has its associated bacteria in the stem also rules out these possibilities. The problems become complicated by the fact that the sugar cane season for growth is very long (12–14 months), and there are no obvious visible indications to measure nitrogen-fixing capacity.

Theoretically it should be possible to use the technique of isotope dilution of ^{15}N derived from the growth medium by atmospheric ^{14}N, for estimating the proportion of plant nitrogen due to fixation. In this method ^{15}N labelled fertilizer is given to the putative nitrogen-fixing plant and to a non-nitrogen-fixing control plant. Then the per cent plant nitrogen due to fixation could be calculated as:

$$\%\ N_2 \text{ fixed} = \left(1 - \frac{\%\ \text{atom excess } ^{15}N \text{ fixing crop}}{\%\ \text{atom excess } ^{15}N \text{ non-fixing crop}}\right) \times 100.$$

One of the key requirements for this method, however, is that the non-fixing control should have the same growth period as the putative nitrogen-fixing crop, and so far a satisfactory non-fixing control crop has not been found in this laboratory. Several factors such as the large size of the sugar cane plant, the need for a large plot for experimentation and the large requirement of ^{15}N fertilizer which in turn necessitates the high ^{15}N dilution rates in tracer studies, render the isotope dilution method for quantifying the amount of nitrogen fixed by sugar cane, rather impracticable [33].

It is possible, on the same principle, to use natural isotopic variation (δ ‰ ^{15}N) to obtain an integrated value for fixation, because high δ ‰ ^{15}N values are usually found in Brazilian soils [34]. This principle was applied in our laboratory as a semi-quantitative test in a nitrogen balance experiment [24] carried out in large containers with 90 kg of soil and the figure obtained was about 17 per cent of plant nitrogen at harvest due to fixation, with a cultivar now known to have one of the poorest nitrogenase activities among the Brazilian cultivars tested. In a field test the δ ‰ ^{15}N values of five sugar cane varieties were compared in our laboratory [34] and significant differences were found both between varieties and between available soil nitrogen, indicating quite a substantial fixation of the order of 30 per cent of plant nitrogen. A study reported [35] δ ‰ ^{15}N values for the CB 45-3 cultivar grown either with fertilizer for seven years or without fertilizer for 7 and 15 years which showed the occurrence of significant isotopic dilution by atmospheric $^{14}N_2$ in plants from all the treatments which implied substantial assimilation of recently fixed nitrogen.

An earlier tentative calculation [8] of fixation rates, based on extrapolation from rates of ^{15}N uptake by non-amended intact root systems, under simulated normal atmosphere conditions, suggested that 3.4 kg N ha^{-1} yr^{-1} is fixed in or on the plant and 50 kg N ha^{-1} yr^{-1} in the rhizosphere. Purchase [25] used acetylene reduction technique to estimate that about 25 kg N ha^{-1} yr^{-1} might be derived from nitrogen fixation by the rattoon crop in South Africa.

FUTURE WORK

It is clear that we still need to know a lot more about the identification and characterization of the bacteria in and on the root surface and in the rhizosphere of sugar cane. It appears that we might have a situation where there is a characteristic group of bacteria commonly associated with sugar cane roots. This group of bacteria seems to include non-nitrogen-fixers, which may contribute something other than fixed nitrogen to the association. Presumably this will be either through growth factors or maybe as a means of reducing local PO_2 and thus provide an oxygen protection mechanism.

It is not known whether bacterial genera commonly found associated with sugar cane are adapted specifically to sugar cane or whether they are ubiquitous types, and this requires further study by serological methods and inoculation experiments. For such studies plants have to be grown aseptically free from bacteria associations. This may not be possible by ordinary propagation methods, but is feasible through tissue culture.

Although sugar cane encourages the growth and prolification of its bacterial system under natural conditions, the levels of bacteria present in the tissues are very erratic. Consequently, in some circumstances inoculation with bacteria may prove useful. For example, planting setts are treated with hot-water (50°C) for disease control and in such instances, inoculation with nitrogen-fixing bacteria could be a useful practice. At present very little information is available on the usefulness of, or appropriate methods for inoculation. In the long term there is the possibility of developing bacterial strains for improved compatibility, for tolerance to mineral fertilizer and for energy conservation.

Relatively short term improvements in the nitrogen-fixing association can be made by appropriate choice of the cultivar, as it is apparent that some can support nitrogen-fixation better than others. Testing of cultivars can be done relatively easily, by comparing acetylene-reduction activity against a standard variety. The development of improved lines by breeding and tissue culture methods is also now considered to be a potential approach.

There is very little knowledge of the plant physiological factors which enhance associative nitrogen fixation e.g., light, temperature, daylength and capacity for sugar formation. Sugar cane provides facility for good gas transport within tissues although this feature does not appear to be a critical factor in nitrogen-fixation. Almost nothing is known about the biochemistry of the fixation process in relation to the potential exudation of growth factors and carbohydrate into the root region in associate nitrogen fixation. Should one be looking for "leaking or non-leaking" root systems? There is also no information on the cost, in terms of sugar consumption for supporting associative nitrogen-fixing systems.

When high levels of nitrogenous fertilizers are used nitrogen fixation will almost certainly be repressed. However, it has been shown in our laboratory that varieties from Hawaii developed under a high nitrogen fertilizer regimes have still retained the capacity for associative nitrogen-fixation. Accurate data on the amount of nitrogen fixed under field conditions, and the cultural and management practices influencing high levels of nitrogen fixation in the field are yet to come forth by carefully planned experiments. Another limitation is that almost all the work so far has been done with 'plant' cane and one knows very little about the situation in rattoon crops.

REFERENCES

1. Malavolta, E., Haag, H.P., Mello, F.A.F. and Brasil Sobr., M.O.C. *Nutricão Mineral de Plantas Cultivadas,* Chapter 5, (1974).
2. Alvarez, R. Segalla, A.L. and Catani, R.A. 1958. *Bragantia 17,* 141-146. Anonymous 1976. Planalsucar, Annual Report for 1975, p. 80.
3. Gomes, F.P. and Cardoso, E.M. Adubacão Mineral da Cana-de-acucar. Editora Aloisi Ltda., Piracicaba, Brasil, Chapter 7 (1958).
4. Arruda, H.V. *Bragantia 19,* 1105-1110 (1960).
5. Takahashi, D.T. *Hawaiian Planters' Record 58,* 95-101 (1970).
6. Dobereiner, J., Day, J. and Dart, P.J. Nitrogenase activity in the rhizosphere of sugar cane and some other tropical grasses, *Plant and Soil 47,* 191 (1972).
7. Ruschel, A.P., Henis, Y. and Salati, E. Nitrogen-15 tracing of N_2-fixation with soil-grown sugar cane seedlings, *Soil Biology and Biochemistry 7,* 181 (1975).
8. Ruschel, A.P., Victoria, R.L., Salati, E., Henis, Y. Nitrogen fixation in sugar cane, *Saccharum officinarum,* in Proc. Symposium Uppsala 1976, *Ecol. Bull.* Stockholm *26,* 297-303 (1978).
9. Vose, P.B., Ruschel, A.P., Victoria, R.L. and Matsui, E. Potential N_2-fixation by sugar cane, *Saccharum* sp., in solution culture. I. Effect of NH_4^+ v NO_3^-, variety and nitrogen level, pp. 119-125, in *Associative N_2-fixation* (Editors, P.B. Vose and A.P. Ruschel), Vol. II, CRC Press, Boca Raton, Florida (1981).

10. Ruschel, A.P., Matsui, E., Vose, P.B. and Salati, E. Potential N_2-fixation by sugar cane, *Saccharum* sp., in solution culture. II. Effects of inoculation and dinitrogen fixation as directly measured by $^{15}N_2$, pp. 128–132, in *Associative N_2-fixation* (Editors, P.B. Vose and A.P. Ruschel), Vol. II, CRC Press, Boca Raton, Florida (1981).
11. Matsui, E., Vose, P.B. Rodrigues, N.S. and Ruschel, A.P. Use of $^{15}N_2$ enriched gas to determine N_2-fixation by undisturbed plants in the field, pp. 153–161, in *Associative N_2-fixation* (Editors, P.B. Vose and A.P. Ruschel), Vol. II, CRC Press, Boca Raton, Florida (1981).
12. Dobereiner, J. Influencia da cane-de-acucar na populacão de de *Beijerinckia* do solo, *Rev. Bras. Biol. 19*, 251 (1959).
13. Freitas, J.R., Ruschel, A.P. and Vose, P.B. Radiorespirometry studies as an indication of soil microbial activity in relation to the root system in sugar cane and comparison with other species, pp. 141–144, in *Associative N_2-fixation* (Editors, P.B. Vose and A.P. Ruschel), Vol. II, CRC Press, Boca Raton, Florida (1981).
14. Dobereiner, J. Nitrogen fixing bacteria of the genus *Beijerinckia* Derx. in the rhizosphere of sugar cane, *Plant and Soil 15*, 211–216 (1961).
15. Arias, D.E., Gatti, I.M., Silva, D.M., Ruschel, A.P. and Vose, P.B. Primeras observaciones al microscopio electronico de bacterias fijadoras de N_2 en las raiz de caña de azücar (*Saccharum officinarum* L.), *Rev. Turrialba 28*, 203 (1978).
16. Hegazi, N.A., Eid, N., Faraq, R.S. and Monib, M. Asymbiotic N_2-fixation in the rhizosphere of sugar cane planted under semi-arid conditions of Egypt, *Rev. Ecol. Biol. Sol. 16*, 23–37 (1979).
17. Patriquin, D.G., Graciolli, L.A. and Ruschel, A.P. Nitrogenase activity of sugar cane propagated from stem cutting in sterile vermiculite, *Soils Biol. Biochem. 12*, 413–417 (1980).
18. Silva, D.M., Ruschel, A.P., Matsui, E., Nogueira, N.L. and Vose, P.B. Determination of the activity of N_2-fixing bacteria in sugar cane roots and bean nodules using tritiated acetylene reduction and electron microautoradiography, pp. 145–151, in *Associative N_2-fixation* (Editors, P.B. Vose and A.P. Ruschel), Vol. II, CRC Press, Boca Raton, Florida (1981).
19. Ruschel, R. and Ruschel, A.P. Inheritance of N_2-fixing ability in sugar cane, pp. 133–140, in *Associative N_2-fixation* (Editors, P.B. Vose and A.P. Ruschel), Vol. II, CRC Press, Boca Raton, Florida (1981).
20. Graciolli, L.A. and Ruschel, A.P. Microorganisms in the phyllosphere and rhizosphere of sugar cane, pp. 91–101, in *Associative N_2-fixation* (Editors, P.B. Vose and A.P. Ruschel), Vol. II, CRC Press, Boca Raton, Florida (1981).
21. Graciolli, L.A. Bacterias fixadoras de nitrogenio em cana-de-acucar (*Saccharum* sp.) em folhas, caules e raizes, Dissertacao ESALQ, Univ. of Sao Paulo (1982).

22. Costa, J.M.F. and Ruschel, A.P. Seasonal variation in the microbial population of sugar cane stalks, p. 109, in *Associative N_2-fixation* (Editors, P.B. Vose and A.P. Ruschel), Vol. II, CRC Press, Boca Raton, Florida (1981).
23. Ruschel, A.P. Associative N_2-fixation by sugar cane, pp. 81-90, in *Associative N_2-fixation* (Editors, P.B. Vose and A.P. Ruschel), Vol. II, CRC Press, Boca Raton, Florida (1981).
24. Ruschel, A.P. and Vose, P.B. Present situation concerning studies on associative nitrogen fixation in sugar cane, CENA Boletim Cinetifico BC-045, July 1977, pp. 27, Piracicaba (Brasil) (1977)
25. Purchase, B.S. Nitrogen fixation associated with sugar cane, *Proc. S. Africa Sugar Tecjnol. Assoc. 1980(6)*, 173-176 (1980).
26. Singh, K. Role of *Azotobacter* in sugar cane culture and the effect of pesticide on its population in soil, pp. 103-107, in *Associative N_2-fixation* (Editors, P.B. Vose and A.P. Ruschel), Vol. II, CRC Press, Boca Raton, Florida (1981).
27. Rennie, R.J. Dinitrogen-fixing bacteria: computer assisted identification of soil isolates, *Can. J. Microbiol. 26*, 1275-1283 (1980).
28. Rennie, R.J., de Frettas, J.R., Ruschel, A.P. and Vose, P.B. Isolation and identification of N_2-fixing bacteria associated with sugar cane (*Saccharum* sp.), *Can. J. Microbiol. 28*, 462-467 (1982).
29. Dart, P.J. and Wani, S.P. Non-symbiotic nitrogen fixation and soil fertility, in 12th Int. Congr. Soil Sci., New Delhi, India, 8-16 Feb., 1982, *Symposium Papers 1*, 3-27 (1982).
30. Ruschel, A.P. and Ruschel, R. Varietal differences affecting nitrogenase activity in rhizosphere of sugar cane, pp. 1941-47, in *Proc. XVI Congr. Int. Soc. Sugar Cane Technol.* (Editors, F.S. Reis and J. Dick), Vol. 2, IMPRES, Sao Paulo, Brazil, (1977).
31. Ruschel, A.P., Orlando, F. J., Zambello, Jr., E. The effect of nitrogen, phosphorus and potassium fertilization and irrigation on nitrogenase activity and yield of sugar cane, p. 1903, in *Proc. XVI Congr. Int. Soc. Sugar Cane Technol.* (Editors, F.S. Reis and J. Dick), Vol. 2, IMPRES, Sao Paulo, Brazil (1977).
32. Van Dillewijin, C. *Botany of Sugarcane*, Chronica Botanica, Waltham, Mass., p. 371 (1952).
33. Vose, P.B., Ruschel, A.P., Victoria, R.L., Tsai Saito, S.M. and Matsui, E. 15-Nitrogen as a tool in biological nitrogen fixation research, pp. 575-592, in *Biological Nitrogen Fixation Technology for Tropical Agriculture* (Editors, P.H. Graham and S.C. Harris), CIAT, Cali, Colombia (1982).
34. Vose, P.B., Ruschel, A.P. and Salati, E. Determination of N_2-fixation, especially in relation to the employment of nitrogen-15 and of natural isotope variation, *2nd Latin American Botanical Congr.*, Brasilia, 89 (1978).

35. Ruschel, A.P. and Vose, P.B. Nitrogen cycling in sugar cane, in Proceedings SCOPE/UNEP Symposium on Nitrogen Cycling in Ecosystems of South America and the Caribbean, Cali, March 1981, *Plant and Soil 64*, 139-146 (1982).

10. Nitrogen Fixation in Wetland Rice Field

I. Watanabe and P.A. Roger

INTRODUCTION

Rice grown in flooded conditions is called wetland rice. Rice grown in moist conditions similar to those used for growing cereals like wheat and maize, is called dryland rice. Because wetland rice can support the highest population per unit area of land [1], rice is mostly grown in densely populated zones and, frequently, in subsistence agricultural systems where crops are grown without synthetic fertilizers. In such cropping systems, wetland rice yields are higher than those of dryland rice. Yield differences can be attributed partly to better water supply and partly to higher fertility of wetland soils. The higher nitrogen fertility of wetland soils is exemplified by the higher dependence of wetland rice on soil nitrogen [2]. Without an external nitrogen supply, wetland rice should deplete soil nitrogen more than dryland rice. Nevertheless, wetland rice has been grown without fertilizer application for a longer time than dryland rice, with little or no decline in yield [3]. It is likely that the higher and more consistent nitrogen fertility of wetland soils can be attributed to biological nitrogen fixation. Nitrogen balance study, described later, indicates this higher nitrogen fertility status of wetland rice cultivation. Because reviews on biological nitrogen fixation in flooded rice soils [4-5] and flooded soils [6] are already available, this chapter focuses on the recent advancement of knowledge.

THE FLOODED RICE FIELD ECOSYSTEM AS A NITROGEN FIXATION SITE

Principal characteristics of wetland rice fields are determined by flooding of the soil and presence of rice plants. The most important change caused by flooding is in the aeration of soil. Because oxygen moves ten thousand times more slowly through a water phase than through a gaseous phase, the capacity of a soil to exchange gases with the atmosphere decreases as it becomes water-saturated. Waterlogging of a soil quickly leads to anaerobic conditions that develop a few millimetres beneath the soil surface in the

Table 1. Major nitrogen-fixing microorganisms in flooded soil-rice ecosystems

Sites	Major nitrogen-fixing microorganisms	Representative genera
Floodwater and surface soil	Free-living blue-green algae	*Nostoc*, *Anabaena*, and others
	Epiphytic blue-green algae	*Nostoc*, *Calothrix*, and others
	Anabaena in symbiosis with *Azolla*	*Anabaena azollae*
	Photosynthetic bacteria	*Rhodopseudomonas*, *Rhodospirillum* and others
	Methane-oxidizing bacteria	*Methylomonas* and others
	Sulphur-oxidizing bacteria	*Thiobacillus*
	Aerobic heterotrophic bacteria	*Azotobacter*, *Derxia*, *Beijerinckia*
	Microaerophilic bacteria	*Azospirillum*
Anaerobic soil	Facultatively anaerobic bacteria	*Bacillus*
	Strictly anaerobic bacteria	*Clostridium*, *Propionibacterium*
	Sulphate-reducing bacteria	*Desulfovibrio*
Plant (mostly root)	Microaerophilic bacteria	*Azospirillum*, *Pseudomonas*, *Alcaligenes*
	Facultatively anaerobic bacteria	*Enterobacter*, *Klebsiella*

reduced layer. After submergence the number of fungi, actinomycetes, and aerobic bacteria is reduced and anaerobic bacteria increase. Surface soil (a few mm) and floodwater remain in an oxidized state and are colonized by aerobic microflora and photosynthetic organisms. Rice plant provides another niche for microbial activity, in the floodwater, including shoots, roots, and the rhizosphere. The subsoil develops beneath the plough pan layer.

The major environments in paddy fields are: floodwater, surface-oxidized soil, reduced soil, rice plants, and subsoil. The reader is referred to reviews on chemistry of the submerged soil [7] and microbiology of flooded rice soil [8-9].

Nitrogen-fixing organisms are distributed in these different sites, where environmental conditions for the growth and nitrogen-fixing activities differ. Major nitrogen-fixing microorganisms in each site are summarized in Table 1.

Floodwater

The floodwater is a photic zone where aquatic communities, including bacteria, prokaryotic and eukaryotic algae, and aquatic weeds, provide organic matter to the soil surface. Little is known about the amount of organic matter contributed by aquatic phototrophs. In a paddy field in the Philippines, the primary production of floodwater communities was equivalent to the productivity values in eutrophic lakes [10]. The productivity of the aquatic photosynthetic biomass probably rarely exceeds 1000 kg dry weight per hectare [11].

In floodwater, aquatic weeds and basal portions of rice shoots are colonized by epiphytic bacteria and algae. Epiphytic nitrogen fixation becomes agronomically significant in deepwater rice where the submerged plant biomass is very high [81].

Surface soil (oxidized layer)

The surface soil is at a redox potential (Eh) higher than 300 mV. The depth of oxidized layer is from 2 to 20 mm and is dependent on the reducing capacity or oxygen-consuming capacity of the soil, owing to microbial respiration and Fe^{+2} oxidation [12]. In the oxidized layer, NO_3^-, Fe^{+3}, SO_4^{-2}, and CO_2 are stable [13] and aerobic bacteria predominate [14-15]. Methane and hydrogen evolved from the anaerobic soil are partly oxidized in the surface soil [16].

Anaerobic soil (reduced layer)

In anaerobic soil, the reduction process predominates. Eh ranges from 300 m V to −300 m V. Takai and Kamura [17] divided the reducing process of paddy soil into two stages: before and after iron reduction is completed. In the first stage, oxygen absorption, nitrate reduction, manganese reduction, and iron reduction proceed in this order, and ammonia and carbon dioxide are liberated. In the second stage, sulphide and methane are produced and the population of anaerobic bacteria increases. The main organic acids detected in reduced paddy soils are acetic, propionic, and butyric acids [18]. Methane formation is accompanied by the decrease of organic acids and carbon dioxide.

Organic matter particles in soil are important microsites for microbial activity. Organic matter is provided to anaerobic soil as crop residues, decomposed material from the aquatic biomass, and organic fertilizers. Wada and Kanazawa [19] developed a technique to fractionate soil particles according to size and density. They found that about 30 per cent of the organic matter in a paddy soil exists in particles larger than 37 μm and that particle size of organic debris decreases during decomposition. Microbial activities are concentrated on the soil aggregates that contain decayed organic debris. The presence of organic debris makes anaerobic soil heterogenous. The activities of soil fauna [20] make microaerophilic sites in anaerobic layers. As Dommergues [21] pointed out, submerged paddy soils are far from being uniformly reduced and should be regarded as complex systems formed by the juxtaposition of microenvironments that are either sites of oxidation reaction or sites of reduction reaction.

Rice root and the rhizosphere

The early concept of rhizosphere suggested that the plant exudes organic substances on which soil microorganisms grow and, consequently, the root in the soil is surrounded by these microorganisms. It now appears that besides the soil adjacent to the root (rhizosphere, in the strict sense), mucilagenous layers on the surface of the epidermis and intercellular spaces among epidermis layers as well as inner tissues of the epidermis and cortex are also inhabited by microbial colonies. These provide more or less continuous media for their activities [22]. Secretion of carbon compounds by roots provides energy sources for microbial growth and is greatly accelerated by the presence of microorganisms [23-24]. Invasion of bacteria, fungi, and protozoa in wetland rice roots was observed at later stages of rice growth [25]. As in other marsh plants, rice roots receive oxygen from aerial parts of the plant and oxidize the rhizosphere. The brownish colour of rice roots indicates oxidation of ferrous iron to ferric iron and its precipitation along the root surface [26].

Rice roots grown in submerged soil have fewer and shorter root hairs and are straighter than those found in dry soil [27]. Redox conditions in the rhizosphere are determined by the balance of oxidizing and reducing capacities of rice roots. Nutrient deficiencies, particularly in nitrogen and potassium, accelerate the reduction of rhizosphere [28-29].

Subsoil

The soil beneath the plough pan is aerobic in well-drained soils and anaerobic in poorly drained soils. Its role in providing nitrogen to rice is sometimes apparent.

Table 2. Nitrogen balance in long-term fertility trials in wetland rice soils (from Watanabe, Craswell, and App [30]) and unpublished data

Site	Cropping per year	Duration years	Treatment	Kg N ha^{-1} yr^{-1}			
				Input	Soil change	Plant uptake*	Balance
Aomori, Japan (41°N)	Wetland rice	21	PK	0	-20	45	+25
			NPK	57	-35	66	-25
Kagawa, Japan (34°N)	Wetland rice and barley	21	PK	0	-42	80 (55)	+38
			NPK	157	-18	154 (96)	-21
Sorachi, Japan (45°N)	Wetland rice	12	PK	0	-44	142	+98
			NPK	39	-51	136	+46
Ishikawa, Japan (36°N)	Wetland rice	22	Unfertilized	0	-34	53	+19
			PK	0	-30	64	+34
			CaPK	0	-34	72	+38
			NPKCa	100	-15	119	+ 4
Shiga, Japan (35°N)	Wetland rice and wheat	40	Unfertilized	0	-1.7	41 (30)	+39
			PK	0	-13.1	67 (51)	+55
			NPK	152	+ 2.2	112 (74)	-37
Los Baños, Philippines (14°N)	2 wetland rices (1st–24th crops)	12	Unfertilized	0	+30	116	+146
	2 wetland rices (24th–33rd crops)	5	Unfertilized	2 (rain)	±20**	71	49–89

*Values in parentheses are N uptakes by rice.

**Soil was analyzed after the 24th and the 33rd crops. No change of soil nitrogen was observed. This value is a standard error of analysis.

QUANTIFICATION OF BIOLOGICAL NITROGEN FIXATION

Balance studies

Higher nitrogen fixation in wetland conditions has long been considered the reason for the higher nitrogen maintenance levels in wetland soils than in dryland soils. The nitrogen balance of long-term fertility experiments gives a quantitative answer to the question of nitrogen gains in wetland conditions. Table 2 indicates the results of experiments in Japan and the Philippines where soil nitrogen changes have been determined and compared with crop nitrogen removal. In non-fertilized plots, net gains of soil nitrogen ranged from 20 to 70 kg N/ha per year (in Japan) or per crop (in the Philippines) except a peat soil at Sorachi. The addition of phosphorus and potassium increased nitrogen gains, but in the most heavily nitrogen-fertilized treatments, net losses of nitrogen occurred.

Nitrogen balance data from long-term fertility plots must be examined with care because of possible errors in soil analysis and unquantified role of subsoils. Most Japanese data on soil nitrogen appear significant, because the experiments were conducted for sufficiently long periods. However, in many nitrogen balance calculations, soil nitrogen changes in the subsoil are not considered. A high positive nitrogen balance in Sorachi, where a peat layer was located below the ploughed layer, suggests that subsoil nitrogen contribution may not always be negligible [31]. It is difficult to quantify the contribution of subsoil, unlike other inputs such as rain and irrigation water which can be quantified.

Yanagisawa and Takahashi [32] reported nitrogen uptake data from fertility trials without nitrogn fertilizer in Japan. Average nitrogen removal from 15 experimental stations was 64 kg N/ha, indicating that a net addition of about 60 kg N/ha per crop is necessary to replace nitrogen used by the crop. However, unless subsoil contribution, soil nitrogen changes, volatilization, and other losses are determined, nitrogen gains estimated from nitrogen uptake in no-nitrogen plots can only give an approximate evaluation of nitrogen fixation.

In pot experiments, sources of nitrogen inputs into flooded rice systems are easily controlled. Data on nitrogen balance in pot experiments indicate that: more net nitrogen gain was obtained when the soil surface was exposed to light than when it was protected from light [34–35], more net nitrogen gain was obtained in planted pots than in unplanted pots [33–35], and wetland conditions yielded more nitrogen than dryland ones [35]. These data substantiate an early study by De [36], who showed the importance of blue-green algae in maintaining soil fertility in flooded soil. The role of rice plant in stimulating nitrogen gains or reducing losses is not well analyzed. It is highly probable that nitrogen gains are stimulated by rhizospheric (associative) nitrogen fixation and that nitrogen losses are mitigated by the continuous absorption of soil nitrogen by the plant, otherwise its nitrogen would be lost.

Nitrogen balance sheets give only the sum of nitrogen gains and losses. To estimate gross nitrogen gain, quantification of nitrogen losses is essential. Experiments by Ventura and Watanabe [35] with $^{15}N_2$ labelled soil showed that differences in nitrogen gains (dark versus light treatments, wetland versus dryland) were caused by. differences in nitrogen fixation but not differences in nitrogen losses.

Acetylene reduction techniques

It is well recognized that acetylene reduction activity (ARA) assay cannot be used as a quantitative tool unless $^{15}N_2$ incorporation experiments are made in identical conditions and experimental ratio of N fixed to reduced acetylene is determined. Because ARA assay is more sensitive than $^{15}N_2$ incorporation technique, it is difficult to utilize both methods for the same period, particularly when the nitrogen-fixing activity is low. However, acetylene reduction technique is useful, especially when comparative studies are made. When this technique is used in the paddy field, the following limitations should be considered.

Acetylene diffusion into flooded soil and back diffusion of formed ethylene are slow. To introduce acetylene into the system, evacuation of the gas phase [37] and mechanical disturbance [38] were proposed. Mechanical disturbance is also necessary to recover ethylene from the soil [38].

Acetylene inhibits nitrogen-fixing activity of methane-oxidizing bacteria, therefore, the contribution of methane-oxidizing bacteria to nitrogen-fixation cannot be determined [39]. Methane oxidation occurs in paddy soils [40].

Acetylene is decomposed in anaerobic conditions. Thus, prolonged incubation under anaerobic condition leads to a high ARA value due to the stimulation of nitrogen-fixation by microorganisms that probably use the decomposition products of acetylene [41].

Spatial variation of activity is large and *in situ* ARA shows a log-normal distribution [42]. High numbers of replicates (more than 6) and composite soil samples are needed for accuracy. In addition, logarithmic transformation of data is necessary for statistical analysis [42]. Methods for algae, plant associative activity, and anaerobic soils are described by Roger et al. [43], Lee and Watanabe [38], and Matsuguchi et al. [37]. Problems in acetylene reduction assay have been extensively reviewed by Knowles [44]. Acetylene methodology has been most frequently used to assess the activity of specific components of the nitrogen-fixing biomass. Some examples of measurements are given in Table 3. ARA values of photodependent nitrogen fixation have been recently summarized by Roger and Kulasooriya [45]. *In situ* photodependent ARA values ranged from 0 to 600 μ mol $C_2H_4 \cdot m^{-2} \cdot h^{-1}$ in Senegal [43]; 0.2 to 3 m mol $C_2H_4 . m^{-2} \cdot day^{-1}$ in the Philippines [48]; 0.1 to 4 m mol $C_2H_4 \cdot m^{-2} \cdot day^{-1}$ in Thailand [50]; 0.03 to 0.9 m mol $C_2H_4 . m^{-2} \cdot day^{-1}$ in Indonesia [51]; and 0.8 to 3.2 m mol $C_2H_4 \cdot m^{-2} . day^{-1}$ in Malaysia [52]. Data from Indonesia and Malaysia may include some surface soil activities.

Maximum value may be near 4 m mol C_2H_4·m $^{-2}$·day^{-1}. ARA associated with rice plant ranged from 10 to 50 μ mol C_2H_4·m^{-2}·plant^{-1}. ARA of anaerobic soil ranged from 0 to 0.5 n mol C_2H_4·g^{-1}·h^{-1} [46].

^{15}N technique

The $^{15}N_2$ incorporation technique gives a direct demonstration of nitrogen-fixation. However, the technique cannot be used in the field to quantify nitrogen-fixation during the growth cycle of rice because of its high cost, point-time measurement, and the need for sophisticated apparatus to control environmental conditions in the closed chamber. In addition, results of laboratory experiments on photodependent nitrogen-fixation are often expressed per unit weight of soil. Because photodependent nitrogen-fixing activities are determined by the surface exposed to light, it is absolutely necessary to express activity per unit of surface area. Data on $^{15}N_2$ incorporation are given in Table 4. If we assume a square metre contains 10^5 g soil and 25 plants, maximum values of heterotrophic, photodependent, and associative (rhizospheric) nitrogen fixation are 30, 43, and 7.2 mg N·m^{-2}·day^{-1}. But no data on the $^{15}N_2$-fixing rates of various agents in the same soil are available.

Recently more attention has been paid to the ^{15}N dilution technique (substrate labelling technique) for quantifying the contribution of nitrogen-fixation in the nitrogen nutrition of organisms or plants [61]. The method is based on the fact that an organism growing at the expense of a substrate labelled with ^{15}N (combined nitrogen) in a system where no nitrogen-fixation occurs accumulates more ^{15}N than a similar organism growing on the same substrate in a similar system where nitrogen-fixation occurs.

The validity of the estimation by this technique depends on the choice of the non-nitrogen-fixing control. Ventura and Watanabe [35] applied the ^{15}N dilution technique to assess the contribution of photodependent nitrogen-fixation in wetland rice using rice plants grown in pots covered by black cloth as control. From this experiment, it was found that two rice crops absorbed 20–30 per cent of nitrogen gain in flooded soil.

No data are available for the application of this technique to assess nitrogen fixation by blue-green algae and *Azolla-Anabaena* symbiosis.

Relative importance of various nitrogen-fixing agents

Nitrogen balance data shown in Table 2 indicate that photodependent nitrogen fixation is more active than heterotrophic nitrogen-fixation in soils tested in the tropics. Watanabe et al. [48] estimated the contribution of nitrogen fixers in floodwater and on the soil surface by measuring ARA before and after the removal of floodwater and surface soil and their subsequent replacement with alga-free water. They concluded that nitrogen-fixing activity by blue-green algae, and perhaps bacteria associated with the algal biomass, was greater than that of other microorga-

Table 3. N_2-fixation measured by acetylene reduction assay and percentage contribution of floodwater, soil, and rhizosphere

Site	Treatment	Fixing rate kg N ha^{-1} $crop^{-1}$*	Contributions (%)			Method	References
			F	S	R**		
Japan	NPK	11	0	60	40	*In vitro*	Matsuguchi [46]
	NPK+compost	17	0	70	30		
	NPK+straw	19	5	80	15		
Japan	No fertilizer	1.1	<5	>85	<10	*In vitro*	Panicksakpatana et al. [47]
	Green manure	3.8	<5	>85	<10		
Philippines	Wet season No fertilizer	24 (163 days)	43	22	35	*In situ*	Watanabe et al. [48]
	Dry season No fertilizer	34 (168 days)	69	17	14		Watanabe et al. [49]

*Assume that 1 mol C_2H_4 reduction is equivalent to 9.3 g N-fixed.

**F=floodwater; S=reduced soil; R=rhizosphere.

nisms associated with wetland plants. Wada et al. [62] and Panichsakpatana et al. [47], on the other hand, reported very little photodependent nitrogen-fixing activity in floodwater and surface soil, and concluded that these sites were not important for nitrogen fixation in rice fields (Table 3).

Field acetylene reduction assays by Lee et al. [63] could not detect nitrogen-fixing activity of the reduced layer, because acetylene does not diffuse into the bulk of the reduced layer. Watanabe et al. [49], who adopted a soil core assay method to measure reduced soil ARA, reported an average daily ARA of 0.30 m mol $C_2H_4 \cdot m^{-2}$ in Maahas soil (pH 7.5) during a rice growing season (Table 3). The relative importance of the various nitrogen-fixing agents depends on environmental conditions and cultural practices. For example, different authors have reported a higher activity of floodwater in unfertilized plots than in fertilized plots [47, 48, 62]. The effects of environmental and agricultural factors are discussed in the following sections.

ECOLOGY OF PHOTODEPENDENT NITROGEN-FIXING ORGANISMS

Photodependent nitrogen-fixing organisms that develop in the photic zone (floodwater and soil-water interface) are free-living and epiphytic blue-green algae (BGA), *Anabaena azollae* in symbiosis with *Azolla*, and photosynthetic bacteria. The ecology of these organisms is summarized in this section. Further information is contained in recent reviews on BGA in rice fields by Roger and Kulasooriya [45] and on *Azolla* by Lumpkin and Plücknett [64-65], Peters and Calvert [66], and Watanabe [67].

Free-living blue-green algae

In paddy fields BGA growth and algal successions are governed by climatic, physicochemical, and biotic factors.

Light intensity is the most important climatic factor. BGA are sensitive to high light and develop protective mechanisms like vertical migrations in the water of submerged soils; preferential growth in shaded zones like embankments, under or inside decaying plant material, or a few millimetres below the surface; photophobotaxis; photokinesis, and stratification of the strains in algal mats where nitrogen-fixing strains grow under a layer of eukaryotic algae [68]. In areas with high incident light intensities, BGA develop later in the crop cycle when the plant cover is dense enough to protect them from excessive light [69]. On the other hand, light deficiency may also be a limiting factor. In Japan, available light under the canopy was below the compensation point of the phytoplankton during the later part of the growth cycle [70]. In the Philippines, during the wet season, under moderate light, ARA was higher in bare soil than in planted soil [48].

Temperature is rarely a limiting factor for BGA in paddy fields because the range of temperature permitting their growth is larger than that required by rice. However, temperature influences the composition of the algal biomass and the productivity. The optimal temperature for BGA is 30-35°C. Low temperatures decrease productivity and favour eukaryotic algae. High temperatures favour BGA and increase algal productivity [45].

pH is the most important soil factor, among the soil properties, determining the composition of algal flora. Under natural conditions BGA grow best in neutral-to-alkaline environments. In rice fields positive correlations occur between water pH and BGA number, soil pH and BGA spores, soil pH and the nitrogen-fixing algal biomass in samples homogenous for stage of rice development, and in fertilization and plant cover density [68]. The beneficial influence of high pH on BGA growth is demonstrated by the addition of lime, which increases BGA growth and N_2 fixation [45]. However, the presence of certain strains of BGA in soils with pH values between 5 and 6 has been reported [71-72].

Phosphorus availability is another important factor determining BGA growth. Okuda and Yamaguchi [73] incubated 117 submerged soils and noted that BGA growth was closely related to the available phosphorus content of the soil.

Biotic factors (organisms) that limit BGA growth are pathogens, antagonistic organisms, and grazers. Of these, only grazers have been documented. The development of zooplankton populations, especially Cladocerans, Copepods, Ostracods, and mosquito larvae prevented the establishment of algal blooms within one or two weeks [74]. Grazing rates and algal diet preferences of Ostracods were studied by Grant and Alexander [75] who estimated the potential consumption of BGA at an average field density of 10,000 Ostracods m^{-2} to be about 120 kg (fresh weight) ha^{-1} day^{-1}. An economic alternative for controlling Ostracod populations is the application of crushed seeds of neem tree (*Azadirachta indica*) [76]. Snails form another group of algal grazers in submerged paddy fields. The biomass of snails can be as much as 1.6 t/ha in some rice fields in the Philippines [45]. Commercial pesticides that can control grazers are expensive and uneconomical [76].

Agricultural practices to encourage BGA growth: The growth of nitrogen-fixing BGA in rice fields is most commonly limited by low pH, phosphorus deficiency, and grazer populations. Application of phosphorus and lime has frequently produced positive results. An increase in algal biomass has also been reported as a secondary effect of insecticide application [45].

Recently, surface application of straw was reported beneficial for BGA growth and photodependent ARA [77-78]. This may be due to an increase of CO_2 in the photic zone, a decrease of mineral nitrogen and O_2 concentration in the floodwater, and the provision of microaerobic microsites by the straw. Increased CO_2 availability and low nitrogen concentration are known to favour the growth of nitrogen-fixing BGA. Low O_2 concentration in the photic zone may have increased their specific nitrogen-fixing activity.

Epiphytic blue-green algae

Nature and distribution of nitrogen-fixing blue-green algae: Epiphytic BGA have been observed on wetland rice [79], deepwater rice [173–174], and on weeds growing in rice fields [80]. A comparison of these different hosts [175] revealed that epiphytism and the associated ARA on wetland rice at seedling, tillering, and heading stages and on the submerged weed *Chara* were predominantly due to colonies of *Gloeotrichia* sp. visible to the naked eye. Epiphytic algae on wetland rice at maturity, on the submerged weed *Najas*, and on deepwater rice could be observed only under a microscope The dominant algae were *Nostoc, Calothrix*, and *Anabaena*. A unique finding was that BGA also exist inside the cavities of senescent rice leaf sheaths. This "endophytism", however, in addition to being not confined to rice, was not present in living healthy tissues. A frequent observation was that older parts of the hosts and plants with rough surfaces supported more epiphytic BGA. It was concluded that epiphytism is possibly related to abiotic effect, of which a mechanical effect in relation to the roughness of the host surface appears to be important.

Nitrogen-fixing activity of epiphytic BGA: Rates of light-dependent ARA on wetland rice gradually diminished from seedling to maturity mainly due to the concomitant decrease of *Gloeotrichia* epiphytism and the reduction of available light. In deepwater rice there was also a decrease in specific ARA (activity per gram of host) from heading to maturity but this was compensated by an increase in the host biomass so that a constant activity per plant was observed at both stages. The results of ARA measurements indicated that nitrogen contribution by nitrogen-fixing microorganisms epiphytic on wetland rice is low but epiphytic BGA play an important role in inoculum conservation because floating algae and soil algae are frequently washed from the field during heavy rains [79–80]. On the other hand, epiphytic nitrogen fixation on deepwater rice makes a substantial nitrogen contribution to this ecosystem (10–20 kg N/ha) mainly due to the greater biomass available for colonization by epiphytic BGA [81].

Fate of epiphytically fixed nitrogen: The importance of epiphytic nitrogen-fixation and the availability of epiphytically fixed nitrogen was evaluated by Watanabe et al. [81] and Watanabe and Ventura [60], using ^{15}N techniques. Direct evidence of nitrogen-fixation associated with deepwater rice was obtained by exposing submerged parts of a plant to $^{15}N_2$ for nine days. There was higher enrichment of $^{15}N_2$ in submerged nodal roots and leaf sheaths where BGA grew epiphytically. During the nine-day-period 8 mg nitrogen was fixed by the plant and at maturity (Table 4) and 40 per cent

Table 4. Evaluation of N_2-fixation in flooded soil and/or wetland rice by $^{15}N_2$ experiments

Systems	Exposure period (days)	Daily N_2-fixing rate	References	Note
Heterotrophic in soil	30	0.133 (μg/g soil)	Rao [53]	
Heterotrophic in soil	30	0.13 ~ 0.3 (μg/g soil)	Kalininskaya et al. [54]	(a)
Photodependent in floodwater +soil	28	3.0 ~ 5.6 (mg/m²)	MacRae and Castro [55]	(b)
Photodependent + heterotrophic floodwater + soil	730	0.15 (μg/g soil)	Reddy and Patrick [56]	
Photodependent in floodwater +soil	30	43 (mg/m²)	"	
Photodependent in floodwater +soil + weed	30	5.8 ~ 7.5 (mg/m²)	"	
Associative in plant	7	195 – 350 (μg/plant)	Ito et al. [57]	
Associative + heterotrophic	7 ~ 13		Yoshida and Yoneyama [58]	(c)
in plant + soil		54 ~ 105 (μg/plant)		
in plant		13 – 20 (μg/plant)		
Associative + heterotrophic	13		Eskew et al. [59]	(d)
in plant + soil		576 (μg/plant)		
in plant		99 (μg/plant)		
Associative + heterotrophic	3		Eskew et al. [59]	
in plant + soil		238 (μg/plant)		
in plant		40 (μg/plant)		
Photodependent epiphytic with deepwater rice	9	900 (μg/plant)	Watanabe and Ventura [60]	(e)

(a) Three kinds of planted paddy soils were used. Soils were taken at tillering stage of rice and after harvest.

(b) Original data which were expressed per g soil were converted to surface area. Ten g soil was placed in 1.1 cm diameter tube. Three kinds of soil were used.

(c) Range of data was obtained by four experiments.

(d) After exposure of plant and soil system to $^{15}N_2$, the plant was grown further to maturity in normal air. $^{15}N_2$ remaining in soil and plant might have been further fixed. Therefore, actual exposure period was longer than 13 days.

(e) Only aquatic parts of plants were exposed to $^{15}N_2$. Plant was grown further in deep water to maturity.

of the fixed nitrogen was found in parts of the plants not directly exposed to $^{15}N_2$ [60].

In the shallow-water rice, epiphytic BGA make only a small contribution to the nitrogen input but in deepwater rice they produce a substantial nitrogen input especially important in a cropping system where nitrogen fertilizer is seldom applied.

Symbiotic blue-green algae and *Azolla*

Anabaena azollae fixes molecular nitrogen in symbiosis with the water fern *Azolla*. Genus *Azolla* comprises two subgenera, Euazolla and Rhizosperma. *A. filiculoides, A. microphylla, A. caroliniana,* and *A. mexicana* belong to Euazolla; *A. pinnata* and *A. nilotica* belong to Rhizosperma. The various species are widely distributed in tropical and temperate fresh water ecosystems throughout the world. *Azolla* is indigenous in some rice fields and can grow in others when inoculated.

Azolla has been used as a green manure for wetland rice culture for centuries in northern Vietnam [83] and southeastern China [82]. Studies of *Azolla* use in paddy fields were initiated at the International Rice Research Institute [102], the Central Rice Research Institute in India [84], and the University of California [85] in the mid-1970s. Farmers recently adopted *Azolla* as green manure on more than 5,000 ha in South Cotabato, Philippines [86]. Previously, only *A. pinnata* was used as green manure for wetland rice in Asia. Various Euazolla species have been recently introduced to Asia and their growth and nitrogen-fixing capabilities are being examined [87-89].

The morphology, life cycle, and physiology of *Azolla-Anabaena* symbiosis are not discussed in this chapter. Reader is referred to other reviews [64, 68, 90, 91]. This section describes environmental factors that affect *Azolla* growth in flooded rice soils: temperature, light intensity, water and wind, pH, mineral nutrition, and pest pressure.

Temperature: Temperature requirement differs among *Azolla* species. The optimum for *A. pinnata, A. mexicana,* and *A. Caroliniana,* grown at constant temperature under artificial light (15 klx), was about 30°C, whereas *A. filiculoides* required 25°C [92]. The response of nitrogenase activity to temperature from 10 to 42°C also showed that *A. filiculoides* grows best under lower temperatures. Although *A. pinnata* is widely distributed in the tropics, it grows best in cooler season [89]. Watanabe and Berja [93] examined the growth of *A. pinnata, A. filiculoides, A. mexicana,* and *A. caroliniana* at average temperatures of 22, 29, 33°C with 8°C difference between day (12 hours) and night (12 hours). Growth rate during the exponential phase was either larger at higher temperature or almost equal at the three temperature levels except in *A. filiculoides* which exhibited a lower growth rate at higher temperature. Maximum biomass at stationary phase decreased as temperature increased. At 22 and 29°C, *A. caroliniana* achieved the highest biomass. At 33°C, strains of *A. pinnata* recorded the highest biomass. This experiment also showed that *Azolla* becomes more sensitive to higher temperature at higher plant density than at lower plant density. Temperature response is related to other factors. The lower the temperature, the lower is the optimum light intensity for growth and nitrogenase activity [94-95]. Because *Azolla* use depends on tolerance for high temperature in the tropics and tolerance for cold in the temperate region, various species were tested for their adaptability to different climatic conditions [87-89]. From trials in Hanchow, China, *Azolla* species were classified into four groups according to temperature response:

1) cold-tolerant, heat-sensitive type: *A. filiculoides* and *A. rubra* (the latter is close to *A. filiculoides*), optimum temperature about 20°C, minimum limit −5 to −8°C, maximum limit 38 to 40°C.

2) heat-tolerant, cold-sensitive type: *A. microphylla* and *A. mexicana*, optimum temperature 25-30°C, minimum limit 5 to 8°, maximum limit 45°C.

3) relatively cold-tolerant and heat-tolerant type: *A. caroliniana* and *A. pinnata* var. *imbricata* (*A. pinnata* species which is common in Asia), optimum temperature 25 to 30°C, minimum limit −3 to −5°C, maximum limit 45°.

4) cold-sensitive and heat-sensitive type: *A. pinnata* var. *africana* (*A. pinnata* which is common in Africa) and *A. nilotica*, optimum temperature about 25°, minimum limit, 8 to 3°C. Maximum limit about 38°C.

Light: Experiments with short periods of exposure to various light intensity showed that light saturation for nitrogen-fixing activity is about 200 $\mu E \cdot m^{-2} \cdot sec^{-1}$ (16 klx) [96], 5 klx [97], and 5 to 10 klx [98]. When growing *in situ*, Azolla requires higher light intensities because the plants overlap each other. Growth increases with light intensity up to values of 400 $\mu E \cdot m^{-2} \cdot sec^{-1}$ or 32 klx [92], 40 klx [98], 49 klx [94]. Further increase in light intensity was reported to retard growth [94, 99].

Although shading not only reduces light intensity but also the temperature of water and air during sunny midday, according to the experience in our laboratory, some shading is beneficial for the growth of the ferns. Further, *A. pinnata, A. mexicana*, and *A. caroliniana* have been observed to turn red in strong sunlight and remain green in shade.

Water and wind: Azolla is sensitive to drought. Ungerminated sporophyte is more tolerant of desiccation than vegetatively growing plants. The fern can grow on the surface of water-saturated soil but it grows more slowly than on water surface because abscission of leaves, which triggers further vegetative propagation, is easier in floating condition. Wind and wave action, as well as other strong turbulence, causes fragmentation and diminishes growth. Therefore, *Azolla* is not found on large lakes or swiftly moving waters [94]. It was often experienced in the Philippines that *Azolla* did not survive after a typhoon.

pH: In buffered liquid medium, *Azolla* produced equal biomass between pH 5 and 8, but its growth was retarded at pH 9 [92]. Singh [100] reported that acid soils (pH 3 to 3.6) did not support growth and the fern died.

Nutrient supply: Except for nitrogen which can be supplied entirely by nitrogen fixation, the macronutrients essential to the symbiosis are the same as those of other photoautotrophs. Deficient levels of macronutrients in *A. pinnata* are 0.08 per cent P, 0.4 per cent K, 0.18 per cent Ca, and 0.016 per cent Fe on dry matter basis [89]. In continuous flow culture, Subudhi, and Watanabe [101] determined that threshold phosphorus concentration is between 0.06 and 0.03 ppm in water and 0.1 per cent in plants. Euazolla species required higher threshold phosphorus in the plant than *A. pinnata.* The growth of *Azolla* floating on floodwater is limited by the release of phosphorus from soil to floodwater which is too slow to meet its requirement. Floating *Azolla* utilizes only water-soluble phosphate (superphosphate) which is easily adsorbed by the soil. Therefore, split application of phosphate fertilizer directly on the leaves of the plant is needed to maximize the efficiency of utilization of phosphate applied to *Azolla* [102].

In some phosphate-rich and light-textured soils in the Philippines, *Azolla* can double its biomass in about three days without external source of phosphate [86]. In some alkaline soils, supplemental iron is required to support the growth of *Azolla* [96].

Pests and predators: Azolla is attacked by many kinds of pests—insects, snails, algae, and fungi. Insects cause the most severe damage. *Azolla* pests have been reviewed by Lumpkin and Plucknett [65]. Among Lepidoptera, *Pyralis, Nymphula* and *Cryptoblabes* are reported in China and South and Southeast Asia. The larvae feed on *Azolla* leaves and construct a tunnel of *Azolla* leaves and sometimes roots, and live inside the tunnels [103]. Damage by Lepidoptera is more severe at higher temperatures. Among Diptera, *Polypedilum, Chrironomus,* and other Chrironomidae are destructive pests. *Bagous* (Coleoptera) and a grasshopper, *Criotettix* (Orthoptera), are also *Azolla* pests. The latter are the most destructive among all insects [104]. Snails (*Lymnaea*) eat young roots and leaves but are less harmful than insects. Some species and strains of *Azolla* are sensitive to fungus attack. This attack is triggered or accompanied by diminished *Azolla* growth under unfavourable conditions (the authors' observation). *Rhizoctonia* and *Sclerotium* were isolated from *A. pinnata* [105]. Chemical pest control is possible, but use of pesticides limits the economical feasibility of *Azolla* as a green manure [86]. Cheap ways of controlling pest damage must be sought.

Photosynthetic bacteria

Photosynthetic bacteria are generally thought to make no significant nitrogen contribution to paddy fields, because they have a low nitrogen-fixing activity that is inhibited by O_2 whereas the floodwater and the soil-water interface are likely to be aerobic. However, these organisms are relatively abundant and as many as 10^5 to 10^7 have been found per ml of floodwater or per gram of soil [106]. Recent results of Habte and Alexander [107] suggested that fixation by photosynthetic bacteria increases when BGA are killed by chemical treatments. Experiment at IRRI (unpublished) showed a 10^5 times increase of photosynthetic bacterial population after surface incorporation of 4 t of straw per hectare; maximum value recorded was 10^8 *Athiorodaceae* per gram of 5 mm top soil. Filters inhibiting specifically the photosynthetic activity of BGA were used in a pot experiment to assess the N_2-fixing activity of photosynthetic bacteria. No nitrogen accumulation was detected in pots covered with filters whereas nitrogen accumulated in the surface soil of the controls (IRRI Annual Report for 1981). This indicates at least a low activity of photosynthetic bacteria compared with BGA.

ECOLOGY OF HETEROTROPHIC NITROGEN-FIXING ORGANISMS

Two types of heterotrophic nitrogen fixation are recognized. One is associated with the rice plant and the other is dependent on organic debris from crop residues and aquatic biomass.

N_2 fixingmicroorganisms associated with rice

As mentioned in the section on nitrogen balance, positive nitrogen gains are higher in the presence of rice. Nitrogen fixation takes place in close association with the rice plant, particularly with the roots. There is evidence that nitrogen fixation is greater in planted soil than in unplanted soil [54, 108]. This section deals with nitrogen fixation by bacteria living on or in rice tissue (associative nitrogen fixation). Recent development on associative nitrogen fixation was reviewed by Patriquin [176].

Evidences of process: Sen [109] discussed the presence of heterotrophic nitrogen-fixing bacteria in rice roots, but the significance of his suggestion was overlooked. Döbereiner and Campelo [110] studied the presence of nitrogen-fixing bacteria in or on the roots of tropical graminaceous plants. Rinaudo and Dommergues [111] and Yoshida and Ancajas [108] found that some nitrogen-fixing (acetylene reduction) activity was associated with wetland rice roots.

Initial suggestions were based on ARA of the excised root [108]. Field assays of less disturbed soil cores with plants developed [63, 113] showed ARA values which were higher in the presence of plants [63].

Field daily ARA values in the Philippines [48, 114–115], Thailand [50], Senegal [116] and the West Indies [117] were in the range of 10–50 μmol C_2H_4 per plant except in acid soils [50, 118].

Incorporation of $^{15}N_2$ into wetland rice plants enclosed in a ^{15}N-enriched atmosphere was demonstrated by Ito et al. [57], Eskew et al. [59], and Yoshida and Yoneyama [58]. Ito et al. [57] transferred field-grown rice plants to water culture and measured ^{15}N uptake for seven days. Water culture avoids the contribution of nitrogen fixation in soil and dilution of $^{15}N_2$ by nitrogen gas in soil solution. The basal portions of shoots and roots were most enriched with ^{15}N. Transfer of the fixed ^{15}N to growing tissue was low during the seven-day exposure to $^{15}N_2$. Eskew et al. [59] exposed the rice plant at heading stage in soil for 13 days under $^{15}N_2$ gas. Immediately after the exposure, one tiller was analyzed for ^{15}N. Only a small amount was detected. The plant was grown to maturity under normal air. During this period, ^{15}N content in panicles and shoots increased. At maturity, 68 per cent of ^{15}N in the plant was recovered in the panicles, indicating that fixed nitrogen is available to rice. Compared with the results of the other authors, the results of Yoshida and Yoneyama [58] showed a much faster transfer of ^{15}N from soil and roots to shoots and panicles. The reason for this apparent discrepancy is unknown. In experiments where rice plants were grown in soil, 75–80 per cent of the fixed atmospheric nitrogen was recovered in the soil. This means that root zone soil is a more active site for nitrogen fixation in flooded-soil rice systems than the rice plant itself.

N_2 fixing bacteria associated with rice root: Various nitrogen-fixing bacteria have been isolated from the rhizosphere, roots, culms and leaf sheaths of rice. The colony forming units (CFU) of aerobic bacteria in the rhizosphere soil or roots were higher than those of anaerobic bacteria [119-120]. Rhizospheric bacteria include *Azotobacter* [121], *Azospirillum* [122-125], *Pseudomonas* [125-127], *Enterobacter* [125-128], *Klebsiella* [125] and *Alcaligenes* [128].

To determine predominant nitrogen-fixing bacteria of the rice root and rhizosphere, the rhizosphere soil (attached to the root), the washing from the root and the macerated root, and the macerated basal portion of shoots were diluted and inoculated to 0.1 per cent tryptic soy agar plates. The colonies that developed were isolated and assayed for nitrogen fixation activity on semisolid medium amended with yeast extract. About 80 per cent of the isolates from rice root gave positive N_2-fixation results [126]. The majority of N_2ase positive bacteria, identified as *Pseudomonas* [127], had uptake type hydrogenase activity [129], and were H_2-dependent chemolithotrophs (unpublished). CFU of *Azospirillum* was about ten or hundred times lower than CFU of *Pseudomonas* [130]. Among *Azospirillum*, *A. lipoferum* was found at higher frequency than *A. brasilense* [124]. Predominance of *Pseudomonas* over *Azospirillum* was also determined by fluorescent antibody staining of root (unpublished).

Thomas Bauzon et al. [125] used a "spermosphere model" to study the composition of the nitrogen-fixing microflora of a rice rhizosphere (rhizosphere soil plus root). Serial dilutions of a rhizosphere sample were inoculated to semisolid inorganic medium supplemented with yeast extract, on which the germinated rice seedling (spermosphere model) was grown in the dark. N_2-fixing bacteria were isolated from the highest dilution which gave nitrogenase activity associated with the spermosphere. This system produced N_2-fixing isolates with a 65 per cent frequency. Densities were 10^5 per gram of dry sample in the initial rice rhizosphere. The bacteria isolated in this system were *Klebsiella oxytoca*, *Enterobacter cloacae*, *Azospirillum* and *Pseudomonas paucimobilis.*

Factors affecting nitrogen-fixing activity and microflora

Rice growth stage: Many investigators have shown that N_2-fixing activity per plant reached its maximum at or near the heading stage [48, 108, 115, 131-132]. A similar trend was observed with ARA of the planted soil [47, 48] and with ARA per gram of dry weight of root [133].

Moisture condition: N_2-fixing activity (ARA) was much higher with wetland rice than with dryland rice [108, 130]. Densities of N_2-fixing bacteria (*Pseudomonas* and *Azospirillum*) per root weight were also higher in wetland fields than in dryland fields [130]. *Pseudomonas*, which was dominant in wetland rice roots, was also found in the roots of aquatic plant *Monochoria vaginalis*, but not in the roots of dryland crops and grasses [134]. The reasons for preference for wetland conditions are not known.

Light intensities: Some data show a higher apparent ARA during daytime than at night [119, 135-136]. A higher N_2-fixing rate during daytime possibly results from an increased exudation of photosynthetate in the rhizosphere. On the other hand, it is also possible that more active transport of gases (C_2H_2 C_2H) during the daytime results in an apparent higher ARA. ARA measurement in a system where gases were forced to circulate [137] did not show a clear-cut trend about diurnal variations. To demonstrate the former possibility, it is necessary to make measurements under conditions where the transport of C_2H_2 to the root and the recovery of C_2H_4 are not limiting.

Oxygen: Because the densities of aerobic bacteria in the rhizosphere were higher than those of anaerobic bacteria and because the aerobically or anaerobically isolated N_2-fixing bacteria were either microaerophilic or facultative anaerobic N_2-fixing bacteria [138], it is believed that the nitrogen-fixing process is microaerophilic. By measuring ARA of rice plant in water culture, Watanabe and Cabrera [137] found that ARA in pots aerated with gas with 20 per cent O_2 almost equalled that of untreated pots where oxygen dissolved in water was less than 0.1 ppm. Whereas denitrification was completely stopped by aeration, ARA was not inhibited by aeration (unpublished). These facts suggest that nitrogen fixation associated with rice roots is protected from oxygen damage. On the other hand, van Berkum and Sloger [136], measuring ARA of intact plants with roots that were directly exposed to the atmosphere, found that 0.25 per cent O_2 completely inhibited ARA. When oxygen transport from leaves was stopped by sealing the cut ends of shoots, 0.25 per cent O_2 in the atmosphere surrounding the roots was optimum for ARA of roots. This seems to indicate a microaerophilic nature of nitrogen-fixation of rice roots.

Considering that N_2-fixation associated with dryland rice is much lower than that associated with wetland rice, it is likely that too much aerobic condition harms N_2-fixation. Oxygen requirement and tolerance mechanisms for oxygen associated with nitrogen fixation in rice are still unsolved.

Combined nitrogen: Experiments with isolated roots [139] and with water-cultured rice [137] showed that nitrogen-fixation associated with rice root was sensitive to NH_4^-N. ARA was inhibited by 70 per cent at 0.33 mM NH_4^+ [137]. N_2-fixation associated with plants grown in the soil is less sensitive to the application of mineral nitrogen fertilizer. In fields where fertilizer nitrogen application did not exceed 100 kg N/ha, mineral nitrogen did not depress ARA associated with rice [49]. At the heading stage, when nitrogen fixation associated with rice root is most active, NH_4^+ content of the soil is as low as that of unfertilized soil [140]. This fact may explain the low sensitivity of associative nitrogen-fixation to mineral nitrogen fertilizer when the whole crop cycle is taken into account.

Nitrogen-fixation sites: Because repeated washing does not remove N_2-fixing microflora from roots [126], it is postulated that nitrogen-fixing flora is located firmly on the surface of or in rice roots. However, information on intracellular and intercellular distribution of diazotrophs in rice roots is insufficient. Huang et al. [141] inoculated Gram negative rods with peritricous flagella into gnotobiotic rice seedlings and observed the presence of bacteria inside the cortex cells using an electron microscope. The basal portion of roots has been reported to be more active than the distal portion [142]

The basal portions of shoots located in the floodwater are also nitrogen fixation sites [143]. The relative importance of the basal portion of shoots is higher at earlier stages of wetland rice growth.

Differences due to species and varieties: It is expected that more nitrogen would be obtained in flooded soil by breeding rice varieties that stimulate greater nitrogen-fixation. Aiming at these goals, differences in associative N_2-fixation between *Oryza* species and *O. sativa* varieties were sought. Many researchers reported this difference by ARA [115, 131, 132, 144-145]. However, because varietal differences change according to growth stage [115, 131], it is not clear if the observed differences reflected the nitrogen fixation rate during the entire growth period of rice. Techniques to detect differences in nitrogen fixation during the entire growth period are needed. It is not known if observed differences among varieties are genetical or physiological.

N_2-fixing microorganisms in anaerobic soil

Heterotrophic N_2-fixation occurs in flooded soil free from living rice roots as shown by $^{15}N_2$ incorporation or by ARA (Tables 3 and 4). Distribution of aerobic N_2-fixing bacteria in rice soil has been extensively studied [5]. CFU of these aerobic nitrogen-fixing bacteria may decrease in flooded conditions. However, some data did not show the decrease of CFU of aerobic N_2-fixing bacteria [146-147]. Dilution or plate counts sometimes give counts of aerobic nitrogen-fixing organisms as high as 10^5 or more [114, 147]. Little is known, however, about predominant aerobic (probably microaerophilic) nitrogen-fixing bacteria in paddy soils.

Recently, the presence of *Azospirillum* in flooded soil has received attention. Charyulu and Rao [148] found that CFU of *Azospirillum* in the flooded soil increased during rice straw decomposition. CFU levels per gram soil were about 10^6 except in acid saline soil ($< 10^1$). This result indicates that *Azospirillum* is one of the active nitrogen-fixing organisms in flooded rice soils.

In laboratory experiments on flooded soils incubated with straw, the increase of nitrogen-fixation in anaerobic conditions was accompanied by the multiplication of *Clostridium* [149].

Sulphate-reducing bacteria are abundant in flooded soils (more than 10^4 CFU/g soil). Nitrogen-fixation associated with straw increased with the addition of sulphate, 1-2 mg N being fixed per gram of sulphate reduced. Obviously sulphate-reducing bacteria play a role in nitrogen fixation in flooded rice soils, but contribute little to nitrogen gain [150].

Many reports show that heterotrophic nitrogen fixation is more active in flooded or anaerobic soils than in aerobic soils [151-154]. Yoneyama et al. [155] reported that lower Eh values (less than -200 mV) favour nitrogen fixation (acetylene reduction) in flooded soils amended with straw. Matsuguchi [46] and Panichsakpatana et al. [47] also reported that decrease of Eh up to -300 mV stimulated acetylene reduction in the reduced plough layer soil. However, some data indicated that more $^{15}N_2$ was fixed in nonflooded than in flooded soils [156-157]. Difference in N_2-fixation between both conditions were dependent on soils and organic matter. No explanation was given by this group of researchers.

In anaerobic plough layer, organic matter supply limits nitrogen-fixation. Reports by Rao [158] showed from 1 to 7 mg nitrogen fixed per gram straw added. From Charyulu and Rao [156] data, nitrogen fixed per gram of added rice straw (5 g/kg soil) in flooded soils ranged from 0 (acid saline soil) to 1.6 mg (acid sulphate soil). From nitrogen balance data, nitrogen gain stimulated by straw application was about 4 mg N/g straw in planted soil (IRRI 1979 Annual Report). Previous unpublished IRRI data indicated that straw incorporation into flood fallow pots did not have a statistically significant effect on nitrogen balance. Recent results (IRRI 1981 Annual Report) also indicated that nitrogen balance in the presence of rice straw was stimulated by rice plants. Charyulu, Nayak, and Rao [159] found that $^{15}N_2$ fixation stimulated by the addition of cellulose was higher in soil from the planted fields than from fallow fields. This is consistent with the hypothesis that nitrogen fixation, which depends on the externally added organic matter, is higher in the root zone than in fallow soil.

INCREASE IN RICE YIELDS BY THE USE OF NITROGEN-FIXING ORGANISMS

From agricultural point of view, the final goal is to increase nitrogen fixation in flooded soil and increase rice yields by replacing or supplementing expensive chemical fertilizer nitrogen. Possible means of increasing nitrogen fixation in flooded soils include addition of materials that stimulate nitrogen-fixation such as organic amendments and phosphorus application, inoculation of nitrogen-fixing organisms, and selection of rice cultivars that stimulate greater nitrogen fixation. Inoculation of nitrogen-fixing organisms has been tried in combination with other treatments like phosphorus application. Here, the effects of inoculation with BGA, *Azolla*, and N_2-fixing bacteria are described and the reasons for the often reported yield increase in rice are discussed.

Effect of algal inoculation

Reviewing literature on BGA and rice, Roger and Kulasooriya [45] concluded the following effects of algal inoculation. Algalization may effect plant size, nitrogen content, and the number of tillers, ears, spikelets, and filled grains per panicle. Grain yield has been the most frequently used criterion for assessing the effects of algalization. Results of field experiments report an average yield increase of about 14 per cent over the control corresponding to about 450 kg grain per hectare per crop where algal inoculation was effective. Results of pot experiments report an average yield increase of about 42 per cent. Because of better BGA growth in pots than *in situ*, pot experiments may be suitable only for quantitative studies.

From field experiments where algalization was done with and without mineral fertilizers (NPK) it appears that yield increase by mineral fertilizers is always higher than that strictly due to algalization. Average yield increase in the presence of nitrogen fertilizers (14.6 per cent) does not significantly differ from that in the absence of nitrogen fertilizers (14.3 per cent). Because biological N_2-fixation is known to be inhibited by inorganic nitrogen, the beneficial effect of algalization in the presence of nitrogen fertilizers was most frequently interpreted as resulting from growth-promoting substances produced by BGA. Such a hypothesis needs to be proven because algalization experiments have been conducted on a "black box" basis, where only the last indirect effect (yield) of an agronomic practice (algalization) was observed. No data on nitrogen fixation and algal biomass measurements in an inoculated paddy field are available. Therefore, the relative importance of nitrogen fixation by inoculated BGA in increasing rice yield, compared with other possible effects like auxinic effects, effects on soil properties, increase of phosphorus availability, etc. is still unknown.

If yield increase is due to nitrogen fixation, part of fixed nitrogen must be utilized by the rice plant. A recent experiment, described below, has quantified the utilization of algal nitrogen by rice.

Availability of algal nitrogen to rice

Uptake by rice of nitrogen fixed by BGA was demonstrated on a qualitative basis by Renaut et al. [160] and Venkataraman [161], using ^{15}N tracer technique. In a quantitative experiment Wilson et al. [162] recovered 37 per cent of the nitrogen from a rice crop from ^{15}N labelled *Aulosira* sp. spread on the soil and 51 per cent of the nitrogen from the same material incorporated into the soil. The study was conducted on a laboratory scale and did not include analysis of ^{15}N remaining in the soil.

Pot and field experiments at IRRI, using ^{15}N labelled *Nostoc* sp., showed that availability of nitrogen from dried BGA incorporated in the soil was between 23 and 28 per cent for the first crop and between 27 and 35 per cent for two crops. Surface application of the algal material reduced the availability to 14-23 per cent for the first crop and 21-27 per cent for two crops [163]. Availability of nitrogen from fresh algal material was similar to that of dried material when surface applied (14 per cent) but much higher (38 per cent) when incorporated [164]. The pot experiment demonstrated that algal nitrogen was less available than ammonium sulphate for the first crop, but for two crops algal nitrogen availability was very similar to that of ammonium sulphate [165]. That indicates a slow-release nature of BGA nitrogen, which reflects the cumulative effects of algal inoculation [45]. The ^{15}N balance in plants and soil after two crops (pot experiment, dried algae) showed that losses from ^{15}N ammonium sulphate were more than twice that from BGA regardless of the mode of application. From these results it was concluded that the BGA material, because of its organic nature, is less susceptible to nitrogen losses than inorganic fertilizer and that its low C/N ratio (5-7) gives it a better nitrogen availability than an organic fertilizer like green manure. The relative availability of algal nitrogen to rice depends on susceptibility to decomposition of the algal material which varies not only with the strains [165], but also with the physiological state as demonstrated by the discrepancy between the values reported by Wilson et al. [162] and those reported by Tirol et al. [163]. Wilson et al. used an algal material collected directly from the flask culture and blended after resuspension in distilled water, whereas Tirol et al. used an algal material dried at room temperature, comprising mainly vegetative cells in dormancy and akinetes, and therefore less susceptible to decomposition.

Effect of *Azolla* inoculation

The effect of *Azolla* on rice yield is principally determined by the biomass and its nitrogen content. Table 5 shows the reported values of maximum biomass and its nitrogen content in the field. However, nitrogen gain in *Azolla* biomass may not be exclusively from atmospheric nitrogen. Soil nitrogen may contribute to a nonnegligible extent. No experimental data on the contribution of soil nitrogen to the nitrogen nutrition of *Azolla* in the field are available. Under favourable experimental conditions,

Table 5. Azolla biomass, its nitrogen and daily N-accumulating rate

Species	Conditions and site	Maximum biomass			Average daily N accumulation	Source
		Dry matter (t/ha)	N content (kg/ha)	Days		
A. filiculoides	Fallow paddy, USA	1.7	52	35	1.5	Talley and Rains [166]
	Shallow pond, USA	1.8	105	—	—	Talley et al. [96]
	Fallow paddy, China	2.2	75	20	3.8	Li et al. [88]
	Rice canopy, China	1.7	66	16	4.1	"
A. mexicana	Pond, USA	0.8	39	39	1.0	Talley et al. [96]
	Fallow paddy, USA	1.1	38	—	—	Talley and Rains [166]
A. caroliniana	Fallow paddy, Philippines	—	40	26	1.5	Watanabe et al. [89]
	Fallow paddy, China	1.8	73	20	3.6	Li et al. [88]
A. pinnata	Fallow, paddy, India	2.3	104	30	3.5	Singh [84]
	Fallow paddy, Philippines	1.1	48	30	1.6	Watanabe et al. [89]
	Rice canopy, Philippines (4 crops)	—	73	91	0.8	Watanabe et al. [89]
	Rice canopy, Vietnam	—	25	60	0.4	Than and Thuyet [83]
	Rice paddy, China	1.8	67	20	3.4	Li et al. [88]
	Rice canopy, China	0.81	36	16	2.2	"

Table 6. Effect of *Azolla* on rice yields in Asia

	Yield (t/ha)				
	Control[a]	N fertilized[b]	*Azolla* plot[c]	Increase due to *Azolla*	Source
China, Zhejiang, 1964	3.7	ND[d]	4.4 w/o I, BT	0.7	Liu [82]
			4.4 I, 1 time, BT	0.7	
			4.9 I, 2 times, BT	1.2	
			5.0 I, 3 times, BT	1.3	
China, Zhejiang	4.7	5.6 (60)	5.2 I, AT	0.5	Li et al.[e] [88]
			5.7 I, BT	1.0	
			5.0 I, BT, AT	1.3	
Vietnam, 1958–67	2.4	ND	2.8 I, BT	0.4	Dao and Do [167]
India, 1976 (Kharif)	4.2	5.5 (40)	4.9 I, BT	0.7	Singh [84]
1977 (Rabi)	1.7	3.2 (40)	2.6 I, BT	0.9	"
Thailand, 1977	2.6	2.9 (37.5)	3.5 I, BT	0.9	Sawatdee et al. [168]
IRRI, 1979–80	4.2	5.2 (77)	5.4 I, BT, AT	1.2	Watanabe et al. [89]
4 Asian countries	2.9	3.5 (30)	3.6 w/o I, AT	0.7	Kikuchi et al.[f] [86]
		4.0 (60)	3.6 I, BT	0.7	
			3.5 I, BT	0.6	
			4.0 I, BT, AT	1.1	

[a] No *Azolla* and no chemical N.
[b] Figures in parentheses are levels of chemical N applied (kg N/ha).
[c] I = incorporated; w/o I = without incorporation; BT = before transplanting; AT = after transplanting.
[d] ND = Data not available.
[e] Means of 4 species of *Azolla*.
[f] Thailand, India, China and Nepal of all sites for 2 years.

Azolla can accumulate 40–120 kg/ha nitrogen within 30 days (Table 5). Gains can be enhanced by growing a second or a third *Azolla* crop after the first crop is incorporated or collected. Annual nitrogen production by *Azolla* thus cultivated throughout the year reaches 450 kg N/ha in the Philippines [102] and 800 kg in India [84] by *A. pinnata*, and 1.2 ton in China by *A. filiculoides* [88, 169].

Azolla is grown either before or atter rice transplanting (dual culture) or both. Practices in *Azolla* use have been reviewed by Watanabe [67]. Table 6 summarizes yield responses to *Azolla*. Results at many sites of the International Network Soil Fertility and Fertilizer Evaluation for Rice (INSFFER), which was participated in by many rice growing countries, showed that one *Azolla* crop and its incorporation increased rice yield to the same extent as application of 30 kg N/ha of urea. Growing four or more *Azolla* crops within the wide row spaced rice canopy and incorporating them in soil gave the same yield as 70–100 kg N/ha of chemical nitrogen fertilizer [89]. The increased yield effect of *Azolla* is primarily due to its nitrogen.

Availability of *Azolla* nitrogen to rice

Azolla's nitrogen becomes available to rice after decomposition. Dried *Azolla* releases 47 per cent of its nitrogen as ammonium for four weeks, while fresh *Azolla* releases 60 per cent for the same period. Availability of nitrogen from dried *Azolla* to one rice crop is about 60 per cent of that of ammonium sulphate [177]. The decomposition rate of *Azolla* depends on its nitrogen content and the presence of other components. The lower the nitrogen content, the less available is its nitrogen [89]. Shi et al. [170] compared nitrogen availability of milk vetch, water hyacinth, and *Azolla* labelled with ^{15}N to wetland rice and the following crop of buckwheat. Although *Azolla* had the highest nitrogen content among the green manures, the availability of nitrogen from *Azolla* to wetland rice was the lowest. This was attributed to a high content in lignin, which is about 30 per cent or more on a dry matter basis [87]. Watanabe et al. [89] studied ^{15}N uptake by rice from *Azolla* in pot and field experiments. In pots, 50 per cent of nitrogen from incorporated *Azolla* was absorbed by the plant, whereas 10 per cent was absorbed when *Azolla* floated on the water. In the field, 28–26 per cent of *Azolla* nitrogen was absorbed by rice when *Azolla* was incorporated 30 to 53 days after transplanting. But when *Azolla* was kept on the surface only 15 per cent was absorbed by the plant. Thus, incorporation of *Azolla* showed higher efficiency, as found for inorganic nitrogen fertilizer.

Inoculation of *Azospirillum*

The positive effects of this bacterium to the yields of dryland crops and grasses were summarized by Boddey and Döbereiner [171]. In India, Subba Rao [172] reported the cases of *Azospirillum* inoculation to rice grown in the fields at five sites. This bacterium was grown in farmyard-soil mixture.

Rice seedlings were dipped in the slurry of the inoculant before planting. In some cases, yield increase by inoculation was observed at no nitrogen-fertilizer or at 40 kg N/ha. It is not known if the yield increase was due to nitrogen fixation or other factors.

CONCLUSION

The flood-soil-rice ecosystem is favourable for biological nitrogen fixation. Nitrogen fixation by photodependent microorganisms in the photic zone and by heterotrophic microorganisms in the reduced layer and in association with wetland rice has been studied extensively. Despite the enormous volume of these studies, the nitrogen-fixing rate in flooded rice soils has not been satisfactorily quantified. Inoculation with blue-green algae, *Azolla*, and some nitrogen-fixing bacteria has been made and positive responses of rice yield have been reported. However, the mechanism of increasing yield, particularly with blue-green algae, is still poorly understood. Despite great agricultural use of *Azolla*, the botany, physiological, and improvement of agronomical properties of *Azolla-Anabaena* symbiosis are much less developed than those of legume-rhizobia symbiosis. Considering the fact that almost half of the world population uses rice as a major source of calories, research on nitrogen fixation in flooded-soil rice systems should be further strengthened, particularly through cooperation among scientists in developed and developing countries.

REFERENCES

1. De Datta, S.K. *Principles and Practices of Rice Production*, John Wiley and Sons, New York (1981).
2. Mitsui, S. *Inorganic Nutrition, Fertilisation and Soil Amelioration for Lowland Rice*, Yokendo, Tokyo (1954).
3. Matsuo, H., Hayase, T., Yokoi, H. and Onikura, Y. Results of long-term fertilizer experiments on paddy rice in Japan, *Annals of Agronomy 27*, 957-968 (1976).
4. Watanabe, I. and Brotonegoro, S. Paddy fields, pp. 241-263, in *Nitrogen Fixation*, Vol. I, *Ecology* (Editor, W.J. Broughton), Clarendon Press, Oxford (1981).
5. Watanabe, I. Biological nitrogen fixation in rice soils, pp. 465-478, in *Soils and Rice*, The International Rice Research Institute, Los Baños, Philippines (1978).
6. Buresh, R.J., Casselman, M.E. and Patrick, W.H. Jr. Nitrogen fixation in flooded soil systems: A review, *Advances in Agronomy 33*, 149-192 (1980).
7. Ponnamperuma, F.N. Chemistry of submerged soils, *Advances in Agronomy 24*, 29-96 (1972).
8. Yoshida, T. Microbial metabolism of flooded soils, pp. 83-122, in

Soil Biochemistry, Vol. 3 (Editors, E.A. Paul and A.D. MacLaren), Marcel Dekker Press, New York (1975).

9. Watanabe, I. and Furusaka, C. Microbial Ecology of flooded rice soils, pp. 125–168, in *Advances in Microbial Ecology*, Vol. 4, (Editor, M. Alexander), Plenum Publishing Co., New York, U.S.A. (1980).
10. Saito, M. and Watanabe, I. Organic matter production in rice field flood-water, *Soil Science and Plant Nutrition 24*, 427–444 (1978).
11. Roger, P.A. and Watanabe, I. Algae and aquatic weeds as a source of organic matter and plant nutrient for wetland rice, in *Organic Matter and Rice*, International Rice Research Institute, Los Baños, Philippines (1983).
12. Howeler, R.H. and Bouldin, D.R. The diffusion and consumption of oxygen in submerged soils, *Soil Science Society of America Proceedings 35*, 202–208 (1971).
13. Pearsall, W.H. and Mortimer, C.H. Oxidation-reduction potentials in waterlogged soils, natural water and muds, *Journal of Ecology 27*, 485–501 (1939).
14. Ishizawa, S. and Toyota, H. Studies on the microflora of Japanese soil, *Bulletin National Institute of Agricultural Sciences* (Japan) *14B*, 204–284 (1960) (in Japanese, English summary).
15. Hayashi, S., Asatsuma, K., Nagatsuka, T. and Furusaka, C. Studies on bacteria in paddy soil, *Report Institute of Agricultural Research* (Tohoku University) *29*, 19–38 (1978).
16. Harrison, W.H. and Aiyer, P.A.S. The gases of swamp rice soils part II. Their utilization for the aeration of the roots of the crop, *Memoir Department of Agriculture of India, Chemistry Series 4*(4), 1–7 (1915).
17. Takai, Y. and Kamura, T. The mechanism of reduction in waterlogged paddy soil, *Folia Microbiologia 11*, 304–313 (1966).
18. Takijima, Y. Studies on behaviour of the growth inhibiting substances in paddy soils with special reference to the occurrence of root damage, *Bulletin, National Institute of Agricultural Sciences* (Japan) *13B*, 117–252 (1963) (in Japanese, English summary).
19. Wada, H. and Kanazawa, S. Method of fractionation of soil organic matter according to its size and density, *Journal Science Soil and Manure* (Japan) *42*, 12–17 (1970) (in Japanese)
20. Kikuchi, E., Furusaka, C. and Kurihara, Y. Effect of tubificids on the nature of a submerged soil ecosystem, *Japanese Journal of Ecology 27*, 163–170 (1917).
21. Dommergues, Y. Microbial activity in different types of microenvironments in paddy soils, pp. 451–466, in *Environmental Biogeochemistry and Geomicrobiology*, Vol. 2 (Editor, W.E. Krumsbein), Ann Arbor Publisher, Michigan (1978).
22. Old, K.M. and Nicolson, T.H. Electron microscopical studies of the microflora of roots of sand dune grasses, *New Phytologist 74*, 51–58 (1975).

23. Martin, J.K. Factors influencing the loss of organic carbon from wheat roots, *Soil Biology Biochemistry 9*, 1-7 (1977).
24. Barber, D.A. and Lynch, J.M. Microbial growth in the rhizosphere, *Soil Biology Biochemistry 9*, 305-308 (1977).
25. Miyashita, K., Wada, H. and Takai, Y. Decomposition process of the intact roots of rice plant. II. Invasion of soil microorganisms, *Journal Science and Manure* (Japan) *48*, 558-564 (1977) (in Japanese).
26. Van Raalte, M.H. On the oxygen supply of rice roots, *Annalus Botanic Garden Buitenzorg 50*, 99-114 (1941).
27. Kawata, S., Ishihara, K. and Shioya, T. Studies on the root hairs of lowland rice plants in the upland field, *Proceedings Japanese Crop Science Society 32*, 250-253 (1964) (in Japanese, English summary).
28. Okajima, H. On the relationship between the nitrogen deficiency of the rice plant roots and the reduction of the medium, *Journal Science of Soil and Manure* (Japan) *29*, 175-180 (1958) (in Japanese).
29. Trolldenier, G. Secondary effect of potassium and nitrogen nutrition of rice. Change in microbial activities and iron reduction in rhizosphere, *Plant and Soil 38*, 267-279 (1973).
30. Watanabe, I., Craswell, E.T. and App, A. Nitrogen cycling in wetland rice fields in Southeast and East Asia, pp. 4-17, in *Nitrogen Cycling in South-East Asian Wet Monsoonal Ecosystems* (Editors, R. Wetselaar, J.R. Simpson and T. Roswell), Australian Academy of Science, Canberra (1981).
31. Sekiya, S. and Shiga, H. A role of subsoil of paddy field in nitrogen supply to rice plants, *Japanese Agriculture Research Quarterly 11*, 95-100 (1977).
32. Yanagisawa, M. and Takahashi, J. Studies on the factors related to the productivity of paddy soil in Japan with special reference to the nutrition of rice plants, *Bulletin National Institute of Agricultural Sciences* (Japan) *14B*, 41-102 (1964) (in Japanese, English summary).
33. Willis, W.H. and Green, V.R. Movement of nitrogen in flooded soil planted to rice, *Soil Science Society of America Proceedings 13*, 229-237 (1948).
34. App, A., Watanabe, I., Alexander, M., Ventura, W., Daez, C., Santiago, T. and De Datta, S.K. Non-symbiotic nitrogen fixation associated with rice plant in flooded soil, *Soil Science 130*, 283-289 (1980).
35. Ventura, W. and Watanabe, I. ^{15}N dilution technique of assessing the contribution of nitrogen fixation to rice plant, *Soil Science and Plant Nutrition 29*, (1983).
36. De, P.K. The problem of the nitrogen supply of rice. I. Fixation of nitrogen in the rice soils under waterlogged conditions, *Indian Journal of Agricultural Science 6*, 1237-1245 (1936).

37. Matsuguchi, T., Shimomura, T. and Lee, S.K. Factors regulating acetylene reduction assay for measuring heterotrophic nitrogen fixation in waterlogged soils, *Soil Science and Plant Nutrition 25*, 323–336 (1979).
38. Lee, K.K. and Watanabe, I. Problems of acetylene reduction technique applied to water saturated paddy soils, *Applied Environmental Microbiology 34*, 654–660 (1977).
39. de Bont, J.A.M. and Mulder, E.G. Inability of acetylene reduction assay in alkane-utilizing nitrogen-fixing bacteria, *Applied Environmental Microbiology 31*, 640–647 (1976).
40. de Bont, J.A.M., Lee, K.K. and Bouldin, D.R. Bacterial oxidation of methane in a rice paddy, pp. 91–96, in *Environmental role of nitrogen-fixing blue green algae and asymbiotic bacteria, Ecology Bulletin 26* (Editor, U. Granhall), Stockholm (1978).
41. Watanabe, I. and de Guzman, M.R. Effect of nitrate on acetylene disappearance from anaerobic soil, *Soil Biology and Biochemistry 12*, 193–194 (1980).
42. Roger, P.A., Reynaud, P.A., Rinaudo, G.E., Ducerf, P.E. and Traore, T.M. Log normal distribution of acetylene reducing activity *in situ*, *Cahiers ORSTOM Serie Biologie 12*, 133–140 (1977) (in French, English summary).
43. Roger, P.A., Reynaud, P.A. Algae enumeration in waterlogged soils: Distributional ecology of microorganisms and density of sampling, *Revue Ecologie Biologie du Sol 15*(2), 229–234 (1978) (in French, English summary).
44. Knowles, R. The measurement of nitrogen fixation, pp. 327–333, in *Current Perspective in Nitrogen Fixation* (Editors, A.H. Gibson and W.E. Newton), Australian Academy of Science, Canberra (1981).
45. Roger, P.A. and Kulasooriya, S.A. *Blue-green Algae and Rice*, The International Rice Research Institute, Los Baños, Philippines (1980).
46. Matsuguchi, T. Factors affecting heterotrophic nitrogen fixation in submerged soils, pp. 207–222, in *Nitrogen and Rice*, The International Rice Research Institute, Los Baños, Laguna, Philippines (1979).
47. Panichsapatana, S., Wada, H., Kimura, M. and Takai, Y. Nitrogen fixation in paddy soils. III. N_2 fixation and its active sites in soil and rhizosphere soil, *Soil Science and Plant Nutrition 25*, 165–171 (1979).
48. Watanabe, I., Lee, K.K. and de Guzman, M.R. Seasonal change of N_2 fixation in lowland rice field assayed by *in situ* acetylene reduction technique. II. Estimate of nitrogen fixation associated with rice plants, *Soil Science and Plant Nutrition 24*, 465–471 (1978).
49. Watanabe, I., de Guzman, M.R. and Cabrera, D. The effect of nitrogen fertilizer on N_2 fixation in paddy field measured by *in situ* acetylene reduction assay, *Plant and Soil 59*, 135–139 (1981).

50. Cholitkul, W., Tangcham, B., Santong, P. and Watanabe, I. Effect of phosphorus on N_2-fixation measured by field acetylene reduction technique in Thailand long term fertility plots, *Soil Science and Plant Nutrition 26*, 291-299 (1980).
51. Brotonegoro, S., Abdulkadir, S., Harmastini-Sukiman and Parthohardjono, J. Nitrogen cycling in lowland rice fields with special attention to N_2 fixation, pp. 36-40, in *Nitrogen Cycling in Southeast Asian Wet Monsoonal Ecosystem* (Editors, R. Wetselaar, J.R. Simpson and T. Roswall), Australian Academy of Science, Canberra (1981).
52. Broughton, W.J., Hong, T.K., Rajoo, M.J. and Ratnasabapathy, M. Nitrogen fixation in some Malaysian paddy fields, pp. 558-584, in *Soil Microbiology and Plant Nutrition* (Editors, W.J. Broughton, C.K. John, J.C. Rajaro and Beda Lim), University Malaya Publisher, Kuala Lumpur (1979).
53. Rajaramamohan, Rao V. Nitrogen fixation as influenced by moisture content, ammonium sulphate and organic sources in a paddy soil, *Soil Biology Biochemistry 8*, 445-449 (1976).
54. Kalininskaya, T.A., Miller, Yu.M., Belov, Yu.M. and Rao, V.R. ^{15}N nitrogen studies of the activity of non-symbiotic nitrogen fixation in rice-field soils of the Krasnodar Territory, *Vestnik Academii Nauk, SSSR, Seria Biologiya* (4), 565-570 (1977) (in Russian, English summary).
55. MacRae, I.C. and Castro, R.F. Nitrogen fixation in some tropical soils, *Soil Science 103*, 277-280 (1967).
56. Reddy, K.R. and Patrick, W.H.Jr. Nitrogen fixation in flooded soil, *Soil Science 128*, 80-85 (1979).
57. Ito, O., Cabrera, D. and Watanabe, I. Fixation of dinitrogen-15 associated with rice plant, *Applied Environmental Microbiology 39*, 554-558 (1980).
58. Yoshida, T. and Yoneyama, T. Atmospheric dinitrogen fixation in the flooded rice rhizosphere as determined by the N-15 isotope technique, *Soil Science and Plant Nutrition 26*, 551-559 (1980).
59. Eskew, D.L., Eaglesham, A.R.J. and App, A.A. Heterotrophic $^{15}N_2$ fixation and distribution of newly fixed nitrogen in a rice flooded soil system, *Plant Physiology 68*, 48-52 (1981).
60. Watanabe, I. and Ventura, W. Nitrogen fixation by blue green algae associated with deep water rice, *Current Science 51*, 462-465 (1982).
61. Rennie, R.J., Rennie, D.A. and Fried, M. Concept of ^{15}N usage in dinitrogen fixation studies, pp. 131-133, in *Isotopes in Biological Dinitrogen Fixation*, International Atomic Energy Agency, Vienna (1978).
62. Wada, H., Panichsakpatana, S., Kimura, M. and Takai, Y. Nitrogen fixation in paddy soils. I. Factors affecting N_2 fixation, *Soil Science and Plant Nutrition 24*, 357-365 (1978).

63. Lee, K.K., Alimagno, B.V. and Yoshida, T. Field technique using the acetylene reduction method to assay nitrogenase activity and its association with rice rhizosphere, *Plant and Soil 47*, 519–526 (1977).
64. Lumpkin, T.A. and Plucknett, D.L. Azolla: Botany, physiology and use as a green manure, *Economic Botany 34*, 111–153 (1980).
65. Lumpkin, T.A. and Plucknett, D.L. *Azolla as a Green Manure: Use and Management in Crop Production*, Westview Tropical Agriculture, Series 5, Westview Press, Boulder, Colorado. USA (1982).
66. Peters, G.A. and Calvert, H.E. The Azolla-Anabaena Symbiosis, pp. 191–218, in *Advances in Agricultural Microbiology* (Editor, N.S. Subba Rao), Oxford & IBH Publishing Co., New Delhi (1982).
67. Watanabe, I. Azolla-Anabaena symbiosis—Its physiology and use in tropical agriculture, pp. 169–185, in *Microbiology of Tropical Soils and Productivity* (Editors, Y. Dommergues and H. Diem), Martinus Niejhoff, Wageningen (1982).
68. Roger, P.A. and Reynaud, P.A. Free-living blue-green algae in tropical soils, pp. 147–168, in *Microbiology of Tropical Soils and Productivity* (Editors, Y. Dommergues and H. Diem), Martinus Niejhoff, Wageningen (1982).
69. Roger, P.A. and Reynaud, P.A. Dynamics of algal population during culture cycle in the rice field in Sahalien, *Revue Ecologie Biologie du Sol 13*, 545–560 (1976) (in French, English summary).
70. Ichimura, S. Ecological studies on the plankton in paddy fields. I. Seasonal fluctuation in the standing crop and productivity of plant, *Japanese Journal of Botany 14*, 269–279 (1954).
71. Aiyer, R.S. Comparative algological studies in rice fields in Kerala State, *Agricultural Research Journal of Kerala 3*(1), 100–104 (1965).
72. Durrel, L.W. Algae in tropical soils, *Transaction of American Microscopic Society 83*, 79–85 (1964).
73. Okuda, A. and Yamaguchi, M. Nitrogen fixing microorganisms in paddy soils (Part 2). Distribution of blue green algae in paddy soil and relationship between the growth of them and soil properties, *Soil Science and Plant Nutrition 2*, 4–7 (1956).
74. Venkataraman, G.S. The role of blue green algae in agriculture, *Science and Culture 27*, 9–13 (1961).
75. Grant, I.F. and Alexander, M. Grazing of blue green algae (Cyanobacteria) in flooded soils by *Cypris* sp. (Ostracoda), *Soil Science Society of America Journal 45*, 773–777 (1981).
76. Grant, F.I., Tirol, A.C., Aziz, T. and Watanabe, I. Regulation of invertebrate grazers as a means to enhance biomass and nitrogen fixation of Cyanophyceae in wetland rice fields, *Soil Science Society of America Journal* (in press).
77. Matsugicji, T. and Yoo, I.D. Stimulation of phototrophic N_2 fixation in paddy field through rice straw application, pp. 18–25, in *Nitrogen*

Cycling in Southeast Asian Wet Monsoonal Ecosystem (Editors, R. Wetselaar, J.R. Simpson and T. Rosswall), Australian Academy of Science, Canberra (1981).

78. Roger, P.A., Tirol, A. and Watanabe, I. Effect of surface application of straw on phototrophic nitrogen fixation, *International Rice Research Newsletter* 7, 16-17 (1982).
79. Roger, P.A., Kulasooriya, S.A., Barraquio, W.L. and Watanabe, I. Epiphytic nitrogen fixation on wetland rice, pp. 62-66, in *Nitrogen Cycling in Southeast Asian Wet Monsoonal Ecosystem* (Editors, R. Wetselaar, J.R. Simpson and T. Roswall), Australian Academy of Science, Canberra (1981).
80. Kulasooriya, S.A. Roger, P.A. Barraquio, W.L. and, Watanabe, I. Epiphytic nitrogen fixation on weeds in a rice field ecosystem, pp. 55-61, in *Nitrogen Cycling in Southeast Asian Wet Monsoonal Ecosystem* (Editor, R. Wetselaar et al.), Australian Academy of Science, Canberra (1981).
81. Watanabe, I., Ventura, W., Cholitkul, W., Roger, P.A. and Kulasooriya, S.A. Potential of biological nitrogen fixation in deep water rice, pp. 191-200, in *Proceedings of the 1981 International Deepwater Rice Workshop*, International Rice Research Institute, Los Baños, Philippines (1982).
82. Liu, C. Use of azolla in rice production in China, pp. 375-394, in *Nitrogen and Rice*, International Rice Research Institute, Los Baños, Philippines (1979).
83. Tuan, D.T. and Thuyet, T.R. The use of azolla in rice production in Vietnam, pp. 395-504, in *Nitrogen and Rice*, International Rice Research Institute, Los Baños, Philippines (1979).
84. Singh, P.K. Use of azolla in rice production in China, pp. 407-418, in *Nitrogen and Rice*, International Rice Research Institute, Los Baños, Philippines (1979).
85. Rains, D.W. and Talley, S.N. Use of azolla in North America, pp. 419-433, in *Nitrogen and Rice*, International Rice Research Institute, Los Baños, Philippines (1979).
86. Kikuchi, M., Watanabe, I. and Haws, L.D. Economic evaluation of Azolla use in rice production, in *Organic Matter and Rice*, International Rice Research Institute, Los Baños, Philippines (1983).
87. Li, Z., Zu, S., Mao, M. and Lumpkin, T.A. Study on the utilization of eight Azolla species in agriculture I, *Zhongguo Nongye Kexue* (*Chinese Agricultural Science*), 19-27 (1982) (in Chinese, English summary).
88. Li, Z., Zu, S., Mao, M. and Lumpkin, T.A. Studies on the utilization of eight Azolla species in agriculture II, *Zhongguo Nongye Kexue* (*Chinese Agricultural Science*), 72-78 (1982) (in Chinese, English summary).

89. Watanabe, I., Bai K., Berja, N.S., Espinas, C.R., Ito, O. and Subudhi, B.P.R. The *Azolla-Anabaena* complex and its use in rice culture, *International Rice Research Paper Series*, 69 (1981).
90. Moore, A.W. Azolla: Biology and agronomic significance, *Botanical Review 35*, 17-34 (1969).
91. Becking, J.H. Environmental requirements of Azolla for use in tropical rice production, pp. 345-373, in *Nitrogen and Rice*, International Rice Research Institute, Los Baños, Philippines (1979).
92. Peters, G.A., Toia, R.E. Jr., Evans, W.R., Christ, D.K., Mayne, B.C. and Poole, R.E. Characterization and comparisons of five N_2-fixing Azolla-Anabaena associations, *Plant Cell Environment 3*, 261-269 (1980).
93. Watanabe, I. and Berja, N.S. The growth of four species of *Azolla* as affected by temperature, *Aquatic Botany 15*, (1982).
94. Ashton, P.J. Effect of some environmental factors of growth of *Azolla filiculoides* LAM, pp. 124-138, in *Orange River Progress Report* (Institute for Environmental Science), Univ. O.F.S. Bloemfontein, South Africa (1974).
95. Talley, S.N. and Rains, D.W. *Azolla filiculoides* LAM as a fallow season green manure for rice in a temperate climate, *Agronomy Journal 72*, 11-18 (1980).
96. Talley, S.N., Talley, B.J. and Rains, D.W. Nitrogen fixation by Azolla in rice fields, pp. 259-281, in *Genetic Engineering in Nitrogen Fixation* (Editor, A. Hollander), Plenum Publishing Company, New York (1977).
97. Peters, G.A. and Mayne, B.C. The *Azolla-Anabaena Azollae* relationship. II. Localization of nitrogenase activity as assayed by acetylene reduction, *Plant Physiology 53*, 820-824 (1974).
98. Watanabe, I. Azolla and its use in lowland rice culture, *Tsuchi to Biseibutsu* (*Soil Microbes*) (Tokyo) *20*, 1-10 (1978).
99. Roger, P.A. and Reynaud, P.A. Preliminary data on the ecology of *Azolla Africana* in Sahelian zone, *Ecologia Plantarum 14*, 75-84 (1979) (in French, English summary).
100. Singh, P.K. Symbiotic algal N_2-fixation and crop productivity, pp. 37-65, in *Annual Review of Plant Science*, Vol. 1, Kalyani Publisher, New Delhi (1979).
101. Subudhi, B.P.R. and Watanabe, I. Differential phosphorus requirement of azolla species and strains in phosphorus limited continuous culture, *Soil Science and Plant Nutrition 27*, 237-247 (1981).
102. Watanabe, I., Berja, N.S., del Rosario, Diana C. Growth of Azolla in paddy field as affected by phosphorus fertilizer, *Soil Science and Plant Nutrition 26*, 301-307 (1980).
103. Salmal, N. and Kulshreshtha, J.P. Biology and control of the pest of *Azolla-Anabaena*, a nitrogen fixing fern, *Oryza 15*, 204-207 (1978).

104. Takara, J. Insect pests on *Azolla pinnata* at Bangkhen, Thailand, *International Rice Research Newsletter 6*, 12-13 (1981).
105. Shahjahan, A.K.M., Miah, S.A., Nahar, M.A. and Majid, M.A. Fungi attack azolla in Bangladesh, *International Rice Research Newsletter 5*, 17-18 (1980).
106. Kobayashi, M., Takahashi, E. and Kawaguchi, K. Distribution of nitrogen-fixing microorganisms in paddy soils of Southeast Asia, *Soil Science 104*, 113-118 (1967).
107. Habte, M. and Alexander, M. Nitrogen fixation by photosynthetic bacteria in lowland rice culture, *Applied Environmental Microbiology 39*, 342-347 (1980).
108. Yoshida, T. and Ancajas, R.R. Nitrogen-fixing activity in upland and flooded rice fields, *Soil Science Society America Proceedings 37*, 42-46 (1973).
109. Sen, M.A. Is bacterial association a factor in nitrogen assimilation by rice plants? *Agricultural Journal India 24*, 229 (1929).
110. Döbereiner, J. and Campelo, A.B. Non-symbiotic nitrogen fixing bacteria in tropical soils, *Plant and Soil*, special volume, 457-470 (1971).
111. Rinaudo, G. and Dommergues, Y. The validity of acetylene reduction method for the determination of nitrogen fixation in rice rhizosphere, *Annals Institute Pasteur 121*, 93-99 (1971) (in French, English summary).
112. Yoshida, T. and Ancajas, R. Nitrogen fixation by bacteria in the root zone of rice, *Soil Science Society of America Proceedings 35*, 156 (1971).
113. Balandreau, J. and Dommergues, Y. Assaying nitrogenase (C_2H_2) activity in the field, *Bulletin Ecological Communication* (Stockholm) *17*, 247-254 (1973).
114. Watanabe, I., Lee, K.K. and Alimagno, B.V. Seasonal change of N_2-fixing rate in rice field assayed by *in situ* acetylene reduction technique. I. Experiments in long term fertility plots, *Soil Science and Plant Nutrition 24*, 1-13 (1978).
115. Watanabe, I. and Barraquio, W. Nitrogen-fixing (acetylene reduction) activity and population of aerobic heterotrophic nitrogen fixing bacteria associated with wetland rice. *Applied Environmental Microbiology 37*, 813-819 (1979).
116. Rinaudo, G. Hamad-Fares, I. and Dommergues, Y. Nitrogen fixation in the rice rhizosphere methods of measurement and practices suggested to enhance the process, pp. 313-322, in *Biological Nitrogen Fixation in Farming Systems of the Tropics* (Editors, A. Ayanaba and P.J. Dart), John Wiley, Chichester (1977).
117. Boddey, R.M., Quilt, P. and Ahmad, N. Acetylene reduction in the rhizosphere of rice: Method of assay, *Plant and Soil 50*, 567-574 (1978).

118. Daroy, Ma.L. and Watanabe, I. Nitrogen tixation by wetland rice grown in acid soil, *Kalikasan Philippine Journal of Biology 11*, (1982).

119. Balandreau, J.P., Rinaudo, G., Oumarov, M.M. and Dommergues, Y. Asymbiotic N_2 fixation in paddy soils, pp. 611-628, in *Proceedings of 1st International Symposium on Nitrogen Fixation*, Vol. 2 (Editors, W.E. Newton and C.J. Nyman), Washington State Univ. Press (1976).

120. Trolldenier, G. Influence of some environmental factors on nitrogen fixation in the rhizosphere of rice, *Plant and Soil 47*, 203-217 (1977).

121. Purushothaman, D., Oblisami, G. and Balasun, C.S. Nitrogen fixation by Azotobacter in rice rhizosphere, *Madras Agricultural Journal 63*, 555-560 (1976).

122. Lakshmi Kumari, M., Kavimandan, S.K. and Subba Rao, N.S. Occurrence of nitrogen fixing *Spirillum* in roots of rice, sorghum, maize and other plants, *Indian Journal of Experimental Biology 14*, 638 (1976).

123. Nayak, D.N. and Rajaramamohan Rao, V. Nitrogen fixation by *Spirillum* sp. in rice roots, *Archives for Microbiology 115*, 359-360 (1977).

124. Ladha, J.K., Barraquio, W. and Watanabe, I. Immunological techniques to identify *Azospirillum* associated with wetland rice, *Canadian Journal of Microbiology 28*, 478-485 (1982).

125. Thomas-Bauzon, D., Weinhard, P., Villecourt, P. and Balandreau, J. The spermosphere model, I: Its use in growing, counting, and isolating N_2-fixing bacteria from the rhizosphere of rice, *Canadian Journal of Microbiology 28*, 922-928 (1982).

126. Watanabe, I. and Barraquio, W. Low levels of fixed nitrogen required for isolation of free-living N_2-fixing organisms from rice roots, *Nature 277*, 565-566 (1979).

127. Barraquio, W., Ladha, J.K. and Watanabe, I. Isolation and identification of dinitrogen fixing *Pseudomonas* associated with wetland rice, *Canadian Journal of Microbiology 29*, (1983).

128. Qiu, Y., Zhou, S., Mo, X., Wang, D. and Hong, J. Study of nitrogen fixing bacteria associated with rice root. I. Isolation and identification of organisms, *Acta Microbiologica Sinica 21*, 468-472 (1981) (in Chinese, English summary).

129. Watanabe, I., Barraquio, W. and Daroy, M.L. Predominance of hydrogen utilizing bacteria among nitrogen-fixing bacteria in wetland rice roots, *Canadian Journal of Microbiology 28*, 1051-1054 (1982).

130. Barraquio, W., Guzman, M.R. de, Barrion, M. and Watanabe, I. Population of aerobic heterotrophic nitrogen fixing bacteria associated with wetland and dryland rice, *Applied Environmental Microbiology 43*, 124-128 (1982).

131. Lee, K.K., Castro, T. and Yoshida, T. Nitrogen fixation throughout growth, and varietal differences in nitrogen fixation by the rhizosphere of rice planted in pots, *Plant and Soil 48*, 613-619 (1977).
132. Sano, Y., Fujii, T., Iyama, S., Hirota, Y. and Komagata, K. Nitrogen fixation in the rhizosphere of cultivated and wild rice strains, *Crop Science 21*, 758-761 (1981).
133. Van Berkum, P. and Sloger, C. Ontogenetic variation of nitrogenase, nitrate reductase, and glutamine synthetase activities in *Oryza sativa*, *Plant Physiology 68*, 722-726 (1981).
134. Barraquio, W.L. and Watanabe, I. Occurrence of aerobic nitrogen fixing bacteria in wetland and dryland plants, *Soil Science and Plant Nutrition 27*, 121-125 (1981).
135. Boddey, R.M. and Ahmad, N. Seasonal variations in nitrogenase activity of various rice varieties measured with an *in situ* acetylene reduction technique in the field, pp. 219-229, in *Associative N_2 Fixation*, Vol. 2 (Editors, Vose et al.), CRC Press, Boca, Raton, Florida (1981).
136. Van Berkum, P. and Sloger, C. Physiology of root associated nitrogenase activity in *Oryza sativa*, *Plant Physiology 69*, 1161-1164 (1982).
137. Watanabe, I. and Cabrera, D. Nitrogen fixation associated with the rice plant grown in water culture, *Applied Environmental Microbiology 37*, 373-378 (1979).
138. Watanabe, I., Lee, K.K., Alimagno, B.V., Sato, M., del Rosario, Diana C. and de Guzman, M.R. *International Rice Research Paper Series* 3 (1977).
139. MacRae, I.C. Effect of applied nitrogen upon acetylene reduction in the rice rhizosphere, *Soil Biology Biochemistry* 7, 337-338 (1975).
140. Shiga, H. and Ventura, W. Nitrogen supplying ability of paddy soils under field conditions in the Philippines, *Soil Science and Plant Nutrition 22*(4), 387-399 (1976).
141. Huang, S., Tang, L., Zhang, W. and Liu, C. Observation on *Flavobacterium oryzae* sp. Nov. M-Sm 1612 in rice root under electron microscope and the characteristics of nitrogen fixation associated with rice, *Acta Microbiologiea Sinica 22*, 156-159 (1982) (in Chinese, English summary).
142. Diem, G., Rougier, M., Hamad-Fares, I., Balandreau, J.P. and Dommergues, Y. pp. 305-311, in *Environmental Role of Nitrogen Fixing Blue Green Algae and Asymbiotic Bacteria, Ecology Bulletin* (Stockholm) *26* (Editor, U. Granhall), Stockholm, Sweden (1978).
143. Watanabe, I., Cabrera, D. and Barraquio, W. Contribution of basal portion of shoot to N_2 fixation associated with wetland rice, *Plant and Soil 59*, 391-398 (1981).
144. Habte, M. and Alexander, M. Effect of rice plants on nitrogenase activity of flooded soils, *Applied Environmental Microbiology 40*, 507-510 (1980).

145. Rinaudo, G. The fixation of nitrogen in the rhizosphere of rice: Importance of varietal type, *Cahiers ORSTOM, Serie Biologie 12*, 117-119 (1977) (in French, English summary).
146. Rouquerol, T. About the phenomenon of nitrogen fixation in the rice field of Camargue, *Annals Agronomy 13*, 325-346 (1962) (in French, English summary).
147. Ishizawa, S. and Toyoda, H. Microflora of Japanese soils, *Bulletin National Institute agricultural Science* (Japan) *14B*, 204-284 (1964) (in Japanese, English summary).
148. Charyulu, P.B.B.N. and Rajaramamohan Rao, V. Influence of various soil factors on nitrogen fixation by *Azospirillum* spp., *Soil Biology Biochemistry 12*, 343-346 (1980).
149. Rice, W.A. and Paul, E.A. The organisms and biological processes involved in asymbiotic nitrogen fixation in waterlogged soil amended with straw, *Canadian Journal of Microbiology 18*, 715-723 (1972).
150. Durbin, K. and Watanabe, I. Sulfate reducing bacteria and nitrogen fixation in flooded rice soil, *Soil Biology and Biochemistry 12*, 11-14 (1980).
151. Brouzes, R., Lasik, J. and Knowles, R. The effect of organic amendment water content and oxygen on the incorporation of $^{15}N_2$ by some agricultural and forest soils, *Canadian Journal of Microbiology 15*, 899-905 (1969).
152. Rice, W.A., Paul, E.A. and Wettler, L.R. The role of anaerobiosis in asymbiotic nitrogen fixation, *Canadian Journal of Microbiology 13*, 829-836 (1967).
153. O'Tool, P. and Knowles, R. Efficiency of acetylene reduction (nitrogen fixation) in soil: Effect of type and concentration of available carbohydrate, *Soil Biology Biochemistry 5*, 789-797 (1973).
154. Chang, P.C. and Knowles, R. Non-symbiotic nitrogen fixation in some Quebec soils, *Canadian Journal of Microbiology 11*, 29-38 (1965).
155. Yoneyama, K., Lee, K.K. and Yoshida, T. Decomposition of rice straw residues in tropical soils, IV. The effect of straw on nitrogen fixation by heterotrophic bacteria in some Philippine soils, *Soil Science and Plant Nutrition 23*, 287-295 (1977).
156. Charyulu, P.B.B.N. and Rajaramamohan Rao, V. Influence of carbon substrates and moisture regime on nitrogen fixation in paddy soils, *Soil Biology Biochemistry 13*, 39-42 (1981).
157. Charyulu, P.B.B.N. and Rajaramamohan Rao, V. Nitrogen fixation in some Indian rice soil, *Soil Science 128*, 86-89 (1979).
158. Rajaramamohan Rao, V. Nitrogen fixation as influenced by moisture content, ammonium sulfate and organic sources in a paddy soil, *Soil Biology Biochemistry 8*, 445-448 (1976).

159. Charyulu, P.B.B.N., Nayak, D.N. and Rajaramamohan Rao, V. Influence of rice variety, organic matter and combined nitrogen, *Plant and Soil 59*, 399-405 (1981).
160. Renaut, J., Sasson, A., Pearson, H.W. and Steward, W.D.P. Nitrogen fixing algae in Morocco, pp. 229-246, in *Nitrogen Fixation by Free-living Microorganisms* (Editor, W.D.P. Stewart), Cambridge University Press, Cambridge (1975).
161. Venkataraman, G.S. Blue-green algae as a biological input in rice production, pp. 132-142, in *Proceedings National Symposium Nitrogen Assimilation and Crop Productivity*, Hissar, India (1977).
162. Wilson, J.T., Eskew, D.L. and Habte, M. Recovery of nitrogen by rice from blue green algae added in a flooded soil, *Soil Science Society of America Journal 44*, 1330-1331 (1980).
163. Tirol, A., Roger, P.A. and Watanabe, I. Fate of nitrogen from a blue green alga in a flooded rice soil, *Soil Science and Plant Nutrition 28*, (1982).
164. Roger, P.A. and Watanabe, I. Research on algae, blue-green algae, nitrogen fixation at the International Rice Research Institute (1963-81), summarization, problems and prospects, *International Rice Research Paper Series 78* (1982).
165. Gunnison, D. and Alexander, M. Resistance and susceptibility of algae to decomposition by various microbial communities, *Limnology and Oceanography 20*, 64-70 (1975).
166. Talley, S.N. and Rains, D.W. Azolla as a nitrogen source for temperate rice, pp. 310-320, in *Nitrogen Fixation*, Vol. 2 (Editors, W.E. Newton and W.H. Oxme-Johnson), Park Press, Baltimore, U.S.A. (1980).
167. Dao, T.T. and Do, A. A new idea on equating different kinds of nitrogenous fertilizer with pure nitrogen, *Agriculture Science Technology* (Vietnam) *93*, 168-174 (1970) (in Vietnamese).
168. Sawatdee, P., Seetanum, W., Chermsiri, C., Kanareugsa, C. and Takahashi, J. Effect of *Azolla* as a green manure crop on rice yield in northern Thailand, *International Rice Research Newsletter 2*, 10 (1978).
169. Li, Z. Nitrogen fixation by *Azolla* in rice fields and its utilization in China, pp. 83-95, in *Non-symbiotic Nitrogen Fixation and Organic Matter in the Tropics*, Proceedings of 12th International Congress of Soil Science, New Delhi (1982).
170. Shi, S., Qui, X., Liao, H. The availability of nitrogen of green manures in relation to their chemical composition, *Acta Pedologica Sinica 17*, 240-246 (1980).
171. Boddey, R.M. and Döbereiner, J. Association of *Azospirillum* and other diazotrophs with tropical Graminaceae, pp. 28-47, in *Non-Symbiotic Nitrogen Fixation and Organic Matter in the Tropics*,

Transactions of the 12th International Congress of Soil Science, Symposia Papers 1, New Delhi (1982).

172. Subba Rao, N.S. *Biofertilizers in Agriculture*, Oxford & IBH Publishing Co., New Delhi (1981).
173. Kulasooriya, S.A., Roger, P.A., Barraquio, W.I. and Watanabe, I. Epiphytic nitrogen fixation on deepwater rice, *Soil Science and Plant Nutrition 27*, 19-27 (1981).
174. Martinez, M.R. and Catling, H.D. Contribution of algae to the nutrition of deep water rice, pp. 201-214, in *Proceedings of the 1981 International Deepwater Rice Workshop*, The International Rice Research Institute, Los Baños, Philippines (1982).
175. Kulasooriya, S.A., Roger, P.A., Barraquio, W.L. and Watanabe, I. Biological nitrogen fixation by epiphytic microorganisms in rice fields, *International Rice Research Paper Series 47* (1980).
176. Patriquin, D.G. Grass-bacteria associations, pp. 139-190, in *Advances in Agricultural Microbiology* (Editor, N.S. Subba Rao), Oxford & IBH Publishing Co., New Delhi (1982).
177. Watanabe, I., Espinas, C.R., Berja, N. and Alimagno, B.V. Utilization of the Azolla-Anabaena complex as a nitrogen fertilizer for rice. *International Rice Research Paper Series 11* (1977).

11. Nitrogen Fixation Associated with Grasses and Cereals

R.M. Boddey and J. Döbereiner

One of the earliest suggestion that nitrogen nutrition of a cereal crop could be supplemented by the activity of associated nitrogen-fixing bacteria was made by Sen [146], from India in 1929 while studying lowland rice. Good evidence to support this idea was not forthcoming although nitrogen balance studies done in India and the U.S.A. did indicate that nitrogen gains in flooded rice could still be obtained even if the autotrophic N_2 fixers (blue-green algae etc.) were deprived of light [56, 93, 170, 179].

Although prior to 1970, there were several reports on the accumulation of total nitrogen attributed to biological nitrogen fixation in non-legume plant/soil systems other than rice [81, 86, 105, 119, 149, 178], the discovery of acetylene reduction assay [82] and its application to crop systems such as rice, sugar cane, many grasses, maize, sorghum and wheat, has triggered off extensive research in diazotroph/gramineae associations [5, 6, 15, 16, 40, 65, 67, 68, 108, 128, 136, 144, 159, 182].

Rice, which is the staple diet of 60 per cent of the world's population [147], has been intensively studied with regard to biological nitrogen fixation (BNF). Three recent studies using growth chamber have demonstrated significant $^{15}N_2$ incorporation through heterotrophic nitrogen fixation associated with the root system of rice [75, 86, 184]. It is difficult to estimate the contribution of BNF in field conditions. It was however, estimated from carefully conducted pot trials, that in four to six crops of flooded rice, the heterotrophic BNF contribution was equivalent to 14 to 18 per cent of the nitrogen accumulated in straw and grain [3]. It should however be stressed that the above estimates did not take into account nitrogen losses by way of denitrification and volatilization and hence great caution ought to be exercised in extrapolating these figures to field situations. Nevertheless nitrogen balance studies [77, 93, 168] of lowland rice have indicated a BNF input of the order of 30-60 kg N ha^{-1} crop^{-1} which is in good general agreement with the nitrogen balance data of App et al. [3].

The significant incorporation of $^{15}N_2$ gas has also been demonstrated to be more in the tetraploid batatais cultivar of *Paspalum notatum* than the diploid pensacola cultivar [58]. The tetraploid cultivar also exhibited greater acetylene reduction activity than the diploid cultivar pensacola [65, 68]. These results confirm the role of associative diazotrophs in the nitrogen nutrition of this tropical grass. The pensacola cultivar was used as a 'non-nitrogen fixing control' [30, 31] in an experiment on the application of isotope dilution technique [102] to measure the contribution of BNF in plants grown in concrete cylinders (60 cm) planted in the field. The results showed that the batatais cultivar had significantly more nitrogen than the pensacola cultivar but recovered the same quantity of applied ^{15}N fertilizer (Table 1). The vertical distribution of roots and the residual nitrogen in the cylinder were not significantly different [31]. It was concluded [102], therefore, that the extra nitrogen accretion of approximately 10 per cent of the total nitrogen or 20 kg N ha^{-1} $year^{-1}$, must have come diazotrophic BNF and selection of a suitable cultivar is an essential step to take advantage of the BNF potential in crop cultivation.

Acetylene reduction assays with *Paspalum* using intact cores indicated a BNF input of less than 7 kg N ha^{-1} $year^{-1}$ which can be taken as an underestimation because hammering the metal cylinders into soil to obtain soil cores causes disturbance and lowers root-associated nitrogenase activity [19, 169]. This partly explains why some workers found comparable results [29, 68, 108] and others noticed that root technique (with eight to 18 hours preincubation at low pO_2) always gave higher values than the soil core assays [15, 22, 74, 92, 159].

Using $^{15}N_2$ gas, 60 to 90-day-old sugar cane plants have been shown to incorporate the label [139] in incubation experiments ranging from 30 to 40 hours. In a rather ambitious field experiment [101], during a five-day incubation, only the rhizosphere soil but not the roots showed ^{15}N enrichment but there was considerable leakage of $^{15}N_2$ gas from the porous cortex tissues. There is therefore, no conclusive data on associated BNF contribution in sugar cane either from pot or field studies although estimates vary from 5 kg N ha^{-1} $year^{-1}$ [124] to 50 kg N ha^{-1} $year^{-1}$ [140].

Rennie [132] used the isotope dilution technique in pots to demonstrate a BNF input of 12 per cent of total nitrogen in small nitrogen-deficient eight-week-old maize plants. Hegazi et al. [83] grew maize plants inoculated with *Azospirillum lipoferum* in 10-litre pots containing vermiculite and ^{15}N labelled $(NH_4)_2SO_4$ and found a BNF input of 46, 51 and 61 per cent of the total nitrogen at harvests made at 8, 10, and 12 weeks, respectively.

Recent work at ICRISAT [51, 52], India has demonstrated very significant nitrogen gains in sorghum and millet plants grown in pots with vermiculite in the green house. Nitrogen balance studies were done with a cultivar of sorghum (CSH-5), showing high acetylene reduction activity in undisturbed pots and soil cores [169], capable of high nitrogen yields without the application of nitrogen fertilizer and not responsive to the application of 40 kg N/ha. The results showed large nitrogen increases

Table 1. Nitrogenase activity, nitrogen accumulation and ^{15}N enrichment and recovery of two *Paspalum notatum* cultivars grown in the field[1]

Cultivar	Mean acetylene[2] reduction activity, μ mole C_2H_4 core^{-1} day^{-1}	Total nitrogen in plant material, g.cylinder^{-1}	Weighted mean[3], % ^{15}N excess in shoot+ rhizomes	Weighted mean[3], % ^{15}N excess in roots[5]	% Recovery of added labelled ammonium sulphate[4]	
					in plant material	in soil[5] (residual)
Batatais	1.67	11.59	0.293	0.340	34.1	30.5
Pensacola	0.44	10.24	0.321	0.419	33.1	30.5
Difference significant at P=	0.001	0.01	0.01	0.05	ns[6]	ns[6]

[1]Plants grown in concrete cylinders (60 cm diameter×50 cm deep) filled with red yellow podzolic soil (total nitrogen 0.12%). Results represent means of three N fertilizer levels (1.6, 3.2 and 6.4 kg nitrogen ha^{-1} addition^{-1}) for 7 harvests over a 22-month period.

[2]Acetylene reduction activity measured on intact cores 10 cm in diameter.

[3]Weighted mean % ^{15}N excess = $\dfrac{\sum(\% \; ^{15}N \text{ excess} \times \text{total nitrogen})}{\sum \text{total nitrogen}}$.

[4]$(NH_4)_2SO_4$ containing 5.104 atom per cent ^{15}N excess applied every 2 to 3 weeks.

[5]Mean of data from the highest fertilizer treatment (6.4 kg nitrogen ha^{-1} addition^{-1}) only.

[6]ns = not significant.

Source: Boddey et al. [31].

Table 2. Nitrogen balance for sorghum cv. CSH-5 grown in pots in unsterilized vermiculite with different inoculations and nitrogen fertilizer levels

	Nitrogen applied (mg.pot^{-1})	Nitrogen in dry matter[1] (mg·pot^{-1})	Nitrogen balance[2] (mg.pot^{-1})
Isolate from *Chloris gayana*	0	86	88
	53	161	155
Isolate from *Sorghum halapense*	0	187	335
	53	173	100
Isolate from *Pennisetum clandestinum*	0	130	234
	53	108	204
Enterobacter spp.	0	133	147
	53	110	60
Napier bajra root extract[3]	0	368	542
	53	140	103
Control (boiled inoculant)	0	152	259
	53	89	243
Unplanted	0	—	45
	53		—8
S.E.M. (0.05)		70	110

[1]Average of four pots with five plants each: plant age at harvest 49 days.

[2]Derived from total nitrogen per pot in plant dry matter plus rooting medium at harvest minus nitrogen in seed, inoculum, and in rooting medium at sowing.

[3]Roots of Napier bajra, a hybrid between *Pennisetum purpureum* × *P. americanum* were ground in sterile tap water, left to stand for 5 hours filtered through muslin and the extract used as inoculant.

Source: Dart and Wani [52].

only in planted pots but inoculation with various undefined bacterial isolates did not prove significantly different from the control (Table 2). Inoculation with root extract of *Pennisetum* roots, however, proportioned a significant increase in total nitrogen (vermiculite plus plant) of 283 mg N/pot in comparison with planted control, and 497 mg N/pot in comparison with unplanted controls. Definitive quantification of all these inputs awaits the results of ^{15}N experiments now in progress (S.P. Wani, personal communication).

De-Polli [58] demonstrated $^{15}N_2$ gas incorporation into the roots, rhizomes and shoots of *Digitaria decumbens.* Several other nitrogen balance studies indicate the importance of associated BNF in other grasses, although good quantitative data are not available except for *Paspalum notatum* [52, 80, 86, 131].

SPECIFIC PLANT BACTERIA ASSOCIATIONS

The occurrence of various nitrogen-fixing heterotrophs in the roots and rhizospheres of several non-leguminous plants have been reviewed earlier [62, 91]. *Bacillus* has been found in Canada to associate with specific genotypes of wheat [106]. These nitrogen-fixing bacteria were found in 10^{-6} dilutions of rhizosphere soil from the chromosome substitution line C-R5D of Cadet wheat but in the parent Cadet, no nitrogen fixers were found even in 10^{-2}dilutions of rhizosphere soil. The bacteria got successfully established with the roots of the same line of wheat in monoxenic culture [96] by invading the root cortex close to the emergence of lateral roots and by embedding on root surface in a polysaccharide, probably of bacterial origin produced as a result of high unfavourable pO_2.

The parents of the C-R5D line are Cadet and Rescue. While Cadet is resistant to the common root rot, Rescue is susceptible [95]. Rescue can be made resistant to root rot if its chromosome pair 5D is replaced by that of Cadet and hence susceptibility or resistance is associated with genes. Susceptibility to root rot is also associated with high bacterial counts in the rhizosphere particularly of those possessing cellulolytic and pectinolytic activities. It is however, the resistant C-R5D having low rhizosphere population which favourably supports the nitrogen-fixing *Bacillus* population in the rhizosphere. It has been shown [57] that C-R5D and other root rot-resistant lines have considerably less senescent cells in the root cortex (as assessed by nuclear staining) than the susceptible lines. The implication of this finding is that C-R5D wheat, by virtue of its viable and healthy roots affords good scope for *Bacillus* association with the root system, especially in the interior of roots. This phenomenon cannot be regarded as a loose association because in monoxenic cultures of certain lines of wheat (including C-R5D), association with *Bacillus* [133] yielded between 40 and 60 per cent more nitrogen than non-inoculated controls, attributable to BNF even when the nutrient solution contained 100 mg KNO_3 L^{-1}.

McClung and Patriquin [103] reported the occurrence of *Campylobacter*, a diazotroph associated with the marine marsh grass, *Spartina alterniflora* based on accretion to total nitrogen and acetylene reduction data [35, 46, 103, 121, 126] from intact plants. The following relevant observations were made: (a) the rate of ethylene accumulation in the plant tops by acetylene reduction in intact plants growing in a solution culture was higher than that which could be accounted for ethylene diffusion from the root exterior *via* the stem aerenchyma tissue, (b) the ethylene accumulation in the shoot enclosure (upper phase) was not affected by the presence of 200 μM NH_4Cl or 0.2 per cent $HgCl_2$ in the solution culture whereas ethylene accumulation in the root chamber was inhibited and (c) addition of glucose and malate to roots, stimulated the accumulation of ethylene in the root phase considerably more than in the upper shoot phase.

In lowland rice 80 per cent of the root interior microflora was nitrogen-fixing bacteria [171] (10^8 per g dry root) whereas in the rhizosphere only 2.4 per cent of the bacteria was nitrogen fixers. *Pseudomonas* spp., *Azospirillum*, *Klebsiella* and *Enterobacter* spp. were common diazotrophs of the rice root interior [17, 175]. $^{15}N_2$ studies, however, have shown [76, 184] that in short-term assays, not more than 19 to 25 per cent of the nitrogen fixed was incorporated into the plant, the remainder being left in the soil. It seems, therefore, that despite very significant enrichment of diazotrophs in the root interior, most N_2 fixation occurs in the rhizosphere soil. Boyle and Patriquin [36], applied the two-phase (shoot and root phases) method of raising *S. alterniflora* [35] to grow rice under dry land field conditions but assayed in non-aerated solution culture. They found: (a) the rate of ethylene diffusion from the site of acetylene reduction through the aerenchyma tissue of the plant to the upper phase could be accounted for by the diffusion of ethylene from the root exterior, and (b) the acetylene reduction activity was quickly inhibited by additions of $HgCl_2$. These observations illustrate the differences between *Spartina* and rice systems.

Azotobacter paspali occurs only in the rhizosphere of *Paspalum notatum* and that too only in association with a few tetraploid varieties of this grass [60, 61] and not other cultivars, indicating the highest degree of specificity between grass/diazotroph associations investigated so far.

AZOSPIRILLUM–GRAMINEAE ASSOCIATIONS

In the last few years a great deal of interest has been given to association of *Azospirillum* spp. with the roots of various grasses and cereals [1, 15, 28, 66, 118, 123, 127, 150].

Plant-bacteria interactions

Significant correlation between nitrogenase activity between individual root pieces and the subsequent activity of enrichment cultures on a semisolid malate medium was first reported by Döbereiner and Day [66]. This was followed by a subsequent observation on similar lines with root pieces of maize surface sterilized with a solution of a quaternary NH_4^+ compound for 30 seconds which suggested that root interior was the most important site of nitrogenase activity [40].

Umali Garcia et al. [163] observed that three strains of *Azospirillum brasilense* were adsorbed to the surface of root hairs of pearl millet (*Pennisetum americanum*) significantly more than *Rhizobium trifolli* and *Pseudomonas fluoresence*. This adsorption was more in plants grown in nitrogen-free media (3.50+10.5 bacteria per root hair) than in those grown in 5 mM KNO_3 - added media (3.15±0.60 bacteria per root hair). Very few cells of *Azotobacter vinelandii, Klebsiella pneumoniae* and *Escherichia coli* were adsorbed to the root hair surface, indicating the preferential attachment of *A. brasilense* to root hairs.

Tissue culture studies

Child and Kurz [47] raised tissue cultures of several legumes and non-legumes for periods of up to two weeks in association with *A. brasilense* nir+ (Strain sp. 7) on a nitrogen-added medium. The nitrogenase activity ot these associations with *Azospirillum* was higher with non-legumes than with legumes or with legumes in association with *Rhizobium* spp. In sugar cane tissue cultures, *A. brasilense* (Strain sp. 7) continued to survive and multiply even after sub-culturing of the tissue cultures for up to 18 months. Similar tissue cultures of tobacco, pearl millet and centipede grass were also grown in association with *A. brasilense* for shorter periods but growth of the callus was inhibited by the bacteria. Electron microscope studies revealed that *A. brasilense* grew profusely in association with sugar cane callus tissue and there were as many as 2×10^8 bacteria mm^{-2} on the surface of the callus but the bacteria did not enter the healthy cells [164]. The most interesting observation was the presence of several large round encapsulated forms of *Azospirillum* (C forms) usually containing considerable quantities of polyhydroxybutyrate [20]. Similar forms of *A. brasilense* have also been observed in pure culture by Eskew et al. [75], who suggested that they were probably cyst-like resting stages incapable of nitrogenase activity. According to Berg [20, 21], who studied the fine structure of these cysts under an electron microscope, they bear no similarity with *Azotobacter* cysts and develop when *Azospirillum* grows on the surface of nitrogen-free solid media under unfavourable pO_2. The optimum pO_2 in solution for *Azospirillum* in liquid culture is 0.005 to 0.007 atm. [116] and vigorously shaken cultures lose their nitrogenase activity when the O_2 input is larger than the rate of uptake by the organism. It is tempting therefore to suggest that the lipopolysaccharide capsule is involved in oxygen protection of the nitrogenase enzyme [20]. These C forms were absent on plates containing 0.25 per cent NH_4Cl but they were found among vibroid cells (V forms) on plates containing nitrogen-free media, which may be the reason for the inability of encapsulated forms to fix nitrogen [21]. Similar encapsulated forms of *A. brasilense* were observed on the surface of inoculated pearl millet roots [163].

Host plant specificity groups

Azospirillum strains differ in their ability to grow on glucose, to denitrify and in their requirement for biotin [112, 142]. Tarrand et al. [155] classified *Azospirillum* into two species—*A. lipoferum* which requires biotin and is able to grow on glucose as a sole carbon source and *A. brasilense* which does not require biotin but is unable to grow on glucose. The two species differed in DNA homology and other physiological characteristics. Neyra [112] found that all strains can dissimilate nitrate as nitrite, while nitrite reduction, the key step to denitrification is characteristic of about half the number of strains. The latter were termed nir^+(nitrate reductase positive) strains. Recently a new acid tolerant species, *A. amazonense* has been proposed [100] (Table 3).

Surface sterilized roots of wheat and rice with one per cent chloramine T solution, were found to be predominantly infected by the nir^- strains of *A. brasilense* whereas maize roots were predominantly infected by *A. lipoferum*. Similarly plants possessing C_4 photosynthetic pathway (*Sorghum*

Table 3. Differences between *Azospirillum amazonense* and the known species

	Azospirillum amazonense	*A. lipoferum*	*A. brasilense*
Colony type on potato agar	white flat with raised margin	pink raised	pink raised
Tolerance to O_2 for nitrogenase activity	very low	low	low
Dissimilation of			
$NO_3 \rightarrow NO_2$	$\mp$	+	+
$NO_2 \rightarrow N_2O$	-	$\mp$	$\mp$
Cell width	0.9-1.0	1.0-1.5[1]	1.0-1.2
Polar flagellum	+	+	+
Lateral flagella on nutrient agar	-	+	+
Polymorph cells in alkaline media	-	+	-
Biotin requirement	-	+	-
Use of sucrose	+	-	-
Generation time for N_2 dependent growth at optimal pO_2	10 hours	5-6 hours	6 hours
DNA base comp. (mol % G+C)	67-68	69-70	69-70

Explanation of signs: $\mp$ positive in more than 90% of the strains; + positive in less than 50% of the strains; — negative.

[1]Cells of *A. lipoferum* may become even wider and up to 10 μm long in older alkaline cultures.

Source: Magalhaes et al. [100].

bicolor, Panicum maximum, Pennisetum purpureum, Digitaria decumbens, Hemarthria altissima, Cynodon dactylon, Brachiaria spp.) were also preferentially infected by *A. lipoferum* whereas C_3 cereals (rye, barley and oats) were infected by *A. brasilense*. In all the cases, the root interior was inhabited by nir$^-$ strains. Sugar cane, unlike all other C_4 plants tested, was however preferentially infected by *A. brasilense* nir$^-$, which may have some connection with the high sucrose content of this plant [10, 11, 138].

Spontaneous mutants of *Azospirillum*, resistant to low levels of streptomycin, were selected from uninoculated as well as *Azospirillum*-inoculated maize plants but this marking technique was not feasible to trace the fate of inoculated strains (Table 4). Streptomycin resistance has been observed in the microflora of rhizosphere of several crops [12, 38], in *Rhizobium* isolates from soybean, *Stylosanthes* [145] and *Phaseolus vulgaris* (Pitard, Boddey and Döbereiner, unpublished), presumably due to the antibiotic-producing *Streptomyces* spp. in the rhizosphere of these crop plants [64].

The above findings suggested that for the inoculation of C_4 plants *A. lipoferum* nir$-$ strains resistant to streptomycin were likely to be most successful. This was tested with maize in a field experiment [79], which included treatments of organic matter (40t/ha) and fertilizer nitrogen (60

Table 4. Selection of low level streptomycin resistant *Azospirillum* spp. during the infection of cereal roots in the field

	No. of isolates	% isolates resistant to 20 μg . ml[-1] streptomycin[2]	
		A. lipoferum	*A. brasilense* nir⁻
MAIZE			
Without inoculation			
Soil	32	6	3
Roots[1]	29	81	0
Inoc. with *A. lipoferum*[3] from maize			
Soil	32	91	0
Roots	30	90	4
Inoc. with *A. brasilense*[3] from wheat			
Soil	31	8	50
Roots	28	84	6
WHEAT			
Without inoculation			
Soil	32	9	6
Roots			
Inoc. with *A. brasilense*[3] from wheat			
Soil	32	12	86
Roots	31	0	94
Inoc. with *A. lipoferum*[3] from maize			
Soil	32	70	12
Roots	32	9	88

[1]Roots of maize and wheat collected in the field and surface sterilized for 60 or 15 min. resp. in 1 per cent Chloramine T.

[2]Data are calculated from approximately two strains each from 4 field plots collected at 4 growth stages (32 strains).

[3]Strains isolated from surface sterilized roots and low level streptomycin resistant (20 μg/ml[-1]).

Source: Baldani and Döbereiner [10].

kg N/ha). The inoculants were *A. brasilense* nir⁻ strain isolated from surface sterilized rice roots and *Azospirillum* sp. nir⁻ (Sp. 242 st) isolated from surface sterilized maize roots, both of which were resistant to 100 μg/ml streptomycin. The results (Table 5) showed that Sp. 242 st strain increased total nitrogen by 26 to 52 per cent (= 40 kg N/ha) while *A. brasilense* nir⁻ strain increased the total nitrogen only to a lower extent. It is relevant to point out that the strain Sp. 242 st used in this experiment was originally classified [155] as *A. lipoferum* nir⁻ but due to storage in culture collection, the strain lost its characteristic ability to grow on glucose and mannitol and hence its taxonomic position has become uncertain.

In a more detailed study [13], *A. brasilense* nir⁻ strain 107 st, isolated from surface sterilized wheat roots increased total nitrogen in tops of wheat significantly at P = 0.05 over uninoculated controls, whereas an *Azospirillum* Sp. nir⁻ strain (Sp. 108 st) from maize roots did not do so. In a second

Table 5. Effect of the inoculation of *Azospirillum* spp. and organic matter on the incorporation of nitrogen by maize grown in the field

Treatment[2]	Inoculation[3]	Dry weight[1–4], $g.plant^{-1}$	Total nitrogen[1–4], $g.plant^{-1}$	% increase due to inoc.	Total Nitrogen[5], $kg.ha^{-1}$			
					In plants	Increase due to inoc.	Increase due to fertilization	Increase due to org. mat.
Org. Mat. + N	*A. lipoferum*	202	3.05	32.0	152	37	—26	—1
	A. brasilense nir⁻	208	3.28	42.0	164	48	18	38
	Control	159	2.31	—	115	—	—26	6
Org. Mat.	*A. lipoferum* nir⁻	247	3.56	25.8	178	36	—	63
	A. brasilense nir⁻	214	2.93	3.5	146	5	—	57
	Control	193	2.83	—	141	—	—	70
N	*A. lipoferum* nir⁻	208	3.06	40.4	153	44	38	—
	A. brasilense nir⁻	168	2.52	15.6	126	17	37	—
	Control	149	2.18	—	109	—	38	—
O	*A. lipoferum* nir⁻	163	2.31	61.5	115	44	—	—
	A. brasilense nir⁻	129	1.78	24.5	89	17	—	—
	Control	103	1.43	—	71	—	—	—

[1]The effects of organic matter, of inoculation and of the interaction of organic matter with nitrogen fertilizer were significant at $P=0.01$.

[2]40 tonnes ha^{-1} of municipal garbage compost incorporated at planting and 60 kg N/ha as NH_4NO_3 applied 10 days before flowering.

[3]Inoculated with 20 ml per plant of a culture grown in NFb with 1 g L^{-1} NH_4Cl.

[4]Means of 4 plots and 4 plants per plot.

[5]Based on 50,000 plants per hectare.

Source: Freitas et al. [79].

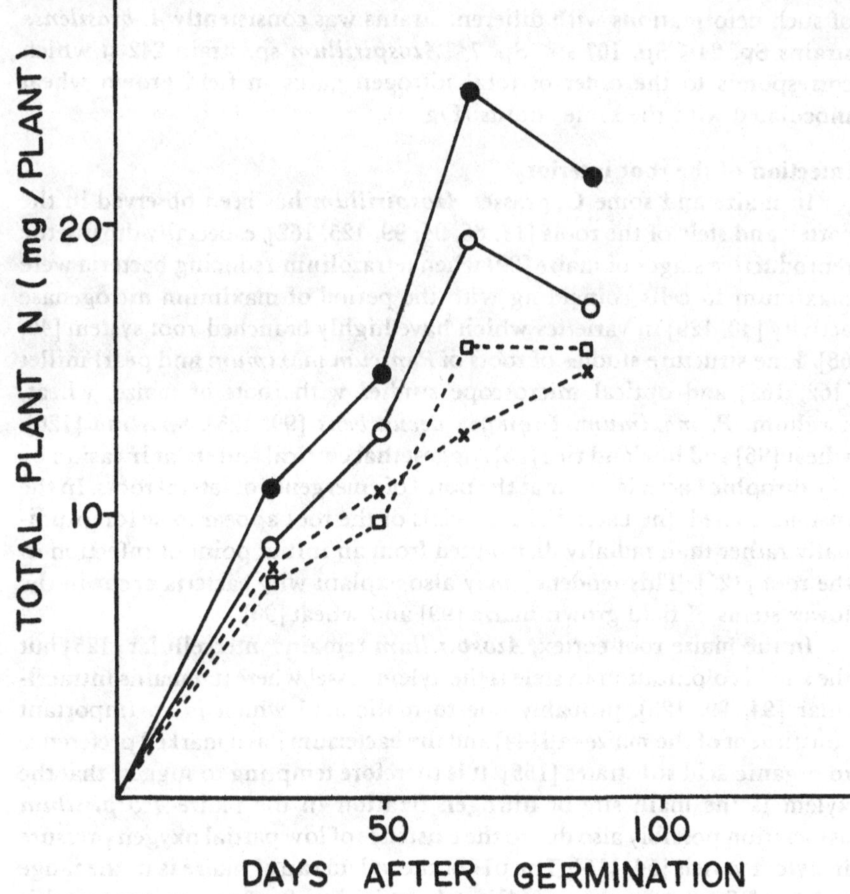

Fig. 1. Total nitrogen accumulation in field grown wheat inoculated with *Azospirillum* spp. Differences between strains were significant at p= 0.01. (● ——— ●) inoculated with *A. brasilense* nir− (Sp 245) isolated from surface sterilized wheat roots collected in the major wheat region (Paraná) of Brazil; (○ ——— ○) inoculated with *A. brasilense* nir− (Sp 107st) isolated from surface sterilized wheat roots collected from a pot experiment in Rio de Janeiro; (□ - - - □) *Azospirillum* spp.(Sp 242st) isolated from surface sterilized maize roots collected in the field in Rio de Janeior; (×- - - ×) uninoculated control treated with an equal amount of heat killed *Azospirillum* cells. Baldani *et al.* [13].

experiment *A. brasilense* nir− strains (Sp. 245 and Sp. 107 st) from wheat roots increased total nitrogen to a greater extent than *A. brasilense* nir− strain Sp. Br 14 from wheat rhizosphere or *Azospirillum* sp. nir− strain (Sp. 242 st) isolated from maize roots (Fig. 1).

Patriquin et al. [127], in agreement with earlier data, showed that monoxenic wheat seedlings inoculated with *Azospirillum* sp. produced root hairs which were equally branched ("tuning fork" deformations) and such hairs were absent in uninoculated controls. The order of frequencies of such deformations with different strains was consistently *A. brasilense* strains Sp. 24< Sp. 107 st< Sp. 7<*Azospirillum* sp. strain 242 st which corresponds to the order of total nitrogen gains in field grown wheat inoculated with the same strains (Fig. 1).

Infection of the root interior

In maize and some C_4 grasses *Azospirillum* has been observed in the cortex and stele of the roots [13, 66, 94, 99, 125, 162], especially during the reproductive stages of maize [99] when tetrazolium-reducing bacteria were maximum in cells coinciding with the period of maximum nitrogenase activity [40, 129] in varieties which have highly branched root system [40, 66]. Fine structure studies of roots of *Panicum maximum* and pearl millet [162, 163] and optical microscope studies with roots of maize, wheat, sorghum, *P. maximum, Diqitaria decumbens* [99, 125], *Spartina* [126], wheat [96] and lowland rice [26] suggest that cortical and stelar invasion by diazotrophic bacteria occur at the point of emergence of lateral roots. In the instances cited, the bacteria in the cells of the root appear to be longitudinally rather than radially distributed from an initial point of infection in the root [125]. This tendency may also explain why bacteria occur in the lower stems of field-grown maize [99] and wheat [90].

In the maize root cortex, *Azospirillum* remains intercellular [125] but the site of colonization in stele is the xylem vessel where it remains intracellular [94, 99, 125], probably due to malic acid which is an important constituent of the maize sap [44] and the bacterium has a marked preference to organic acid substrates [155]. It is therefore tempting to suggest that the xylem is the main site of nitrogen fixation in the maize-*Azospirillum* association possibly also due to the existence of low partial oxygen pressure in xylem vessels [54, 116]. The pH of the xylem sap of maize is in the range of 5.3 to 5.6 in young shoots [44] but *Azospirillum* Sp. 7 grows poorly at this pH, the optimum pH for its growth being 6.8 to 7.8 [66]. Recent preliminary experiments conducted by these authors have however shown that isolates of *Azospirillum* from surface-sterilized maize and wheat roots grow at pH 5.5 or even lower.

The role of malic acid in the transport of potassium in the plant is well documented [104, 158]. It is possible that other organic acids in the xylem may be important substrates in maize roots. In extracts of roots from 30-day-old maize plants, the quantity of trans-aconitic acid was high, followed by fumaric, citric and malic acid in descending order (D.K. Stumpf, personal communication). Trans-aconitate is an important constituent of forage grasses [41, 153] and maize shoots. In maize, this acid can account for 50 per cent of the total organic acid content of the shoot system [48, 49]. There are no data on trans-aconitic acid content in xylem sap of Gramineae.

Several experiments have shown that certain strains of microorganisms can use trans-aconitate as an energy and carbon source which can be

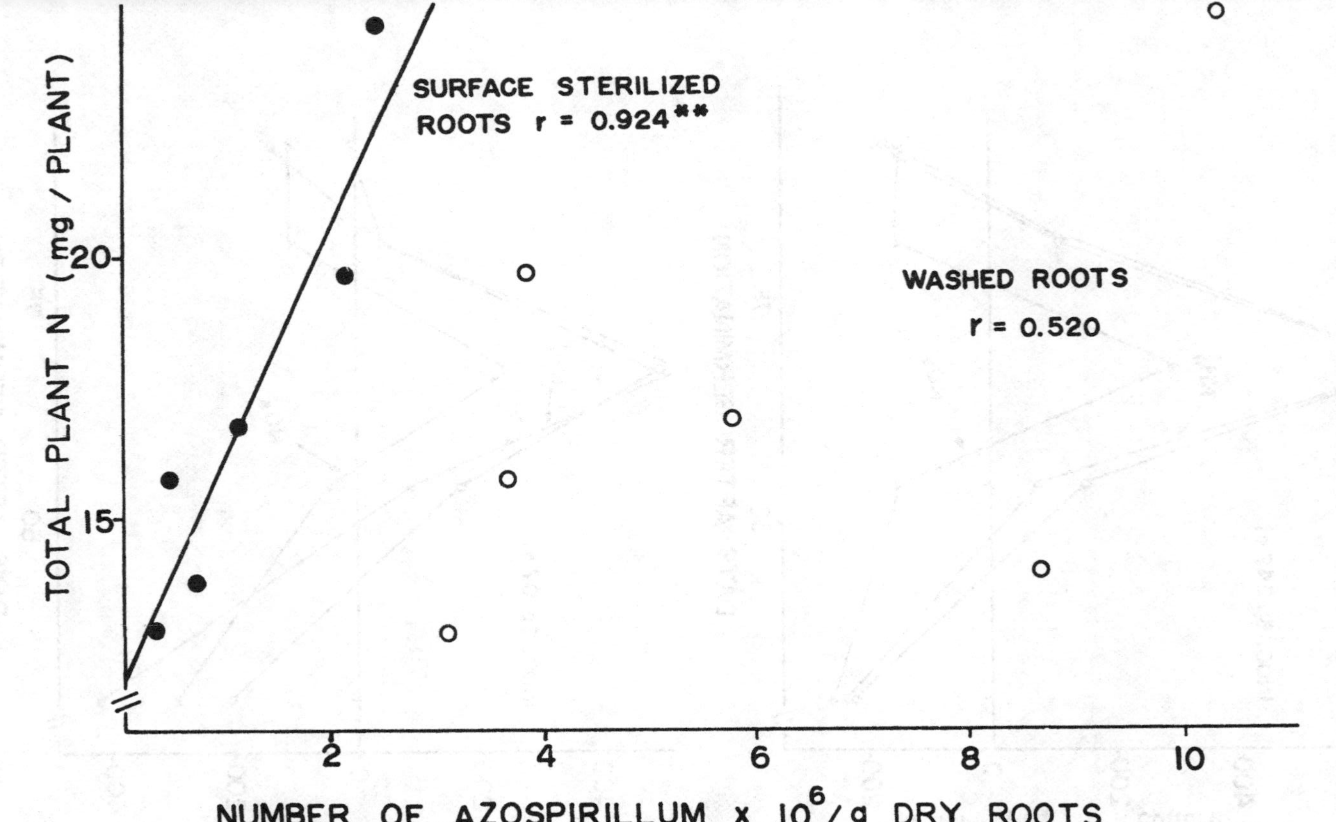

Fig. 2. Correlation of numbers (MPN) of *Azospirillum* spp. in washed (○) and surface sterilized (●) wheat roots with total plant nitrogen incorporated into wheat plants. The data are from the harvest at 64 days from the same field experiment as those in Fig. 1 but include additional treatments (one strain from rhizosphere soil and one strain mixture), Baldani *et al.* [13].

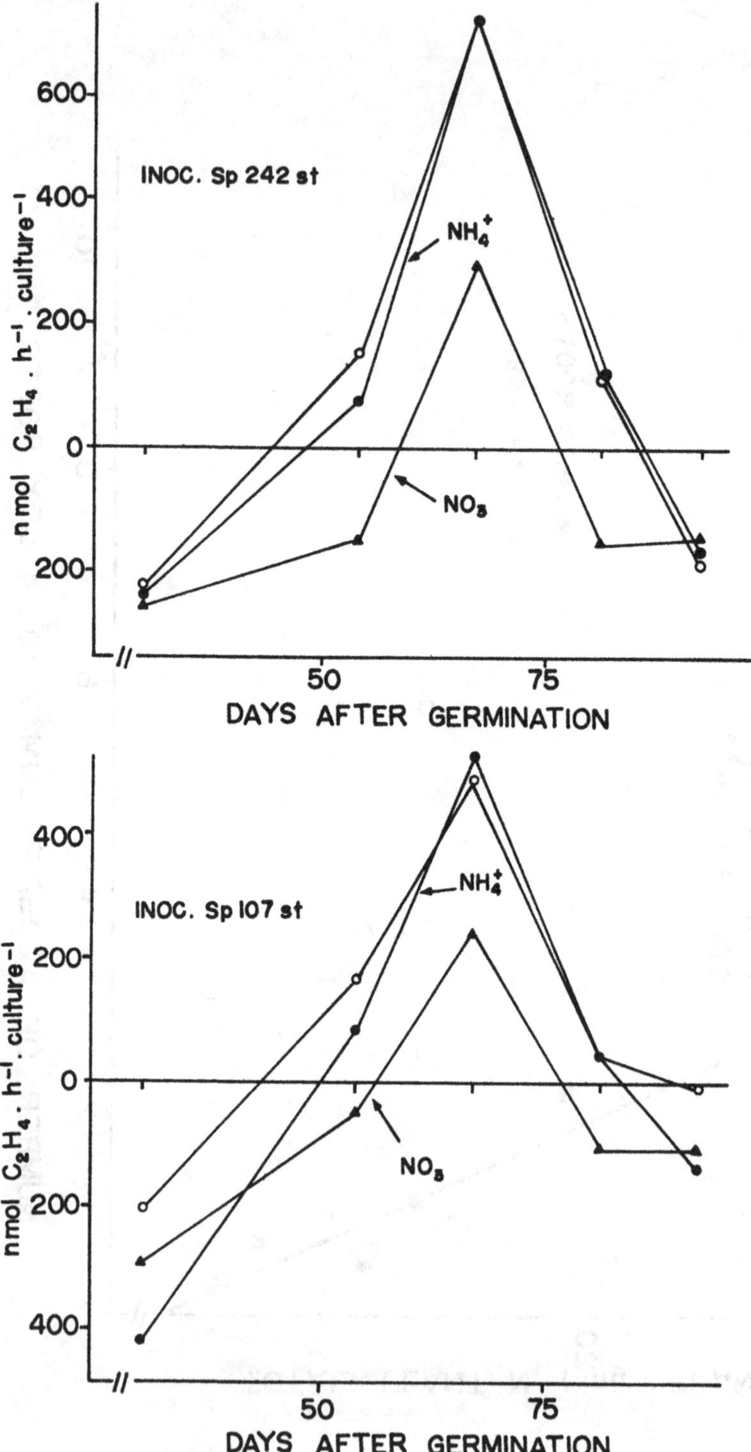

600
400
200
0
200
INOC. Sp 242 st
NH_4^+
NO_3
nmol $C_2H_4 \cdot h^{-1} \cdot$ culture^{-1}
50
75
DAYS AFTER GERMINATION
400
200
0
200
400
INOC. Sp 107 st
NH_4^+
NO_3
nmol $C_2H_4 \cdot h^{-1} \cdot$ culture^{-1}
50
75
DAYS AFTER GERMINATION

summarized as follows: four out of six strains of *Spirillum* sp. [39]; *A. brasilense* strains Sp. 81 and Sp. 82 but not Sp. 7 [14]; 13 out of 17 strains of *A. brasilense* excepting Sp.7 and Cd, 13 out of 14 strains of *A. lipoferum* and one of *Azotobacter paspali* tested in Döbereiner's laboratory and eight out of nine strains of acid-tolerant *A. amazonense* [100]. On the other hand, in Döbereiner's laboratory, one each of *Derxia, Beijerinckia indica, B. camargense, Xanthobacter autotrophicum, X. flavum, Klebsiella pneumoniae, Escherichia coli* and several strains of *Rhizobium* spp. did not grow on trans-aconitate containing medium. An increase in root-associated acetylene reduction activity with the onset of the reproductive phase of growth has been noticed in excised roots of maize [240, 129], rice [183], barley [166], wheat [108] as well as intact plants of rice [25, 27, 143, 173, 174] and sorghum [169]. This fact has also been shown with $^{15}N_2$ in lowland rice [76]. The increased nitrogenase activity during the reproductive phase may have to be related to changes in xylem sap by further investigations.

The cause for the inhibition of nitrogenase activity during the 'lag period' in excised root assays has been ascribed to the accumulation of NO_2^- in the medium [130]. Non-denitrifying (nir⁻) strains of *Azospirillum* spp. accumulate NO_2^- in the culture medium due to high reduction of NO_3^- [110, 112]. The inhibition of nitrogen-dependent growth of the two isolates of *Azospirillum* (Sp. 242 and 107 st) by extracts of roots of young maize plants fed with KNO_3 [181] and the enhancement of growth by root extracts of plants at the grain filling stage can be seen from Fig. 3. The reason for these effects cannot be explained at present although it has been found in a similar experiment that NO_2^- levels in root extracts decreased with age i.e., from 3.2 mM in 18-day-old plants to 1.8 mM in 59-day-old plants (D.F. Xavier, personal communication).

Plant genotype effects

The highly specific *A. paspali/P. Notatum* [61] and *Bacillus* spp./wheat [96] lines associations have already been mentioned. Differences in ARA values of intact plants of rice, which could be correlated ($r=0.605$, $p=0.001$) with root weight have been noticed [98] which suggested the possibility that the differences in ARA values could really be due to rooting density in the small pots used for the experiment. Significant differences in ARA values were reported among the seedlings of rice cultivars grown in small tubes [136] and even among mutants of single strain of rice where values of 30 mutants [71] ranged from 861±456 to 7348+1971

Fig. 3. Effect of root extracts on N_2 dependent growth of *Azospirillum* spp. (Xavier *et al*. [181]). Roots from field grown maize were disintegrated in a blender and a 10^{-1} dilution after filter sterilization mixed in the proportion 1:1 with double strength NFb medium (semi-solid). Data are n moles C_2H_4 produced h^{-1}.culture^{-1} in comparison with medium prepared without root extract. Plants were fertilized with 3 applications of 40 kg N/ha of KNO_3 (▲—▲), $(NH_4)_2SO_4$ (●———●) or no N fertilizer (○—○).

nmole C_2H_4 $h^{-1}g$ dry $root^{-1}$. In a field experiment, however, no significant differences in values from *in situ* ARA assays were noticed between two traditional indica and two modern IRRI cultivars of rice having different origins [27].

In maize, as revealed by excised root assays, considerable significant difference between mean activity of 17 lines of maize during reproductive stages of growth were noticeable [40] and this observation supports the contention that it is more meaningful to take ARA values at flowering and pod stages of the plant when the soil nitrogen is depleted thei eby creating a

Table 6. Differences in nitrogenase activity associated with roots of 30 cultivars of wheat assayed on the intact plant soil system

Cultivar	Nitrogenase activity[1] $n \cdot mol \cdot C_2H_4 \cdot h^{-1} \cdot g \cdot root^{-1}$
Tobari	3 523 a[2]
Lagoa Vermelha	2 171 ab
Cotiporan	2 002 ab
Sonora 63	1 870 ab
CNT 1	1 372 ab
Sonora 64	1 363 ab
Paraguai 214	1 249 ab
MR 72k4	1 240 ab
IAS 62	1 163 ab
CNT 6	1 146 ab
IAS 57	1 087 ab
CNT 5	1 052 ab
IAS 54	923 ab
IAS-Maringá	892 ab
IAS 58	723 ab
CNT 7	628 ab
C 51	611 ab
IAS 53	610 ab
CNT 4	582 ab
IAS 20	569 ab
Inia	563 ab
S 76	563 ab
IAS 53	467 b
IAS 61	381 b
IAS 59	368 b
CNT 3	343 b
Ciano	260 b
Tanori	245 b
BH 1146	160 b
Londrina	111 b

[1]Plants assayed using soil cores (10 cm diameter). Values are means of four assays during grain filling.

[2]Values followed by the same letter are not significantly different at P=0.05 (Tukey test). Least significant difference 3053.

Source: Nery et al. [108].

condition for better BNF output. Ela et al. [73] assayed 27-day-old seedlings of maize and found differences in ARA values and later confirmed the findings through $^{15}N_2$ gas incorporation studies.

In pearl millet, Bouton and Brooks [32] screened cultivars by *in situ* acetylene reduction assays. At ICRISAT, India [52], 334 sorghum and 284 pearl millet lines were screened by soil core ARA assays in field-grown plants to select lines which exhibited consistently high associated ARA. More recent studies, however, suggested variability from plant to plant, partly due to disturbances in the process of the removal and transportation of the soil cores [169].

Using intact soil cores, the ARA of 30 field-grown wheat cultivars during the grain filling stage showed significant differences between the mean ARA values [108] (Table 6). A significant correlation ($r=0.83$, $p=0.01$) between the ARA of intact potted plants with ARA of excised roots of the same plants under one per cent O_2 (Fig. 4), was observable although excised roots on an average gave lower values than the intact systems.

In interpreting the above results, due attention must be paid to differences arising out of mechanical disruption of soil cores [169], slow rates of gas diffusion in soil cores [23, 180] and blue-green algae [29, 173].

Environmental and edaphic effects

High soil moisture has been shown to be positively related to ARA in grasses in Canada, England, the U.S.A., Australia and Brazil [55, 148, 151, 160, 167, 175, 177]. The ARA values of roots from dryland rice had lower ARA than those of wetland rice [182]. In maize and sorghum soil moisture was also related to ARA [9, 83, 169]. It is likely that roots and rhizosphere have more microaerophilic sites in which diazotrophs can be active under high moisture levels [55] which is in consonance with the idea that low pO_2 encourages maximal nitrogenase activity as has been demonstrated in several crops [29, 45, 53, 69, 122, 152, 161].

Maximum ARA activity of rice was recorded at 35°C and insignificant activity at 10° and 15°C [25]. Similar results were reported for maize from *in situ* ARA assay, the activity increasing from 18° to 35°C [9].

Diurnal fluctuations in ARA associated with intact cereals and grasses have been recorded [5, 7, 8, 23, 27, 65, 108, 111, 137, 161, 173] which suggest that there is a close link between photosynthesis, translocation and BNF [7, 22, 27, 65]. At constant temperature very little differences in ARA values were reported for rice [27, 172] and ARA of intact rice plants decreased [25] within 15 minutes after transferring the plants into dark even though root temperature was maintained at 35°C. Decapitation of rice plants had little effect on ARA [173] but it was stressed that in soil cores containing sorghum plants, ARA measurements ought to be made early to avoid underestimations of *in situ* ARA [169]. *S. alterniflora* incubated continuously in the dark exhibited the same ARA as those plants maintained in the

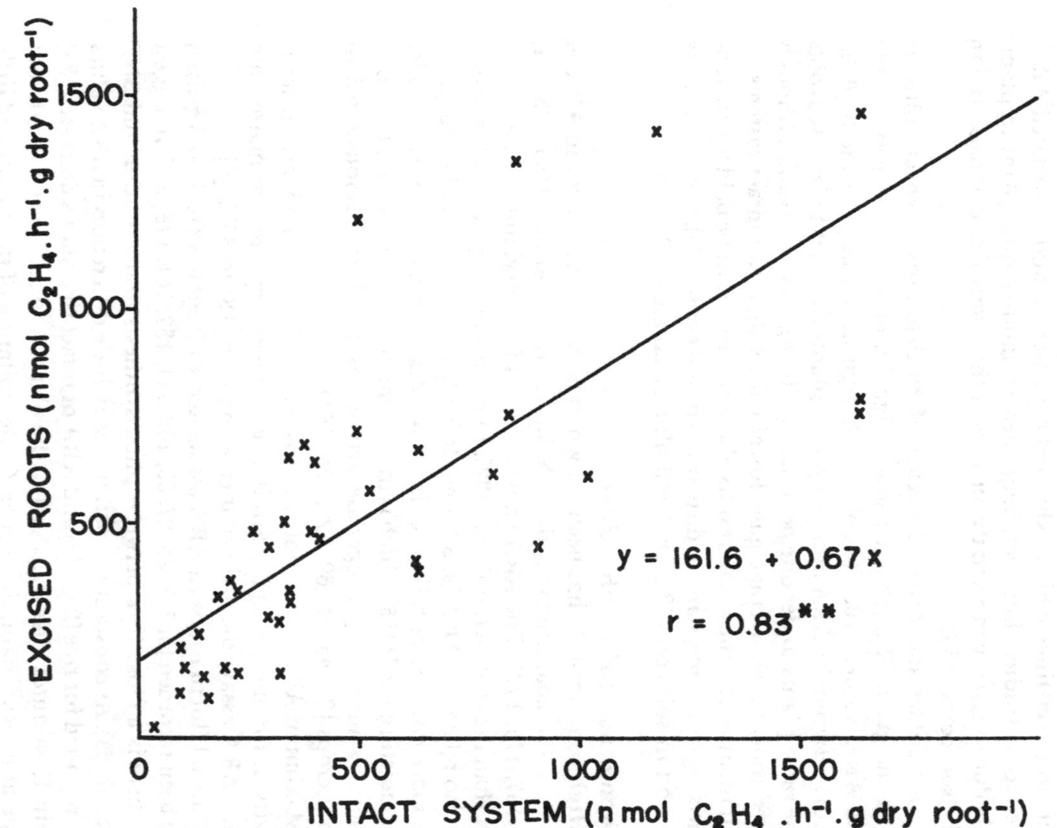

Fig. 4. Correlation of acetylene reduction activity of intact wheat plants grown in pots (500 g soil per pot) with the subsequent activity of excised roots (pre-incubated at pO_2=0.01 atm for 16 h) of the same plants (after Nery *et al.* [108]).

normal day/night cycle for the first 13 days of the experiment, suggesting that there could be a large supply of carbon in the root system from a photosynthate carbon, a pool which was demonstrated in experiments using $^{14}CO_2$[37]. The size of this pool will depend, on the rate of photosynthesis depending on light, temperature and other factors controlling carbon assimilation and the contradictory results so far described can be explained on this basis.

The addition of 10 μg ml^{-1} of urea nitrogen dramatically reduced the ARA of excised roots [161] as well as intact plants [172]. This kind of inhibition was seen by the application 2.8 μg ml^{-1} of ammoniacal nitrogen (200 μm NH_4Cl) with regard to the activity of root surface or the rhizosphere soil but not that of root interior [35]. Inhibition of ARA of excised sorghum roots was noticed to the extent of 50 per cent with 1.4 μg ml^{-1} (100 μm) of ammonium or nitrate nitrogen and the levels of combined nitrogen required for complete inhibition was 140 μg N ml^{-1} [109]. In one instance roots of very young rice plants showed nitrogenase activity [72] but in another instance no ARA was detected in excised roots until 42 days after germination which was attributed to the inability of diazotrophs to fix N_2 [26] in the presence of NH_4^+ levels above 20 μg N g dry soil^{-1} (Figs. 5 and 6). Thus it becomes apparent that if plants are provided with sufficient nitrogen, they will not develop associated ARA. Significant negative correlation ($r = -0.58$, $p = 0.01$) has been seen between the percentage of nitrogen in root and associated ARA [24], indicating that the availability of nitrogen in soil regulates nitrogenase activity [2, 63] (Fig. 7). Fertilizer nitrogen application (20 kg N ha^{-1} of NH_4NO_3) did not influence ARA in field grown *Pennisetum purpureum* and *Digitaria decumbens* [65] whereas heavy applications of nitrogen fertilizers inhibited ARA [129] indicating that reasonable starter dose of fertilizer nitrogen is not harmful to BNF activities.

Effects of *Azospirillum* inoculation on plant yield

While strain selections for rhizobial effects on legumes can be easily done by standard plant tests [165], pot and field trials have to be resorted to for selecting strains of bacteria for associative BNF. Many reports show statistically significant increases in yield and plant nitrogen content of wheat, maize and forage grasses [4, 14, 33, 50, 113, 114, 132]. Clearcut and repeatable strain selections have, however, not been possible [70] and in most experiments, *A. brasilense* type strain Sp. 7 (ATCC 29145) or two similar nir$^+$ strains (Sp. 13 and cd) have been used which were not isolated from respective roots. These nir$^+$ strains have been shown to get established on the root surface whereas nir$^-$ strains obtained from surface-sterilized roots got established within roots.

With all the aforesaid limitations, surprisingly, *Azospirillum* inoculation field trials have proved successful as those initial trials of Florida and Bahamas [34, 150, 156]. Under temperate conditions variable but signifi-

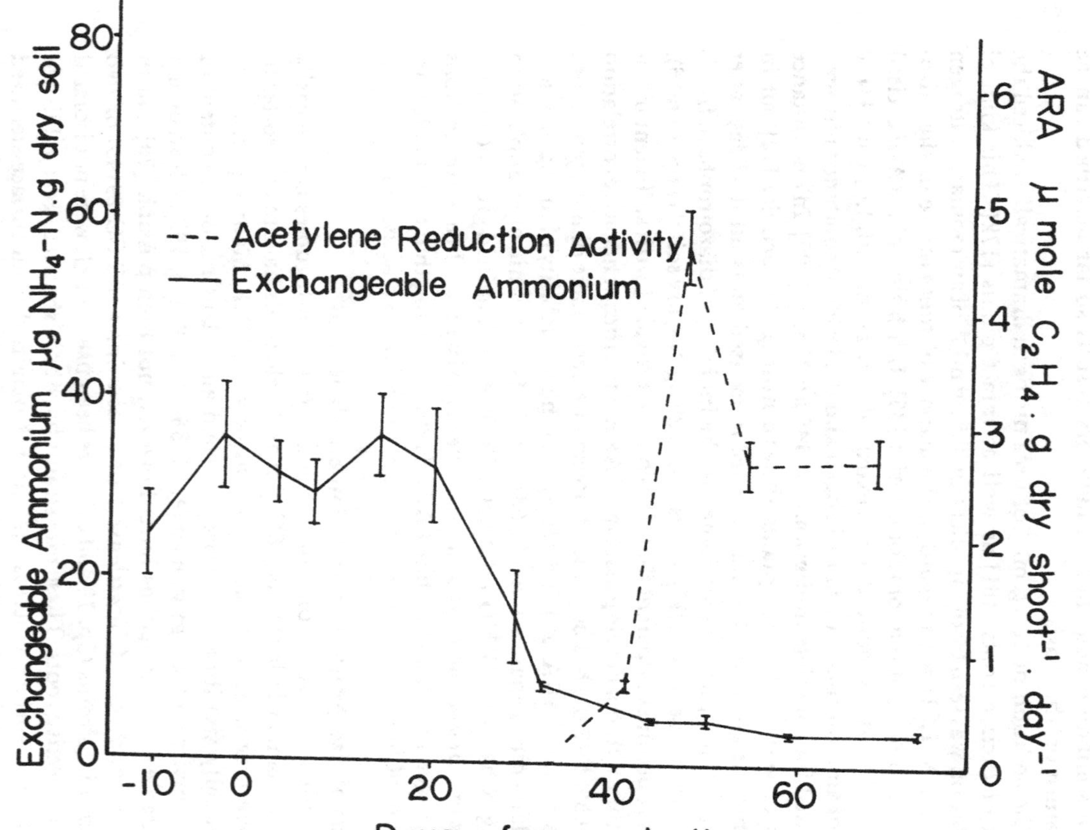

Fig. 5. Soil exchangeable ammonium levels and the acetylene reduction activity of excised roots of rice grown in flooded pots of 2.5 kg of Cunupia fine sandy clay soil (Aquic Eutropept). After Boddey [26].

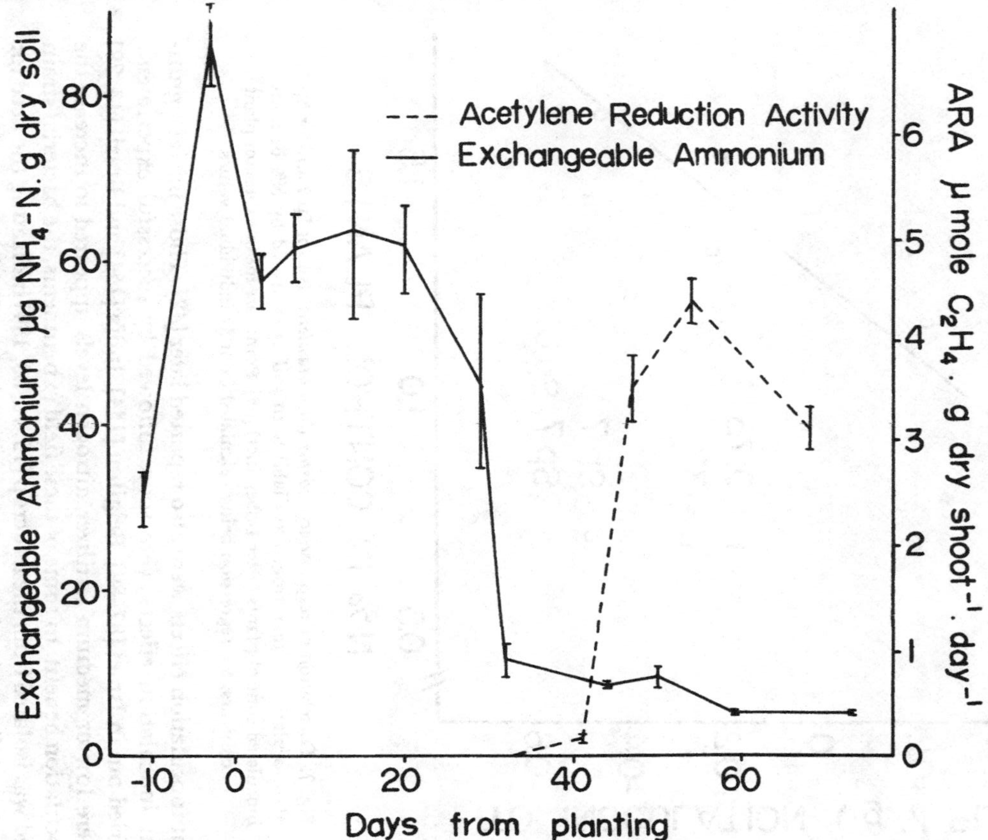

Fig. 6. Soil exchangeable ammonium levels and the acetylene reduction activity of excised roots of rice grown in flooded pots of 2.5 kg of Cacandee clay soil (Typic Pelludert). After Boddey [26].

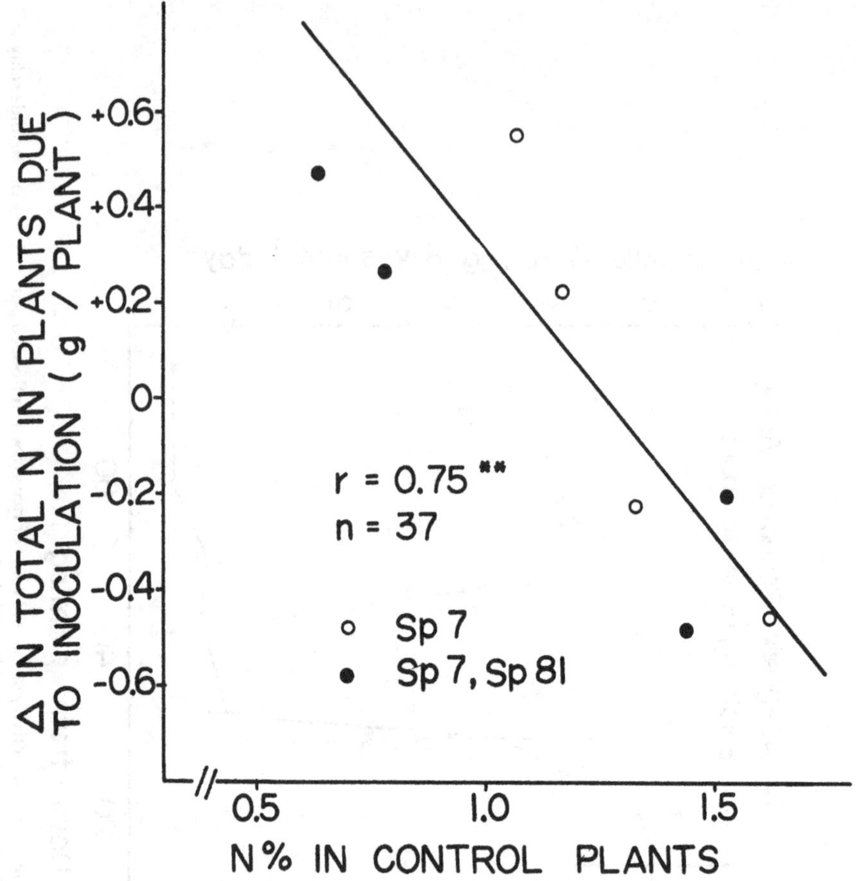

Fig. 7. Correlation of response to *Azospirillum brasilense* (strain Sp 7 and/or Sp 81) inoculation with nitrogen availability in soil as evaluated by N% in non-inoculated control plants (after Döbereiner) [63]. Points are mean values for plant cultivars, the regression being calculated from 37 individual values.

cant inoculation effects were also reported later [42]. Those initial significant inoculation effects were soon followed by successful experiments carried out in Israel [87–89], Belgium [134], India [154] and Brazil [13, 79] where low or medium fertilizer nitrogen levels appeared to increase the inoculation benefit. In some of these field experiments, the bacterial strain used was isolated from the test crop [13, 79, 83, 134] indicating the need for using specific strains for good response (also see Table 5).

Apart from nitrogen, growth substances produced by *Azospirillum* [157] are believed to enhance root growth leading to the better growth of plants. Harmonal effects or the effects of pectinolytic enzymes are believed to produce the so-called 'sponge effect' [115] which enhances the nutrient uptake by plants. No significant effects of inoculation were observed on the vegetative parts of wheat due to inoculation with *A. brasilense* but there was a significant effect of inoculation by way of increased total nitrogen and labelled fertilizer (^{15}N) recovery in the head which may be attributed to an efficient assimilation and translocation of fertilizer nitrogen to the grains (unpublished results from Dobereiner's laboratory). The reduction of NO_3^- to NO_2^- is considered as a limiting factor in NO_3^- assimilation in maize [18] whereas in legumes reduced nitrogen is more efficiently translocated to the grain than NO_3 [120]. This nitrate reductase activity (NO_3^- to NO_2^-) of *Azospirillum* spp. [110] could be responsible for the surprising results in Table 7 if it is assumed that plant enzymes are available to reduce the NO_2^- at the site of its formation.

Table 7. Nitrogen accumulation and ^{15}N incorporation in wheat[1] inoculated with various *Azospirillum* strains

Inoculum	Dry weight, g	Heads		% ^{15}N excess	Excess ^{15}N recovered, mg
		% nitrogen	Total nitrogen mg		
IN GRAIN					
Sp 245[2]	46.7	2.72a	1271a	0.171	2.20a
Sp 107st[2]	42.3	2.83a	1195a	0.190	2.30a
Sp 7[3]	45.2	2.81a	1276a	0.159	2.00a
Control	39.3	2.21b	866b	0.156	1.33b
LSD[4]	9.4	0.33	318	0.044	0.67
P=	n.s.	0.001	0.01	n.s.	0.01
IN STRAW					
Sp 245[2]	89.6	0.985	889	0.163	1.41
Sp 107st[2]	82.0	1.024	843	0.169	1.44
Sp 7[3]	84.1	1.084	910	0.142	1.30
Control	77.0	1.116	855	0.162	1.38
LSD[4]	17.0	0.159	243	0.032	0.52
P=	n.s.	n.s.	n.s.	n.s.	n.s.

[1]Wheat plants (cv. BH 1146) grown in the field in concrete cylinders (60 cm diameter × 50 cm deep) filled with red-yellow podzolic soil. ^{15}N labelled ammonium sulphate (2.10 atom % ^{15}N excess) added at a rate of 20 kg nitrogen ha^{-1} 30 days after seeding.

[2]*Azospirillum brasilense* nir^- isolated from surface sterilized wheat roots.

[3]*A. brasilense* nir+ (ATCC 29145) isolated from rhizosphere soil of *Digitaria decumbens*.

[4]LSD = Least significant difference (Tukey) at P = 0.05.

Source: R.M. Boddey, V.L.D. Baldani and J. Döbereiner, unpublished.

CONCLUSIONS

Root-associated BNF is a major source of nitrogen for lowland rice [93, 168], estimated at 30–60 kg N ha^{-1} $crop^{-1}$, probably between 20 and 50 per cent of the total plant nitrogen. The evidences in other cereal crops for BNF input are not substantial and the data are scanty and inconsistent.

Azospirillum does not show specificity like the wheat-*Bacillus* or *P. notatum*-*A. paspali* associations but the preferential infection of C_3 cereals by *A. brasilense* nir$^-$ strains and C_4 plants by *A. lipoferum* nir$^-$ strains exist although with the exception of sugar cane. Rice has been shown to select for *Pseudomonas*-like N_2- fixing bacteria and *Spartina* roots to *Camphylobacter* which may also be cited as instances of host-bacteria specificity.

Inoculation of cereals with *Azospirillum* spp. has been shown to increase growth and total nitrogen of plants in many field and pot trials in soils with low native *Azospirillum* (as in Israel) or in places where selected strains isolated from surface sterilized roots of the same host were used as inoculants. These effects may be attributed to N_2 fixation and improvement of root growth by growth factor production by the inoculant which indirectly helps in better uptake of plant nutrients from soil.

Whether or not technologies which enhance the efficiency of graminae/diazotroph associations will be exploited for practical agriculture in the foreseeable future will depend on their economic viability and adaptability to the skills and needs of the farmers. It does however take the equivalents of seven barrels of crude oil to produce one ton of fertilizer nitrogen. In 1981, developing countries like Brazil and India, paid 30 per cent of their import bills towards petroleum products which is really significant. Part of this imported petroleum is used in the manufacture of fertilizer nitrogen and therefore, developing countries should aim at supplementing or replacing fertilizer nitrogen for crops with biologically fixed nitrogen. This requires high-priority research on BNF technology, especially in relation to associative symbiosis.

ACKNOWLEDGEMENTS

Part of the research at EMBRAPA-PNPBS had financial support from FINEP and from the Brazilian Research Council (NPq).

REFERENCES

1. Albrecht, S.L., Okon, Y. and Burris, R.H. Effects of light and temperature on association between *Zea mays* and *Spirillum lipoferum*, *Plant Physiol. 60*, 528-531 (1977).
2. Albrecht, S.L., Okon, Y., Lonnquist, J. and Burris, R.H. Nitrogen fixation by corn-*Azospirillum* associations in a temperate climate, *Crop Sci. 21*, 301-306 (1981).
3. App, A.A., Watanabe, I., Alexander, M., Ventura, W., Daez, C., Santiago, T. and De-Datta, S.K. Non-symbiotic nitrogen-fixation associated with the rice plant in flooded soils, *Soil Sci. 130*, 283-289 (1980).
4. Avivi, Y. and Feldman, M. The response of wheat to bacteria of the genus *Azospirillum*, *Israel J. Bot. 31*, 237-245 (1982).
5. Balandreau, J.P. Mesure de l'activité nitrogénasique des microorga-

nismes fixateurs libres d'azote de la rhizosphère de quelques graminées, *Rev. Écol. Biol. Sol. 12*, 273-290 (1975).

6. Balandreau, J.P. and Villemin, G. Fixation biologique de l'azote moleculaire en savane de Lamto (basse Cote-d'Ivoire) Resultats preliminares, *Rev. Écol. Biol. Sol. 10*, 25-33 (1973).
7. Balandreau, J.P., Millier, C.R. and Dommergues, Y.R. Diurnal variations of nitrogenase activity in the field, *Appl. Microbiol. 27*, 662-665 (1974).
8. Balandreau, J.P., Millier, C.R., Weinhard, P., Ducerf, P. and Dommergues, Y.R. A modelling approach of acetylene reducing activity of plant-rhizosphere diazotroph systems, pp. 523-529, in *Recent Developments in Nitrogen Fixation* (Editors, W. Newton, J.R. Postgate and C. Rodriguez-Barrueco) (1977).
9. Balandreau, J.P., Ducerf, P., Hamad-Fares, I., Weinhard, P. Rinaudo, G., Millier, C. and Dommergues, Y. Limiting factors in grass nitrogen fixation, pp. 275-302, in *Limitations and Potentials for Biological Nitrogen Fixation in the Tropics* (Editors, J. Döbereiner et al.), Plenum Press (1978).
10. Baldani, V.L.D. and Döbereiner J. Host-plant specificity in the infection of cereals with *Azospirillum* spp., *Soil Biol. Biochem. 12*, 433-440 (1980).
11. Baldani, J.I., Pereira, P.A.A., Rocha, R.E.M. da and Döbereiner, J. Especificidade na infeccao de raízes por *Azospirillum* spp. em plantas com via fotossintética C_3eC_4*Pesq. Agropec. Bras. 16*, 325-330 (1981).
12. Baldani, J.I., Baldani, V.L.D., Xavier, D.F., Boddey, R.M. and Döbereiner, J. Efeito de calagem no número de actinomicetos e na porcentagem de bactérias resistentes a estreptomicina na rizosfera de milho, trigo e feijão, *Rev. Microbiol. 13*, 250-263 (1982).
13. Baldani, V.L.D., Baldani, J.I. and Döbereiner, J. Effects of *Azospirillum* inoculation on root infection and nitrogen incorporation in wheat, *Can. J. Microbiol.* (in press) (1983).
14. Baltensperger, A.A., Schank, S.C., Smith, R.L., Litell, R.C., Bouton, J.H. and Dudeck, A.E. Effect of inoculation with *Azospirillum* and *Azotobacter* on turf-type Bermuda genotypes, *Crop Sci. 18*, 1043-1045 (1978).
15. Barber, L.E., Tjepkema, J.D., Russell, S.A. and Evans, H.J. Acetylene reduction (nitrogen fixation) associated with corn inoculated with *Spirillum, Appl. Environ. Microbiol. 32*, 108-113 (1976).
16. Barber, L.E., Tjepkema, J.D. and Evans, H.J. Acetylene reduction in the root environment of some grasses and other plants in Oregon, in *Environmental Role of Nitrogen fixing Blue-green Algae and Asymbiotic Bacteria, Bull. Ecol.* (Stockholm) *26*, 366-372 (1978).
17. Barraquio, W.L. and Watanabe, I. Occurrence of aerobic nitrogen fixing bacteria in wetland and dry land plants, *Soil Sci. Plant Nutrit. 27*, 121-125 (1981).

18. Beevers, L. and Hageman, R.H. Nitrate reduction in higher plants, *Ann. Rev. Plant Physiol. 20*, 495–522 (1969).
19. Behling-Miranda, C.H., Boddey, R.M. and Döbereiner, J. The effects of disturbance and soil moisture on intact core acetylene reduction measurements of nitrogenase activity associated with grasses, *An. Acad. Bras. Ciênc. 54*, 760–761 (1982).
20. Berg, R.H., Vasil, V. and Vasil, I.K. The biology of *Azospirillum*-sugarcane association. II. Ultrastructure, *Protoplasma 101*, 143–163 (1979).
21. Berg, R.H., Tyler, M.E., Novick, N.J., Vasil, V. and Vasil, I.K. Biology of *Azospirillum*-sugarcane association enhancement of nitrogenase activity, *Appl. Environ. Microbiol. 39*, 642–649 (1980).
22. Berkum, P. van and Bohlool, B.B. Evaluation of nitrogen-fixation by bacteria in association with roots of tropical grasses, *Microbiol. Rev. 44*, 491–517 (1980).
23. Berkum, P. van and Day, J.M. Nitrogenase activity associated with soil cores of grasses in Brazil, *Soil Biol. Biochem. 12*, 137–140 (1980).
24. Berkum, P. van and Sloger, C. Ontogenetic variation of nitrogenase, nitrate reductase, and glutamine synthetase activities in *Oryza sativa, Plant Physiol. 68*, 722–726 (1981).
25. Berkum, P. van and Sloger, C. Physiology of root-associated nitrogenase activity in *Oryza sativa, Plant Physiol. 69*, 1161–1164 (1982).
26. Boddey, R.M. Biological nitrogen fixation in the rhizosphere of lowland rice, Ph.D. thesis, Faculty Agric., Univ. West Indies, St. Augustine, Trinidad (1980).
27. Boddey, R.M. and Ahmad, N. Seasonal variations in nitrogenase activity of various rice varieties measured with an *in situ* acetylene reduction technique in the field, pp. 219–229, in *Associative N_2 Fixation*, Vol. II (Editors, P.B. Vose and A.P. Ruschel), CRC Press Inc., Palm Beach, Florida, USA (1981).
28. Boddey, R.M. and Döbereiner, J. Associations of *Azospirillum* and other diazotrophs with tropical Gramineae, *Trans. 12th Int. Congr. Soil Sci.* New Delhi, India. Symp. Papers 1, Non-Symbiotic Nitrogen Fixation and Organic Matter in the Tropics, pp. 28–47 (1982).
29. Boddey, R.M., Quilt, P. and Ahmad, N. Acetylene reduction in the rhizosphere of rice: Methods of assay, *Plant Soil 50*, 567–574 (1978).
30. Boddey, R.M., Chalk, P.M., Victoria, R. and Matsui, E. The ^{15}N isotope dilution technique applied to the estimation of biological nitrogen fixation associated with *Paspalum notatum* cv. batatais in the field, *Soil Biol. Biochem. 15*, 25–32 (1983a).
31. Boddey, R.M., Chalk, P.M., Victoria, R.L., Matsui, E. and Döbereiner, J. The use of the ^{15}N isotope dilution technique to estimate the contribution of associated biological nitrogen fixation to the nitrogen nutrition of *Paspalum notatum* cv. batatais, *Can. J. Microbiol.* (in press) (1983b).

32. Bouton, J.H. and Brooks, C.O. Screening pearl millet for variability in supporting bacterial acetylene reduction activity, *Crop Sci. 22*, 680-681 (1982).
33. Bouton, J.H. and Zuberer, D.A. Response of *Panicum maximum* Jacq. to inoculation with *Azospirillum brasilense, Plant Soil 52*, 585-590 (1979).
34. Bouton, J.H., Smith, R.L., Schank, S.C., Burton, G.W., Tyler, M.E., Littell, R.C., Gallaher, R.N. and Quesenberry, K.H. Response of pearl millet inbreds and hybrids to inoculation with *Azospirillum brasilense, Crop Sci. 19*, 12-16 (1979).
35. Boyle, C.D. and Patriquin, D.G. Endorhizal and exorhizal acetylene-reducing activity in a grass (*Spartina alterniflora* Loisel)—diazotroph association, *Plant Physiol. 66*, 276-280 (1980).
36. Boyle, C.D. and Patriquin, D.G. Acetylene reducing activity (ARA) by endorhizosphere diazotrophs, pp. 27-31, in *Associative* N_2 *Fixation*, Vol. II (Editors, P.B. Vose and A.P. Ruschel), CRC Press Inc., Palm Beach, Florida, USA (1981a).
37. Boyle, C.D. and Patriquin, D.G. Carbon metabolism of *Spartina alterniflora* Loisel in relation to that of associated nitrogen-fixing bacteria, *New Phytol. 89*, 275-288 (1981b).
38. Brown, M.E. Stimulation of streptomycin resistant bacteria in the rhizosphere of leguminous plants, *J. Gen. Microbiol. 24*, 369-377 (1961).
39. Büchanan, R.E. and Gibbons, N.E. Bergey's Manual of Determinative Bacteriology, 8th edition, Williams and Wilkins, Baltimore, USA (1974).
40. Bülow, J.F.W. von and Döbereiner, J. Potential for nitrogen fixation in maize genotypes in Brazil, *Proc. Nat. Acad. Sci. 72*, 2389-2393 (1975).
41. Burau, R.G. and Stout, P. Trans-aconitic acid in range grasses in early spring, *Science 150*, 766-767 (1965).
42. Burris, R.H., Albrecht, S.L. and Okon, Y. Physiology and biochemistry of *Spirillum lipoferum*, pp. 303-315, in *Limitations and Potentials for Biological Nitrogen Fixation in the Tropics* (Editors, J. Döbereiner et al.), Plenum Press, New York (1978).
43. Butler, N. and Calverley, J. At the back of every cloud the silver lining, *South, The Third World Magazine*, April, pp. 13-14 (1983).
44. Butz, R.G. and Long, R.C. L-malate as an essential component of the xylem fluid of corn seedling roots, *Plant Physiol. 64*, 684-689 (1979).
45. Capone, D.G. and Budin, J.M. Nitrogen fixation associated with rinsed roots and rhizomes of eelgrass *Zostera marina, Plant Physiol. 70*, 1601-1604 (1982).
46. Casselman, M.E., Patrick, W.H. Jr. and De Laune, R.D. Nitrogen fixation in a Gulf coast salt marsh, *Soil Sci. Soc. Amer. J. 45*, 51-56 (1981).

47. Child, J.J. and Kurz, W.G.W. Inducing effect of plant cells on nitrogenase activity by *Spirillum* and *Rhizobium* in vitro, *Can. J. Microbiol. 24*, 143-148 (1978).
48. Clark, R.B. Organic acids of maize (*Zea mays* L.) as influenced by mineral deficiency, *Crop Sci. 8*, 165-167 (1968).
49. Clark, R.B. Organic acids and mineral cations of corn plant parts with age, *Commun. Soil Sci. Plant Anal. 7*, 585-600 (1976).
50. Cohen, E., Okon, Y., Kigel, J., Nur, I. and Henis, Y. Increases in dry weight and total nitrogen in *Zea mays* and *Setaria italica* associated with nitrogen-fixing *Azospirillum* spp. *Plant Physiol. 66*, 746-749 (1980).
51. Dart, P.J. and Subba Rao, R.V. Nitrogen fixation associated with sorghum and millet, pp. 169-177, in *Associative N_2 Fixation, Vol. I* (Editors, P.B. Vose and A.P. Ruschel), CRC Press Inc., Palm Beach, Florida, USA (1981).
52. Dart, P.J. and Wani, S.P. Non-symbiotic nitrogen fixation and soil fertility, *Trans. 12th Int. Congr. Soil Sci.*, New Delhi, India Symp. Papers 1. Non-symbiotic nitrogen fixation and organic matter in the tropics, pp. 3-27 (1982).
53. Day, J.M. Nitrogen-fixing associations between bacteria and tropical grass roots, pp. 273-288, in *Biological Nitrogen Fixation in Farming Systems of the Tropics* (Editors, A. Ayanaba and P.J. Dart), John Wiley & Sons (1977).
54. Day, J.M. and Döbereiner, J. Physiological aspects of N_2- fixation by a *Spirillum* from *Digitaria* roots, *Soil Biol. Biochem. 8*, 45-50 (1976).
55. Day, J.M., Harris, D., Dart, P.J. and Berkum, P. van. The Broadbalk experiment. An investigation of nitrogen gains from non-symbiotic nitrogen fixation, pp. 71-84, in *Nitrogen Fixation by Free-living Micro-organisms* (Editor, W.D.P. Stewart) (1975).
56. De, P.K. and Sulaiman, M. Influence of algal growth in the rice fields on the yield of crop, *Ind. J. Agr. Sci. 20*, 327-342 (1950).
57. Deacon, J.W. and Lewis, S.J. Natural senescense of the root cortex of spring wheat in relation to susceptibility to common root rot (*Cochliobolus sativus*) and growth of a free-living nitrogen-fixing bacterium, *Plant Soil 66*, 13-20 (1982).
58. De-Polli, H., Matsui, E., Döbereiner, J. and Salati, E. Confirmation of nitrogen fixation in two tropical grasses by $^{15}N_2$ incorporation, *Soil Biol. Biochem. 9*, 119-123 (1977).
59. De-Polli, H., Boyer, C.D. and Neyra, C.A. Nitrogenase activity associated with roots and stems of field grown corn (*Zea mays* L.) plants, *Plant Physiol. 70*, 1609-1613 (1982).
60. Döbereiner, J. *Azotobacter paspali* sp. n. uma bactéria fixadora de nitrogênio na rizosfera de *Paspalum*, *Pesq. Agropec. Bras. 1*, 357-365 (1966).

61. Döbereiner, J. Further research on *Azotobacter paspali* and its variety specific occurrence in the rhizosphere of *Paspalum notatum* Flugge, *Zentralbl. Bakteriol. Parasitenkd (Abteilung 2) 124*, 224–230 (1970).
62. Döbereiner, J. Nitrogen-fixing bacteria in the rhizosphere, pp. 86–120, in *The Biology of Nitrogen Fixation* (Editor, A. Quispel), North Holland, Amsterdam (1974).
63. Döbereiner, J. Nitrogen fixation in grass-bacteria associations in the tropics, pp. 51–68, in *Isotopes in Biological Dinitrogen Fixation*, IAEA, Vienna (1978).
64. Döbereiner, J. and Baldani, V.L.D. Selective infection of maize roots by streptomycin-resistant *Azospirillum lipoferum* and other bacteria, *Can. J. Microbiol. 25*, 1264–1269 (1979).
65. Döbereiner, J. and Day, J.M. Nitrogen fixation in the rhizosphere of tropical grasses, pp. 39–56, in *Nitrogen Fixation by Free-living Micro-organisms* (Editor, W.D.P. Stewart), Cambridge Univ. Press, Cambridge, England (1975).
66. Döbereiner, J. and Day, J.M. Associative symbioses in tropical grasses: characterization of microorganisms and dinitrogen-fixing sites, pp. 518–538, in *Proceeding of the 1st International Symposium on Nitrogen Fixation* (Editors, W.E. Newton and C.J. Nyman), Washington State University Press, Pullman (1976).
67. Döbereiner, J., Day, J.M. and Dart, P.J. Nitrogenase activity in the rhizosphere of sugar cane and some other tropical grasses, *Plant Soil 37*, 191–196 (1972a).
68. Döbereiner, J., Day, J.M. and Dart, P.J. Nitrogenase activity and oxygen sensitivity of the *Paspalum notatum-Azotobacter paspali* association, *J. Gen. Microbiol. 71*, 103–116 (1972b).
69. Döbereiner, J., Day, J.M. and Dart, P.J. Rhizosphere association between grasses and nitrogen fixing bacteria: Effect of O_2 on nitrogenase activity in the rhizosphere of *Paspalum notatum, Soil Biol. Biochem. 5*, 157–159 (1973).
70. Döbereiner, J., Nery, M. and Marriel, I.E. Ecological distribution of *Spirillum lipoferum*. Beijerinck, *Can. J. Microbiol. 22*, 1464–1473 (1976).
71. Dömmergues, Y.R. and Rinaudo, G. Factors affecting N_2 fixation in the rice rhizosphere, pp. 241–260, in *Nitrogen and Rice*. International Rice Research Institute, Laguna, Philippines (1979).
72. Dömmergues, Y.R., Balandreau, J.P., Rinaudo, G. and Weinhard, P. Non-symbiotic nitrogen fixation in the rhizospheres of rice, maize and different tropical grasses, *Soil Biol. Biochem. 5*, 83–89 (1973).
73. Ela, S.W., Anderson, M.A. and Brill, W.J. Screening and selection of maize to enhance associative bacterial nitrogen fixation, *Plant Physiol. 70*, 1564–1567 (1982).
74. Eskew, D.L. and Ting, I.P. Comparison of intact plant and excised

root assays for acetylene reduction in grass rhizospheres, *Plant Sci. Lett. 8*, 327-331 (1977).

75. Eskew, D.L., Focht, D.D. and Ting, I.P. Nitrogen fixation, denitrification and pleomorphic growth in a highly pigmented *Spirillum lipoferum, Appl. Environ. Microbiol. 34*, 582-585 (1977).
76. Eskew, D.L., Eaglesham, A.R.J. and App, A.A. Heterotrophic $^{15}N_2$ fixation and distribution of newly fixed nitrogen in a rice-flooded soil system, *Plant Physiol. 68*, 48-52 (1981).
77. Firth, P., Thitipoca, H., Suthipradit, S., Wetselaar, R. and Beech, D.F. Nitrogen balance studies in the Central Plain of Thailand, *Soil Biol. Biochem. 5*, 41-46 (1973).
78. Flett, R.J., Hamilton, R.D. and Campbell, N.E.R. Aquatic acetylene reduction techniques: Solutions to several problems, *Can. J. Microbiol. 22*, 43-51 (1976).
79. Freitas, J.L.M. de, Rocha, R.E.M. da, Pereira, P.A.A. and Döbereiner, J. Matéria orgânica e inoculação com *Azospirillum* na incorporacão de N pelo milho, *Pesq. Agropec. Bras. 17*, 1423-1432 (1982).
80. Greenland, D.J. Contribution of microorganisms to the nitrogen status of tropical soils, pp. 13-25, in *Biological Nitrogen Fixation in Farming Systems of the Tropics* (Editors, A. Ayanaba and P.J. Dart), John Wiley & Sons (1977).
81. Greenland, D.J. and Nye, P.H. Increases in the carbon and nitrogen contents of tropical soils under natural fallows, *J. Soil Sci. 10*, 284-299 (1959).
82. Hardy, R.W.F., Holsten, R.D., Jackson, E.K. and Burns, R.C. The acetylene-ethylene assay for N_2-fixation: Laboratory and field evaluations, *Plant Physiol. 43*, 1185-1207 (1968).
83. Hegazi, N.A., Monib, M., Amer, H.A. and Shokr, E.S. Response of maize plants to inoculation with Azospirilla and/or straw amendment under Egyptian conditions as measured by ATP, acetylene reduction and ^{15}N-isotope dilution analysis. Paper presented at the 2nd Int. Symp. N_2 Fixation with Non-Legumes, Banff, Alberta, Canada, 5-10 September (1982).
84. Hill, S., Drozd, J.W. and Postgate, J.R. Environmental effects on the growth of nitrogen fixing bacteria, *J. Appl. Chem. Biotechnol. 22*, 541 (1972).
85. Ito, O., Cabrera, D. and Watanabe, I. Fixation of dinitrogen-15 associated with rice plants, *Appl. Environ. Microbiol. 39*, 554-558 (1980).
86. Jaibeyo, E.O. and Moore, A.W. Soil nitrogen accretion under different covers in a tropical rain-forest environment, *Nature 197*, 317-318 (1963).
87. Kapulnik, Y., Sarig, S., Nur, I., Okon, Y., Kigel, J. and Henis, Y. Yield increases in summer cereal crops of Israel in fields inoculated with *Azospirillum, Exp. Agric. 17*, 179-187 (1981a).

88. Kapulnik, Y., Kigel, J., Okon, Y., Nur, I. and Henis, Y. Effect of *Azospirillum* inoculation on some growth parameters and N-content of wheat, sorghum and *Panicum, Plant Soil 61*, 65–70 (1981b).
89. Kapulnik, Y., Sarig, S., Nur, I., Okon, Y. and Henis, Y. The effect of *Azospirillum* inoculation on growth and yield of corn, *Israel J. Bot. 31*, 247–256 (1982).
90. Kavimandan, S.K., Lakshmi-Kumari, M. and Subba Rao, N.S. Non-symbiotic nitrogen fixing bacteria in the rhizosphere of wheat, maize and sorghum, *Proc. Ind. Acad. Sci. Sect. B 87*, 299–302 (1978).
91. Knowles, R. The significance of asymbiotic dinitrogen fixation by bacteria, pp. 33–83, in *A Treatise on Dinitrogen Fixation* (Editors, F.W. Hardy and A.H. Gibson), John Wiley & Sons, New York (1977).
92. Koch, B.L. Associative nitrogenase activity by some Hawaiian grass roots, *Plant Soil 47*, 703–706 (1977).
93. Koyama, T. and App, A. Nitrogen balance in flooded rice soils, pp. 95–104, in *Nitrogen and Rice*, International Rice Research Institute, Laguna, Philippines (1979).
94. Lakshmi, V., Satyanarayana Rao, A., Vijayalakshmi, K., Lakshmi-Kumari, M., Tilak, K.V.B.R. and Subba Rao, N.S. Establishment and survival of *Spirillum lipoferum, Proc. Indian Acad. Sci. Section B 86*, 397–405 (1977).
95. Larson, R.I. and Atkinson, T.G. A cytogenetic analysis of reaction to common root rot in some hand red spring wheats, *Can. J. Bot. 48*, 2059–2067 (1970).
96. Larson, R.I. and Neal, J.L. Selective colonization of the rhizosphere of wheat by nitrogen-fixing bacteria, in *Environmental Role of Nitrogen-fixing Blue-green Algae and Asymbiotic Bacteria, Bull. Ecol.* (Stockholm) *26*, 331–342 (1978).
97. Lee, K.K. and Watanabe, I. Problems of the acetylene reduction technique applied to water saturated paddy soils, *Appl. Environ. Microbiol. 34*, 654–660 (1977).
98. Lee, K.K., Castro, T. and Yoshida, T. Nitrogen fixation throughout growth, and varietal differences in nitrogen fixation by the rhizosphere of rice planted in pots, *Plant Soil 48*, 613–619 (1977).
99. Magalhães, F.M.M., Patriquin, D. and Döbereiner, J. Infection of field grown maize with *Azospirillum* spp., *Rev. Brasil. Biol. 39*, 587–596 (1979).
100. Magalhães, F.M.M., Baldani, J.I. and Döbereiner, J. A new acid-tolerant *Azospirillum* species, *Arch. Microbiol.* (in press) (1983).
101. Matsui, E., Vose, P.B., Rodrigues, N.S. and Ruschel, A.P. Use of $^{15}N_2$ enriched gas to determine N_2 fixation by undisturbed sugar-cane plant in the field, pp. 153–161, in *Associative N_2 Fixation*, Vol. II (Editors, P.B. Vose and A.P. Ruschel), CRC Press Inc., Palm Beach, Florida, USA (1981).

102. McAuliffe, C., Chamblee, D.S., Uribe-Arango, H. and Woodhouse, W.W. Influence of inorganic nitrogen on nitrogen fixation by legumes as revealed by ^{15}N, *Agron. J. 50*, 334-347 (1958).

103. McClung, C.R. and Patriquin, D.G. Isolation of a nitrogen fixing *Campylobacter* species from the roots of *Spartina alterniflora* Loisel, *Can. J. Microbiol. 26*, 881-886 (1980).

104. Mengel, K. and Kirkby, E.A. Principles of plant nutrition, Int. Potash Institute, Bern, Switzerland (1979).

105. Moore, A.W. Non-Symbiotic Nitrogen Fixation in Soil and Soil Plant Systems, *Soils e Fert. 29*, 113-128 (1966).

106. Neal, J.L. Jr. and Larson, R.I. Acetylene reduction by bacteria isolated from the rhizosphere of wheat, *Soil Biol. Biochem. 8*, 151-155 (1976).

107. Neal, J.L. Jr., Larson, R.I. and Atkinson, T.G. Changes in rhizosphere populations of selected physiological groups of bacteria related to substitution of specific pairs of chromosomes in spring wheat, *Plant Soil 39*, 209-212 (1973).

108. Nery, M., Abrantes, G.T.V., Santos, D. dos and Döbereiner, J. Fixação de nitrogênio em trigo, *Rev. Bras. Ci. Solo. 1*, 15-20 (1977).

109. Neyra, C.A. and van Berkum, P. Nitrogenase activity in isolated sorghum roots. Effect of bicarbonate and inorganic nitrogen, *Plant Physiol. 57*, Suppl. Abs. 533 (1976).

110. Neyra, C.A. and van Berkum, P. Nitrate reduction and nitrogenase activity in *Spirillum lipoferum, Can. J. Microbiol. 23*, 306-310 (1977).

111. Neyra, C.A. and Döbereiner, J. Nitrogen fixation in grasses, *Adv. Agron. 29*, 1-38 (1977).

112. Neyra, C.A., Döbereiner, J., Lalande, R. and Knowles, R. Denitrification by N_2-fixing *Spirillum lipoferum, Can. J. Microbiol. 23*, 300-305 (1977).

113. Nur, I., Okon, Y. and Henis, Y. An increase in nitrogen content of *Setaria italica* and *Zea mays* inoculated with *Azospirillum, Can. J. Microbiol. 26*, 482-485 (1980).

114. O'Hara, G.W., Davey, M.R. and Lucas, J.A. Effect of inoculation of *Zea mays* with *Azospirillum brasilense* under temperate conditions, *Can. J. Microbiol. 27*, 871-877 (1981).

115. Okon, Y. *Azospirillum*, physiological properties, mode of association with roots and its application for the benefit of cereal and forage grass crops, *Israel J. Bot. 31*, 214-220 (1982).

116. Okon, Y., Albrecht, S.L. and Burris, R.H. Factors affecting growth and nitrogen fixation of *Spirillum lipoferum, J. Bacteriol. 127*, 1248-1254 (1976).

117. Okon, Y., Albrecht, S.L. and Burris, R.H. Methods for growing *Spirillum lipoferum* and for counting it in pure culture and in

association with plants, *Appl. Environ. Microbiol. 33*, 85-88 (1977).

118. Owens, L. Use of ^{15}N enriched soil to study N_2 fixation in grasses, p. 473, in *Genetic Engineering for Nitrogen Fixation* (Editor, A. Hollaender), Plenum Press, New York (1977).
119. Parker, C.A. Non-symbiotic nitrogen-fixing bacteria in soil. III. Total nitrogen changes in a field soil, *J. Soil. Sci. 8*, 48-59 (1957).
120. Pate, J.S. Transport and partitioning of nitrogenous solutes, *Ann. Rev. Plant Physiol. 31*, 313-340 (1980).
121. Patriquin, D.G. Nitrogen fixation (acetylene reduction) associated with cord grass, *Spartina alterniflora* Loisel, in *Environmental Role of Nitrogen-fixing Blue-green Algae and Asymbiotic Bacteria* (Editor, U. Granhall), *Ecol. Bull.* (Stockholm) *26*, 20-27 (1978a).
122. Patriquin, D.G. Factors affecting nitrogenase activity (acetylene reducing activity) associated with excised roots of the emergent halophyte *Spartina alterniflora* Loisel, *Aquatic Bot. 4*, 193-210 (1978b).
123. Patriquin, D.G. New developments in grass-bacteria associations, in *Advances in Agricultural Microbiology* (Editor, N.S. Subba Rao), Oxford & IBH Publishing Co., New Delhi (1981).
124. Patriquin, D.G. Nitrogen fixation in sugar cane litter, *Biol. Agric. Hort. 1*, 39-64 (1982).
125. Patriquin, D.G. and Döbereiner, J. Light microscopy observations of tetrazolium-reducing bacteria in the endorhizosphere of maize and other grasses in Brazil, *Can. J. Microbiol. 24*, 734-742 (1978).
126. Patriquin, D.G. and McClung, C.R. Nitrogen accretion and the nature and possible significance of N_2 fixation (acetylene reduction) in a Nova Scotian *Spartina alterniflora* stand, *Marine Biol. 47*, 227-242 (1978).
127. Patriquin, D.G., Döbereiner, J. and Jain, D.K. Sites and processes of association between diazotrophs and grasses, *Can. J. Mikrobiol.* (in press) (1983).
128. Pedersen, W.L., Chakrabarty, K., Klucas, R.V. and Vidaver, A.K. Nitrogen fixation associated with root of winter wheat and sorghum in Nebraska, *Appl. Environ. Microbiol. 35*, 129-135 (1978).
129. Pereira, P.A.A., Bülow, J.F.W. von and Neyra, C.A. Atividade da nitrogenase, nitrato redutase e acumulação de nitrogênio em milho braquítico *Zea mays* L. (cv. Piranão) em dois níveis de adubação nitrogenada, *Rev. Bras. Ciê. Solo 2*, 28-33 (1978).
130. Pereira, P.A.A., Baldani, J.I., Döbereiner, J. and Neyra, C.A. Nitrate reduction and nitrogenase activity in excised corn roots, *Can. J. Bot. 59*, 2445-2449 (1981).
131. Purchase, B.S. Nitrogen fixation associated with grasses. A potential source of nitrogen for Rhodesian agriculture, *Rhodesia Agric. J. 75*, 99-104 (1978).
132. Rennie, R.J. [15]N isotope dilution as a measure of dinitrogen fixation

by *Azospirillum brasilense* associated with maize, *Can. J. Bot. 58*, 21-24 (1980).

133. Rennie, R.J. and Larson, R.I. Dinitrogen fixation associated with disomic chromosome substitution lines of spring wheat, *Can. J. Bot. 57*, 2771-2775 (1979).

134. Reynders, L. and Vlassak, K. Use of *Azospirillum brasilense* as biofertilizer in intensive wheat cropping, *Plant Soil 66*, 217-223 (1982).

135. Rinaudo, G. Fixation biologique de l'azote dans trois types de sols de rizières de Côte d'Ivoire, *Thèse de Docteur Ingénieur*, Faculté des Sciences, Montpellier, France (1970).

136. Rinaudo, G. La fixation d'azote dans la rhizosphère du riz: Importance du type variétal, *Cah. ORSTOM, sér. Biol. 12*, 117-119 (1977).

137. Rinaudo, G., Hamad-Fares, I. and Dommergues, Y.R. Nitrogen fixation in the rice rhizosphere: Methods of measurement and practices suggested to enhance the process, pp. 313-322, in *Biological Nitrogen Fixation in Farming Systems of the Tropics* (Editors, A. Ayanaba and P.J. Dart), John Wiley & Sons (1977).

138. Rocha, R.E.M., Baldani, J.I. and Döbereiner, J. Specificity of infection by *Azospirillum* spp. in plants with C_4 photosynthetic pathway, pp. 67-69, in *Associative N_2 Fixation*, Vol. II (Editors, P.B. Vose and A.P. Ruschel), CRC Press Inc., Palm Beach, Florida, USA. (1981).

139. Ruschel, A.P., Henis, Y. and Salati, E. Nitrogen-15 tracing of N-fixation with soil grown sugar cane seedlings, *Soil Biol. Biochem. 7*, 181-182 (1975).

140. Ruschel, A.P., Victoria, R.L., Salati, E. and Henis, Y. Nitrogen fixation in sugar cane (*Saccharum officinarum* L.), in *Environmental Role of Nitrogen-fixing Blue-green Algae and Asymbiotic Bacteria* (Editor, U. Granhall), *Bull. Ecol.* (Stockholm) *26*, 297-303 (1978).

141. Ruscoe, A.W., Newcomb, E.H. and Burris, R.H. Pleomorphic forms in strains of *Spirillum lipoferum* by light and electron microscopy, Poster presented at Steenbock Kettering Int. Symp. Nitrogen Fixation, Madison, Wisconsin, USA, June 12-16 (1978).

142. Sampaio, M.J.A., Vasconcelos, L. de and Döbereiner, J. Characterisation of three groups within *Spirillum lipoferum* Beijerinck, in *Environmental Role of Nitrogen-fixing Blue-green Algae and Asymbiotic Bacteria* (Editor, U. Granhall), *Bull. Ecol.* (Stockholm) *26*, 364-365 (1978).

143. Sano, Y., Fujii, T., Iyama, S., Hirota, Y. and Komagata, K. Nitrogen fixation in the rhizosphere of cultivated and wild rice strains, *Crop Sci. 21*, 758-760 (1981).

144. Schank, S.C., Day, J.M. and De Lucas, E.O. Nitrogenase activity, nitrogen content, *in vitro* digestibility and yield of 30 tropical forage grasses in Brazil, *Trop. Agric. 54*, 119-125 (1977).

145. Scotti, M.R.M.M.L., Sá, N.M.H., Vargas, M.A.T. and Döbereiner, J.

Streptomycin resistance of *Rhizobium* isolates from Brazilian cerrados, *An. Acad. Brasil. Ciê. 54*, 733-738 (1982).

146. Sen, J. Is bacterial association a factor in nitrogen assimilation by rice plants, *Agric. J. India 24*, 229-231 (1929).

147. Singh, P.K. Use of *Azolla* and blue-green algae in rice cultivation in India, pp. 183-196, in *Associative N_2 Fixation*, Vol. II (Editors, P.B. Vose and A.P. Ruschel), CRC Press Inc., Palm Beach, Florida, U.S.A. (1981).

148. Smith, D. and Patriquin, D.G. Survey of angiosperms in Nova Scotia for rhizosphere nitrogenase (acetylene-reduction) activity, *Can. J. Bot. 56*, 2218-2223 (1978).

149. Smith, R.M., Thompson, D.O., Collier, J.W. and Hervey, R.J. Soil organic matter, crop yields, and land use in the Texas Blackland, *Soil Science 77*, 377-388 (1954).

150. Smith, R.L., Bouton, J.H., Schank, S.C., Quesenberry, K.H., Tyler, M.E., Gaskins, M.H. and Littell, C. Nitrogen fixation in grasses inoculated with *Spirillum lipoferum*, *Science 193*, 1003-1005 (1976).

151. Souto, S.M. and Döbereiner, J. Variação estacional da fixação de N_2 e assimilação de nitrato em gramíneas forrageiras tropicais, *Pesq. Agropec. Bras.* (in press) (1983a).

152. Souto, S.M. and Döbereiner, J. Novo método de determinação de fixação de N_2 em raízes extraídas de gramíneas forrageiras tropicais, *Pesq. Agropec. Bras.* (in press) (1983b)

153. Stout, P.R., Brownell, J. and Burau, R.G. Occurrence of trans-aconitate in range forage species, *Agron. J. 59*, 21-24 (1967).

154. Subba Rao, N.S. Response of crops to *Azospirillum* inoculation in India, pp. 137-144, in *Associative N_2 Fixation*, Vol. I (Editors, P.B. Vose and A.P. Ruschel), CRC Press, Palm Beach, Florida, USA (1981).

155. Tarrand, J.J., Krieg, N.R. and Döbereiner, J. A taxonomic study of the *Spirillum lipoferum* group, with descriptions of a new genus, *Azospirillum* gen. nov. and two species *Azospirillum lipoferum* (Beijerinck) comb. nov. and *Azospirillum brasilense* sp. nov., *Can. J. Microbiol. 24*, 967-980 (1978).

156. Taylor, R.W. Response of two grasses to inoculation with *Azospirillum* spp. in a Bahamian soil, *Trop. Agric. 56*, 361-365 (1979).

157. Tien, T.M., Gaskins, M.H. and Hubbell, D.H. Plant growth substances produced by *Azospirillum brasilense* and their effect on the growth of pearl millet (*Pennisetum americanum* L.), *Appl. Environ. Microbiol. 37*, 1016-1024 (1979).

158. Ting, I.P. Towards a model for malate accumulation in plant tissues, *Plant Sci. Letters 21*, 215-221 (1981).

159. Tjepkema, J. and van Berkum, P. Acetylene reduction by soil cores of maize and sorghum in Brazil, *Appl. Environ. Microbiol. 33*, 626-629 (1977).

160. Tjepkema, J.D. and Burris, R.H. Nitrogenase activity associated with some Wisconsin prairie grasses, *Plant Soil 45*, 81–94 (1976).

161. Trolldenier, G. Influence of some environmental factors on nitrogen fixation in the rhizosphere of rice, *Plant Soil 47*, 203–217 (1977).

162. Umali-Garcia, M., Hubbell, D.H. and Gaskins, M.H. Process of infection of *Panicum maximum* and *Spirillum lipoferum*, pp. 373–379, in *Environmental Role of Nitrogen-fixing Blue-green Algae and Asymbiotic Bacteria* (Editor, U. Granhall), *Bull. Ecol.* (Stockholm) *26*, 373–379 (1978).

163. Umali-Garcia, M., Hubbell, D.H., Gaskins, M.H. and Dazzo, F.B. Association of *Azospirillum* with grass roots, *Appl. Environ. Microbiol. 39*, 219–226 (1980).

164. Vasil, V., Vasil, I.K., Zuberer, D.A. and Hubell, D.H. The biology of *Azospirillum*-sugarcane association. I. Establishment of the association, *Z. Pflanzenphysiol. Bd. 95*, 141–147 (1979).

165. Vincent, J.M. A manual for the practical study of root-nodule bacteria, *IBP Handbook No. 15*, Blackwell Scientific Publications, Oxford (1970).

166. Vinther, F.P. Nitrogenase activity (acetylene reduction) during the growth cycle of spring barley (*Hordeum vulgare* L.), *Z. Pflanzenernaehr. Bodenk. 145*, 356–362 (1982).

167. Vlassak, K., Paul, E.A. and Harris, R.E. Assessment of biological nitrogen fixation in grassland and associated sites, *Plant Soil 38*, 637–649 (1973).

168. Walcott, J.J., Chauviroj, M., Chinchest, A., Choticheuy, P., Ferraris, R. and Norman, B.W. Long-term productivity of intensive rice cropping systems on the central plain of Thailand, *Exp. Agric. 13*, 305–316 (1977).

169. Wani, S.P., Dart, P.J. and Upadhyaya, M.N. Factors affecting nitrogenase activity (C_2H_2 reduction) associated with sorghum and millet estimated using the soil core assay, *Can. J. Microbiol.* (in press) (1983).

170. Watanabe, I. Biological nitrogen fixation in rice soils, pp. 465–478, in *Soil and Rice*, International Rice Research Institute, Laguna, Philippines (1978).

171. Watanabe, I. and Barraquio, W.L. Low levels of fixed nitrogen required for isolation of free-living N_2-fixing organisms from rice roots, *Nature* (London) *277*, 565–566 (1979).

172. Watanabe, I. and Cabrera, D.R. Nitrogen fixation associated with the rice plant grown in water culture, *Appl. Environ. Microbiol. 37*, 373–378 (1979).

173. Watanabe, I. and Lee, K.K. Non-symbiotic nitrogen fixation in rice and rice fields, pp. 289–305, in *Biological Nitrogen Fixation in Farming Systems of the Tropics* (Editors, A. Ayanaba and P.J. Dart), John Wiley & Sons (1977).

174. Watanabe, I., Lee, K.K. and Guzman, M. de. Seasonal change of N_2 fixing rate in rice field assayed by *in situ* acetylene reduction technique. II. Estimate of nitrogen fixation associated with rice plants, *Soil. Sci. Plant Nutrit. 24*, 465–471 (1978).

175. Watanabe, I., Barraquio, W.L. and Ladha, J.K. Isolation and identification of dinitrogen-fixing bacteria associated with rice, Paper presented at 2nd Int. Symp. N_2 Fixation with Non-Legumes, Bnaff, Alberta, Canada, 5–10 September (1982).

176. Weier, K.L. Nitrogenase activity associated with three tropical grasses growing in undisturbed soil cores, *Soil Biol. Biochem. 12*, 131–136 (1980).

177. Weier, K.L., Macrae, I.C. and Whittle, J. Seasonal variation in the nitrogenase activity of a *Panicum maximum* var. Trichoglume pasture and identification of associated bacteria, *Plant Soil 63*, 189–198 (1981).

178. White, J.W., Holben, F.J. and Richer, A.C. Maintenance level of nitrogen and organic matter in grassland and cultivated soils over periods of 54 and 72 years, *J. Amer. Soc. Agron. 37*, 21–31 (1945).

179. Willis, W.H. and Green, V.E. Movement of nitrogen in flooded soil planted to rice, *Soil Sci. Soc. Amer. Proc. 13*, 229–237 (1948).

180. Witt, W.W. and Weber, J.B. Ethylene adsorption and movement in soils and adsorption by soil constituents, *Weed Sci. 23*, 302–307 (1975).

181. Xavier, D.F., Baldani, J.I., Baldani, V.L.D. and Döbereiner, J. Effect of maize root extracts on growth and nitrogenase activity of *Azospirillum* spp., *An. Acad. Brasil. Ciê. 54*, 761 (1982).

182. Yoshida, T. and Ancajas, R.R. Application of the acetylene reduction method in nitrogen fixation studies, *Soil Science Plant Nutrit. 16*, 234–237 (1970).

183. Yoshida, T. and Ancajas, R.R. Nitrogen fixing activity in upland and flooded rice fields, *Soil Sci. Soc. Amer. Proc. 37*, 42–46 (1973).

184. Yoshida, T. and Yoneyama, T. Atmospheric dinitrogen fixation in the flooded rice rhizosphere as determined by the N-15 isotope technique, *Soil Sci. Plant Nutrit. 26*, 551–560 (1980).

12. *Azotobacter* and *Azospirillum* Genetics and Molecular Biology

C. Elmerich

INTRODUCTION

The family Azotobacteraceae comprises four genera: *Azotobacter, Azomonas, Beijerinckia* and *Derxia*. All species are aerobic, heterotrophic bacteria. The cells are large, contain granules of poly-β-hydroxybutyrate and the average G+C per cent ranges from 53 to 70. In addition, all species are nitrogen fixers in the free living state in air or under microaerobic conditions [1]. An additional genus, *Azospirillum*, was recently defined [2]. Initially the bacteria were named *Spirillum lipoferum* but it became evident that they could not be classified in the *Spirillum* genus and were relatively closer to *Azomonas* and *Derxia* [2-3]. This is why the genus *Azospirillum*, according to Becking [4] is more likely to belong to the Azotobacteraceae family. In this family, attention was focused mainly on *Azotobacter* defined by Beijerinck in 1901 [1]. Early genetics and biochemistry of nitrogen fixation were performed with *A. vinelandii* [5-8]. In the case of *Azospirillum*, the bacterium was discovered in 1922 by Beijerinck [9], and was forgotten until its rediscovery by Becking [10] and by Döbereiner and Day [11]. In particular, Döbereiner and Day reported the association of these bacteria with the rhizosphere of numerous grasses and suggested their possible agronomic importance. This chapter deals with the actual or potential tools for genetic analysis and cloning in *Azospirillum* and *Azotobacter* and with the molecular biology of nitrogen fixation. Associations with plants and agronomic importance in particular of *Azospirillum* have been reviewed previously [12-14] and in chapter 11 of this book.

TOOLS FOR GENETIC ANALYSIS

Bacteria and growth properties

Azotobacter

Three major *A. vinelandii* strains are used: OP (ATCC13705),

ATCC478 and ATCC12837. Strain OP, which is also named UW [8], is a non-capsulated mutant derivative of strain O (ATCC12518) [15]. Most reported Nif⁻ mutants are derivatives of strain UW. *A. chroococcum* is often the NCIB8003 strain. To the author's knowledge there are no genetic or biochemical reports on *A. beijerinckii* or *A. paspali.* Growth medium is usually the modified Burk's medium as described by Strandberg and Wilson [7] or by Dalton and Postgate [16]. In this medium the carbon source is mannitol, but it can be replaced by various sugars including glucose or sucrose and by organic acids [17]. Rhamnose is a specific growth substrate for *A. vinelandii* [18]. *A. vinelandii* and *A. chroococcum* can be differentiated on the basis of pigmentation. *A. vinelandii* contains a green fluorescent pigment whose production is increased in iron-limited media [19] and which is identified as made of two phenolic compounds [20]. *A. chroococcum* produces a dark-brown pigment [1, 4]. In both species nitrogenase activity is expressed in nitrogen-free medium under aerobic conditions [7, 16]. Resistant forms [21], called cysts, are differentiated in the presence of some carbon sources, such as butanol [22], β-hydroxybutyrate or crotonate [23].

Azospirillum

In the last decade, all the strains of *Azospirillum* studied were isolated, from the roots of grasses in various parts of the world [2, 12–14, 24–27]. Two major *A. brasilense* strains are used: strain Sp 7 (ATCC 29145) isolated from *Digitaria decumbens* in Brazil [28], also named 7000 [29], and strain Cd (ATCC29279) isolated from *Cynodon dactylon* in California [30] and which hyperproduces a pink pigment identified as a carotenoïd [31]. In the case of *A. lipoferum*, the most studied strain is strain Br17 (ATCC29709) isolated from *Zea mays* in Brazil [2]. The two species are easily differentiated on the basis of their nutritional properties. *A. lipoferum* strains grow on various sugars including glucose as well as organic acids and require biotin. *A. brasilense* strains do not grow on glucose nor on most sugars, but they utilize organic acids, such as malate and they are prototrophs [2–3, 32]. Composition of the minimal growth media described in literature is quite variable. The buffered salt base from the Kalininskaia medium [33] is recommended (see composition in [34]). Nutrient broth is a good complete medium [29]. In nitrogen-free minimal medium, nitrogenase is expressed under microaerobic conditions [28, 32, 35–37]. *Azospirillum* forms cysts in old cultures [38–40]. All *A. lipoferum* strains and most of the *A. brasilense* are denitrifiers under anaerobic conditions [3, 12, 30, 41–43].

Mutagenesis

There is no large collection of mutants either of *Azotobacter* or of *Azospirillum.* In *Azotobacter* it is very difficult to obtain mutants [44] except for a Nif⁻ phenotype [8]. The situation is the opposite in *Azospirillum* where auxotrophs can be obtained in large number [45] but Nif⁻

mutants genetically impaired in nitrogen fixation are rather tedious to isolate [46]. The reasons for this failure are not easy to explain since many factors could interfere with a successful isolation. In the case of *Azotobacter*, Sadoff et al. [44] argued that the difficulty to obtain mutants is due to the large number of chromosomes per cell. In the exponential phase of growth, each cell contains four nuclear bodies as observed by nuclear colouration [5, 23]. However, the number of nuclear bodies per cyst is 1, which is in agreement with a nuclear segregation at the end of the exponential growth [23]. DNA renaturation studies showed that the average length of the *Azotobacter* chromosome was similar to *E. coli* [44]. Measurement of DNA content per cell gave a value of 3.4×10^{-14} g per nuclear body, i.e., ten times more DNA than in *E. coli* [23]. A mean of 40 chromosomes per cell in the exponential phase was therefore proposed by Sadoff et al. [44]. Terzaghi [47] examined the survival of *A. vinelandii* after U.V. irradiation. The fact that the rate of survival was similar to that of *E. coli* was not in agreement with ten times increment of DNA except on the assumption that most of the chromosome copies are not biologically functional. In *Azospirillum* the difficulty to obtain *nif* mutants is not explained and might be artefactual. For both genera, however, various types of mutants have been isolated.

Azotobacter

In early reports, Nif⁻ mutants of *A. vinelandii* were described but could not be characterized due to the lack of a methodology [5, 48]. Pigmentation mutants, a leucine auxotroph and an energy metabolism mutant were obtained from *Azotobacter agilis* [49] after X Ray mutagenesis. *Azotobacter agilis* was an early name for *Azomonas agilis* [1], although the strain used by Karlsson and Barker [49] corresponds to *Azotobacter agile*, which is an *A. vinelandii* strain [4]. More recently using N-methyl-N'-nitro-N-nitrosoguanidine (NTG) Nif⁻ mutants were isolated [8, 50–51]. With the same mutagen, Page and Sadoff [52] isolated purine and pyrimidine auxotrophs and a rifampicin (Rif^R) resistant mutant. An adenine-requiring mutant was also described [53]. After U.V. irradiation Terzaghi [54] isolated Nif derepressed mutants. Isolation of Nif⁻ mutants from *A. chroococcum* was also reported [55].

Azospirillum

Classical techniques of mutagenesis previously described for *E. coli* [56], such as U.V. irradiation or treatment by ethylmethane sulphonate (EMS) or NTG, followed by penicillin or D-cycloserine enrichment, can be successfully applied to *Azospirillum*. Auxotrophs for various amino-acids and bases were obtained from at least four different strains: *A. brasilense* Sp 7 [45], *A. brasilense* 13t [57], *A. brasilense* Sp 6 [58], and *A. lipoferum* Br17. When an enrichment procedure for auxotrophs was used the mutation frequency was roughly 1 to 5×10^{-3}. Antibiotic-resistant mutants can be isolated without mutagenesis [59], but the spontaneous rate is very low and mutagenesis is often required (author's unpublished results). Mutants of nitrogen metabolism [29, 46, 60-61] and antimetabolite-resistant mutants [62-64] were described.

Transposon Tn*5* which confers resistance to kanamycin (Km^R) was utilized as a mutagen in *A. brasilense* Sp 7 to isolate auxotrophs [65]. The protocol for mutagenesis was based on the use of the suicide plasmid pJB4JI (Incp-1, Mu, Gm, Tn5) [66]. This plasmid is very stable in *Azospirillum* and the suicide phenomenon described in *Rhizobium* [66] was not observed [65]. However, the introduction of another IncP-1 plasmid into Sp 7 derivatives containing pJB4JI, forced segregation of the resident plasmid, and Km^R auxotrophs were isolated [65]. The mechanism which led to the transposition event was rather obscure, and it could not be demonstrated with certainty that the auxotrophs were due to the physical insertion of the transposon in the corresponding gene. Thus, an alternative procedure for transposition mutagenesis should be looked for. The use of plasmids containing Tn*5* (constructed by Simon et al. [67]), which cannot replicate in *Azospirillum* but can be mobilized into this bacterium, appears to be a promising approach (author's unpublished results).

Deoxyribonucleic acid mediated transformation

The first DNA mediated transformation in *Azotobacter* was reported by Sen and Sen [68], who described interspecific transformation for pigment production between *A. chroococcum* and *A. vinelandii.* Using more reliable genetic markers, optimal conditions for transformation in *A. vinelandii* were developed. In early experiments, Page and Sadoff [52] developed a method of transformation on agar plates where crude DNA preparations (cell lysates) were spotted onto lawns of recipient bacteria. In this method, non-competent bacteria were in contact with the DNA until the stage of competence was reached. Transformations to prototrophy of several auxotrophic and Nif mutants and to rifampicin resistance (Rif^R) were successful [52, 69]. Genetic linkage between auxotrophy and Rif^R [52] and between *nif* mutations [69] was established. The method was also used to transform *Azotobacter* with *Rhizobium* DNA [51, 70-72].

Factors influencing competence were examined, using the same plate assay, but the time of exposure to DNA was limited by the addition of deoxyribonuclease [52, 73-74]. A growth medium for maximum competence was defined by Page and von Tigerstrom [74]. It corresponded to Burk medium [8, 52] supplemented with glucose and ammonia, but without iron addition. Iron limitation was concomitant with the development of the fluorescent pigment characteristic of *A. vinelandii* [20]. However, pigment production was not always associated with competence [74]. Maximal competence was observed in the iron limited medium when at least 0.5 mM calcium was present [75]. Glucose could be replaced by other good carbon sources such as mannitol, glycerol or sucrose [74]. Competence in nitrogen-fixing cultures, first reported by Page and Sadoff [73], was denied by Page and von Tigerstrom [74], until optimal conditions for competence were defined [76].

After the iron-limited medium for competent cells was described, a transformation technique using a liquid assay and purified DNA was developed [77]. The incubation medium for transformation required at least 8 mM Mg^{++} which could not be replaced by Ca^{++}. Frequency of transformation on plates reached 10^{-4} [74] and in liquid 10^{-2} to 10^{-3} [77].

No data are available on the mechanism of transformation itself. Proteins involved in competence were concentrated from supernatants and distilled water washes of cells grown in the iron-limited competence induction medium [75]. Competence could be restored to an iron-calcium-limited culture by the addition of 0.5 mM Ca^{++}. The recovery of competence was mediated by a 60,000 dalton glycoprotein [75], an envelope protein of *A. vinelandii* [78]. Another protein, of pI 5.19, essential in competence was detected from an iron-calcium-limited culture [75]. Though the mechanisms of competence and transformation are not yet understood in *Azotobacter*, transformation appears to be an efficient means of gene transfer in this organism.

David et al. [79] developed a method of transformation with plasmid DNA adapted from the $CaCl_2$ treatment technique used in *E. coli* [80]. Plasmids RP4 (IncP1 group plasmid) [81], RSF1010 (IncQ group plasmid) [82] and derivatives were transformed into *A. vinelandii*. The transformants were stable and all the drug resistances of the plasmids were expressed (see below under Cloning Systems). The maximal frequency of transformation was 10^{-5} when the recipient cells [79] were treated with 200 mM $CaCl_2$.

In *Azospirillum,* transformation by chromosomal DNA was reported [59]. It is likely that in this organism, as in *Azotobacter*, optimal conditions for competence are difficult to achieve. This perhaps explains why no other report on transformation has been published.

Bacteriophages

Azotophages

Azotophages are distributed in five serological unrelated groups: four were identified by Duff and Wyss [83] and the last one by Chuml et al. [84]. All the phages isolated formed clear plaques with turbid halos, the respective sizes of the centre and of the halo being different from one phage to the other [83]. A phage representative for each serological group was studied: A14 (group I), A21 (group II), A31 (Group III), A41 (group IV) and PAV-1 (group V). The A phages were all isolated on *A. vinelandii* strain O [83, 85]. These phages had a host range limited to *A. vinelandii* and *A. chroococcum* and could not plate on strains belonging to *Azomonas* or *Beijerinckia* genera [83]. Therefore, they were useful as taxonomic tools. Moreover, the A series of phages did not plate on non-capsulated strains, such as strain

OP [84]. However, phages titrating on strain OP have been isolated [50, 86]. Among this group is phage PAV-1 [50] which titrates on both strains O and OP [84].

The five groups of phages contained double stranded linear DNA molecules with a molecular weight of 160×10^6 daltons for A14 [87]; 42×10^6 daltons for A21; 47×10^6 daltons for A31 [88] and 29×10^6 for PAV-1 [84]. The molecular weight of phage A41 DNA was not determined because the phage stocks were low and difficult to purify due to the instability in CsCl [88]. The phages belonged to three of the morphological types described by Bradley [89]. All groups had polyhedral heads, likely icosahedral; phages A21, A41 and PAV-1 had a short non-contractile tail [84, 88, 90]; A31 had a long non-contractile tail [88, 90] and A14 had the same morphology as All [85] from the same serological group and presented a contractile tail [87, 90]. The other phages mentioned by Fisher and Brill [8] and Bishop et al. [86] were not characterized.

The five groups of phages did not give stable lysogens but were responsible for the pseudolysogenic conversion of their hosts [91-92]. It was observed, after infection of *A. vinelandii* strain O and plating the survivors, that a large proportion of the colonies were rough, nongummy, yellow pigmented and that the cells lost their polysaccharide coat and became flagellated and motile [85, 91-92]. In particular the phenotypically converted cells were very similar to the acapsulated mutant strain OP [15]. The phenotypically converted cells produced phage and were immune to superinfection by any of the other phages except by PAV-1. In addition, when grown in the presence of phage antiserum, three types of cells were rescued: (i) host cells, capsulated and still sensitive to the five groups of phages; (ii) pseudolysogens, phenotypically converted, producing phages and capable of segregating wild type capsulated cells; and (iii) permanently converted cells, that were resistant to the A phages but still sensitive to PAV-1, unable to produce phage and unable to segregate wild type capsulated cells [91]. Sensitivity to PAV-1 was not surprising since the phage can infect both capsulated and uncapsulated cells [84]. The necessity of the capsule for infection by Group A phages was further confirmed. After digestion of the capsule by a purified polysaccharide depolymerase, no adsorption of phage A21 on the treated cells could be detected [91]. Four copies of phage DNA were present in the pseudolysogen and less than 0.2 copies in the permanently converted cell [92]. The molecular events which led to the permanently converted cells could not be explained [92].

Transduction with Azotophages was mentioned in an early report [93] but no further confirmation has been published. Moreover, Bishop et al. [86] isolated 21 phages that formed plaques on *A. vinelandii* OP and eight on the *A. vinelandii* strain ATCC12877 but did not find transducing ability among them.

***Azospirillum* bacteriophages**

Phages infecting *Azospirillum* were isolated only in Brazil. A phage able to form plaques on *A. brasilense* Sp 7 was reported but not characterized [94]. A temperate phage, named Al-1, forming plaques on *A. lipoferum* Br 17 was studied [95]. Its host range is limited to a few *A. brasilense* and *A. lipoferum* strains from the Brazilian origin. Strain Sp 7 is resistant to Al-1. Bacteriophage Al-1 has an icosahedral head and a long non-contractile tail to which five or six spikes are attached. The phage genome is a double-stranded linear DNA molecule of 22×10^6 daltons with cohesive ends. No transducing properties were detected.

Al-1 formed homogeneous turbid plaques on its hosts, and stable lysogens were obtained with all the sensitive strains available. In the lysogens, the prophage DNA was maintained as a plasmid. This is a new example of extrachromosomal lysogeny described initially for the *E. coli* phage P1 [96]. Two types of spontaneous mutant phages with altered plaque morphology were isolated: totally clear plaque and semi-turbid plaque mutants. Phages from the two types of mutants gave unstable lysogens which could be compared to the pseudolysogens described in *Azotobacter* [91], though no phenotypic conversion was observed [95].

Among 16 *Azospirillum* strains tested, 11 were found lysogenic for a defective prophage. Mitomycin C induced lysis of the cultures but no plaque production was detected on any host [34, 95]. For *A. brasilense* Sp 7, lysis occurred at drug concentrations ranging from 0.1 to 1 μg/ml whatever the growth conditions. Icosahedric phage-like particles were purified from a lysate of strain Sp 7. The particles had an apparent density of 1.3 g/cm^3 and were devoid of DNA [34]. No relationship was established between the lysogenic state and sensitivity to Al-1 [95].

Endogenous plasmids

Plasmid DNA was detected in all *Azospirillum* strains and at least one to six molecular species ranging from 4 to over 300×10^6 daltons were found [34, 57, 97–98]. Characterization of the plasmids was performed by electrophoretic migration of DNA from cell lysates on agarose gels using techniques described by Meyers et al. [99], Eckardt [100] or Casse et al. [101]. In some cases, plasmids were purified by CsCl-ethidium bromide gradients and restriction analysis was performed [34, 98]. The two taxonomic groups *A. brasilense* and *A. lipoferum* cannot be differentiated on the basis of their plasmid content. Spontaneous loss of some plasmids [34] or curing by high temperature [97] or by acridine orange treatment [57] was observed. Phenotypic changes, in particular in glucose utilization [97] and in heavy metals resistance [57], concomitant with plasmid loss were reported. However, no specific feature was demonstrated to be plasmid borne.

In *Rhizobium*, where large plasmids have been identified, some nitrogen fixation genes as well as genes involved in symbiosis are carried on plasmids [102-105]. No homology between *Klebsiella pneumoniae nif* genes and plasmid DNA of several *Azospirillum* strains (including Sp 7 and Br 17) was detected (author's unpublished results). Other investigators, however, reported a plasmid location for *nif* genes in some *Azospirillum* strains [106-107].

In *Azotobacter,* plasmids were detected in *A. chroococcum* [108]. No plasmid was found in *A. vinelandii* strain UW [108] but some were found in other strains. In particular strain AVY5 was found to carry genes homologous to *nifHDK* of *K. pneumoniae* on a 120×10^6 daltons plasmid [109].

Chromosome mobilization

Chromosome mobilization was demonstrated in *A. brasilense* Sp 7 using the R68-45 plasmid as a mobilizing agent [45]. Plasmid R68-45 [110-111] has the property to promote gene transfer in various Gram negative bacteria including *Rhizobium* [112]. It is an IncP-1 plasmid derivative of R68 [110] which confers resistance to kanamycin (Km), tetracycline (Tc) and carbenicillin (Cb). Plasmid R68-45 was introduced in *Azospirillum* by conjugation. Matings should be performed on the surface of solid agar, and the plates should be incubated in humid atmosphere to avoid drying of the *Azospirillum* culture. It was not possible to select for Cb transfer, because *Azospirillum* contains a β-lactamase [34], but selection for either Tc or Km was successful. Crosses were performed between an *Azospirillum* donor and series of multiple auxotrophs [45]. The frequency of transfer of R68-45 between *Azospirillum* mutants was 10^{-2} transconjugants per recipient. Marker transfer occurred at a frequency close to 10^{-6} per recipient, regardless of the marker tested. The ratio between recombinants and transconjugants that received R68-45 was roughly 10^{-5}. Inheritance of R68-45 was observed in 90 per cent of recombinants. The marker transfer promoted by R68-45 appeared to be non-polar which suggested the existence of multiple origins of transfer in *Azospirillum* as previously reported in *Pseudomonas aeruginosa* [111]. Linkage data between several pairs of markers was established [45]. Similar results were obtained with strain *A. brasilense* Sp 6, and in addition, a method of genetic exchange using spheroplast fusion with polyethylene glycol was developed [113].

Cloning systems

Construction of gene banks in bacteriophage λ vectors or in *E. Coli* plasmid vectors is convenient when: (i) a specific probe is available to identify the clones by *in situ* plaque or colony hybridization; (ii) complementation of *E. coli* markers is possible, or (iii) the products of the cloned genes can be identified by their own function or by using a specific antiserum. For example, with the *nifHDK* genes of *K. pneumoniae* as a probe, homologous genes from *A. brasilense* Sp7 were detected in a bacteriophage λgene bank [114] (see also section on Genetics of Nitrogen Fixation on p.

326). However, in order to perform genetic analysis by complementation in the original bacteria, subcloning into a broad host range vector is necessary. For this purpose IncP-1 or IncQ plasmids such as RP4, RK2 (IncP-1) and RSF1010 (IncQ) or their derivatives can be used [67, 115–117].

IncP-1 plasmids are self-transmissible and can be introduced by conjugation into *Azotobacter* [79] and *Azospirillum* [45, 58, 65, 118]. In addition RP4 was introduced by transformation into *Azotobacter* [79]. A 20 kb derivative of RK2 designated pRK290 was constructed [115] which confers resistance to Tc and contains unique restriction sites for *Eco*RI and *Bgl*II. It is not self-transmissible but it can be mobilized with the aid of plasmid pRK2013 [115]. Plasmid pRK290 can be transferred into *Azospirillum*. An *Azospirillum nif* fragment was cloned in pRK290 and *nif* partial diploids were constructed which appeared to be stable [46].

IncQ plasmid are not self-transmissible [82]. Plasmid RSF1010 and its derivatives can be mobilized in *E. coli*, using various sex factors [82], and in other Gram negative bacteria including *Pseudomonas* [119], *Azotobacter* [79], *Rhizobium* [120] and *Azospirillum* (author's unpublished results) using pRK2013 [115], RP4 or derivatives [67]. Recently, Kennedy and Robson [55] used a RSF1010 derivative, pKT230, to clone the *nifA* gene of *K. pneumoniae*. Plasmid pKT230 is 12 kb, it confers resistance to streptomycin and Km and it has unique *Hind*III and *Xho*I sites in the Km^R gene [116]. A *Sal*I fragment carrying the entire *nifA* gene coding region and a part of *nifL* [121] was cloned at the *Xho*I site of pKT230 to yield plasmid pCK1 [55]. The pCK1 plasmid was then introduced into *A. vinelandii* and *A. chroococcum* wild type and *nif* mutants to study the effect of the *nifA* gene product on *Azotobacter* nitrogen fixation [55] (see section on Genetics of Nitrogen Fixation on p. 326). Plasmids RSF1010 and derivatives can be also introduced by transformation in *Azotobacter* [79].

Plasmids pRK290 [115], pLAFR1 [117], or derivatives of RSF1010 such as pKT230 [116] and pSUP106 [67] would indeed be useful tools to construct gene banks of *Azotobacter* and *Azospirillum* DNA in *E. coli*. The clones obtained could be analyzed by complementation after conjugation en masse with a series of *Azotobacter* and *Azospirillum* mutants. Particularly useful should be the cosmid vectors pLAFR1 [117] and pSUP106 [116] which allow cloning of large fragments of DNA due to size selection at the step of *in vitro* packaging in λ heads [122]. Both cosmids were used to obtain gene banks of *R. meliloti* [67, 117]. An alternative procedure was investigated by David et al. [79] who developed a technique to directly transfer plasmid DNA into *Azotobacter* by transformation. These workers suggested that *Azotobacter* could be used as a cloning recipient for *Rhizobium* DNA since it had previously been shown that *Rhizobium* DNA transformed Nif *Azotobacter* mutants to Nif [51] (see section on Transformation on p. 318). Some transformants also produced *Rhizobium* surface antigens [71, 72].

NITROGEN FIXATION

Biochemistry and physiology

Research on *Azotobacter* has contributed substantially to knowledge of the physiology of nitrogen fixation and of the structure and functioning of the nitrogenase complex. Several reviews deal with various aspects of nitrogen fixation [123-128] and only some essential and recent data are discussed here.

The nitrogenase complex

Activity of nitrogenase in *A. vinelandii* [6], as is the case for every other diazotroph studied later [124, 129], requires two nonheme iron proteins: component I and component II [130]. Component I, also designated the MoFe protein, has been crystallized [131-132]. It is a tetramer of 245,000 daltons [133-134] containing 2 Mo atoms per molecule [130, 134], made of two different molecular species of 61,000 daltons ($\alpha_2\beta_2$) [133]. Molybdenum was found associated with component I as a cofactor [135], the FeMoCo, which contained Fe, S^{--} and Mo in the ratio 8, 6, 1 [134] but other ratios were proposed [136]. Component II, also designated the Fe protein, is a dimer formed of two identical subunits of 31,200 daltons, containing 289 amino acids [137]. The enzyme complex of *A. chroococcum* was also purified and characterized [138-139]. As compared to *Azotobacter* little is known about *Azospirillum* nitrogenase. The enzyme of *A. brasilense* strain Sp7 was purified and an activating factor was found to be required for component II activity [36, 140], as has been demonstrated for *Rhodospirillum rubrum* [141]. Using antisera prepared against *K. pneumoniae* MoFe protein and Fe protein, cross-reacting material was precipitated from nitrogen-fixing cultures of *A. brasilense* Sp7. Recently in this laboratory, the molecular weights of the polypeptides after SDS gel electrophoresis were estimated at 60,000 and 64,000 for the MoFe protein and 33,000 and 36,000 for the Fe protein. The two bands observed with the Fe protein might correspond to the presence of active and covalently modified inactive polypeptides in the extracts.

ROLE OF OXYGEN, NITROGEN SOURCES AND MOLYBDENUM IN NITROGEN FIXATION

Oxygen: The nitrogenase complex of all diazotrophs is extremely sensitive to oxygen, both components being rapidly and irreversibly inactivated upon exposure to air [123-128]. However, nitrogenase in cell-free extracts of *Azotobacter* [6, 138] is relatively oxygen stable. It was first suggested that oxygen tolerance was due to a particular state of the enzyme, but more recent experiments suggested an alternative mechanism [142-143]. Oppenheim et al. [142] proposed that protection against oxygen was due to the important cytoplasmic membrane that appears in organisms grown in nitrogen-free medium [144]. Haaker and Veeger [143] demonstrated the presence of a protein factor that forms an oxygen stable complex with the nitrogenase components. This factor was purified as a 24,000 dalton iron-sulphur protein [145]. A 14,000 dalton protein with similar functions was purified from *A. chroococcum* [146]. *Azotobacter* fixes nitrogen in air, though excess aeration inhibits nitrogenase activity [7, 16, 147–150]. Sensitivity to oxygen inhibition has been correlated to the highly adaptative respiratory activity of the cells [16, 147–149]. When cells of low respiratory activity were subjected to high aeration (oxygen stress) nitrogenase activity was immediately switched off and returned when aeration was lowered. Two protection mechanisms were proposed to account for this observation: a respiratory protection and a conformational change of the nitrogenase structure [16, 147–148]. Conformational protection was likely due to the oxygen stabilizing factor [127, 145–146]. In addition, it was demonstrated in *A. chroococcum* that the biosynthesis of the nitrogenase complex is repressed by oxygen [151], as in *K. pneumoniae* [152].

In nitrogen-free semi-solid media, *Azospirillum* behaves as a typical microaerobic organism and forms a pellicle below the surface, which moves with the oxygen availability [28, 37]. In liquid cultures maintained in fermentors, dissolved oxygen tensions between 0.003 and 0.007 atm were found to be optimal for growth under conditions of nitrogen fixation and the doubling time was 5 to 7 hours [32, 36, 153–154]. No protective mechanism of the nitrogenase complex, as observed in *Azotobacter*, was detected at increased oxygen tensions [43, 154].

Nitrogen sources: Molecular nitrogen is not an inducer of nitrogenase biosynthesis and ammonia appears to be a repressor [7, 155]. However, ammonia is not a repressor *per se* and needs to be metabolized since repression is relieved by methionine sulphoximine, an inhibitor of ammonia assimilation. This was shown initially in *Azotobacter* [156] and later in *Azospirillum* [35]. Many papers deal with ammonia or nitrogen source effects on nitrogenase biosynthesis in batch or continuous cultures (see [7, 35, 42–43, 150, 153, 162]. No nitrogenase was synthesized when ammonia or easily assimilated nitrogen sources were present in the growth medium. Excretion of ammonia [163] and of other nitrogenous compounds [154] during growth under nitrogen-fixation conditions was reported for *Azospirillum*. The fact that nitrate respiration is compatible with nitrogen fixation in *Azospirillum* was controversial [42–43]. It was finally established that nitrate repressed nitrogenase biosynthesis, but that nitrogen fixation occurred at the early stage of growth for a short period of time before the assimilation of ammonia resulting from nitrate reduction [162]. In *Azotobacter*, when ammonia was added to a nitrogen-fixing culture, nitrogenase biosynthesis stopped and the rate of decay of activity was much higher than the expected dilution rate [157, 160]. Loss of activity was not due to a rapid enzyme turn-over and was concomitant to inactivation of component I as shown by electron paramagnetic resonance (EPR) signal

[158]. In addition, even at very low concentration, ammonia caused a partial inhibition of the nitrogenase activity [160, 164] which was explained by a change in the energization of the membrane [165]. This inhibition did not result in covalent modification of the Fe protein as observed in photosynthetic bacteria [128, 166]. In contrast, covalent modification of nitrogenase Fe protein occurred in *Azospirillum* as demonstrated by the necessity of an activating factor in crude extracts [140] and by the *in vivo* reversible switch-off of nitrogenase activity upon addition of 50 μM ammonia (author's unpublished results). It should be also pointed out that both genera contained an ammonia permease [164, 167-168] which might play a role in the inhibition and/or repression of nitrogenase.

Molybdenum: It has been known for a long time that Mo is essential for nitrogen fixation [128, 169]. When grown in Mo-deficient medium containing tungstate, *A. vinelandii* produced an inactive component I that could be reactivated *in vivo* by the addition of molybdate to the growth culture [169]. In cell-free extract of cells grown in the presence of tungstate, the addition of FeMoCo restored component I activity [135]. Reactivation by MoO_4^{--} was also achieved. ATP and an additional factor of low molecular weight (between 2,000 and 4,000) was required for the reactivation [170]. A molybdenum storage protein was detected in both N_2 and ammonia grown cells [171]. This protein contains two types of subunits of 21,000 and 24,000 daltons and is likely to have a tetramer organization with a binding capacity of at least 14.5 atoms of Mo per molecule. This storage protein is not the source of a preformed FeMoCo for nitrogenase [171].

Alternative system of nitrogen fixation

Bishop et al. [172] obtained Nif^+ pseudorevertants of the Nif^- strains UW6 and UW10 [173] that retained the original *nif6* and *nif10* mutations of the parent strains. The pseudorevertants displayed growth on nitrogen-free medium at a lower growth rate than the wild type and reduced acetylene at a rate of 3 to 4 per cent of the Nif^+ control. They produced four ammonia-repressible proteins detected by two-dimensional gel electrophoresis. In addition phenotypic reversion to Nif^+ of a series of Nif^- mutants when grown in media devoid of Mo, but containing tungsten, vanadium or rhenium salts, was also observed [172, 174]. Under those conditions the mutants reduced acetylene at a low rate, though the cells lacked the typical EPR signal of the MoFe protein. These observations led Bishop et al. [172-174] to propose the existence in *A. vinelandii* of an alternative pathway for nitrogen fixation whose functioning was independent of Mo.

Genetics of nitrogen fixation

Though genetics of nitrogen fixation was initiated in *Azotobacter* [8, 50], our current knowledge on the nitrogen fixation (*nif*) genes comes from the study of *K. pneumoniae*. In this species the existence of a good transduction system [175] and the isolation of plasmid pRD1 [176] that carried the entire *nif* cluster of *K. pneumoniae* were most useful in the development of *nif* genetics [126, 177].

Homology with K. pneumoniae (nif) genes and cloning of nif DNA

In *K. pneumoniae*, 17 *nif* genes organized in seven transcriptional units have been identified [126, 177–181]. *NifHDK* are the structural genes for the nitrogenase complex polypeptides; *nifH* codes for the single subunit of component 2 (Fe protein) and *nifDK* for the two subunits of component 1 (MoFe protein). The three genes and the *nifY* gene are transcribed in the order *nifHDKY*. This operon and a part of *nifE* are carried by a 6.2 kb *EcoRI* fragment which was cloned in plasmid pSA30 [182]. Using this fragment as a hybridization probe, homology was detected with total DNA from a large number of diazotrophs [183–184] including *A. vinelandii* [183] and several strains of *Azospirillum* [65, 114].

In early experiments with *A. vinelandii* 5 *EcoRI* restriction fragments of 12.5; 8.1; 3.5; 2.2 and 0.8 kb hybridizing with the *nif* probe were detected [183]. Homology was limited to *nifH* and *nifD* genes of *K. pneumoniae*. It was not clear why so many *EcoRI* restriction fragments were found to hybridize with the probe. This could account for the existence of pseudogene (s) (as reported in *Anabaena* [185]) or for *nif* gene reiteration (as reported in *Rhizobium phaseoli* [186]) or could support the hypothesis of an alternative pathway for nitrogen fixation as proposed by Bishop et al. [172]. However, more recently, Bishop and Bott [187] detected only three *EcoRI* restriction fragments of 1.4, 2.6 and 4.1 kb which shared homology with the pSA30 *nif* probe. The 4.1 kb fragments were cloned in plasmid pBR325 to yield plasmid pBL1 and pBL3. Plasmid pBL1 shared homology with *nifK* and *nifD* gene and plasmid pBL2 shared homology with *nifD* and possibly with *nifH*. The two *nif* fragments were likely to be adjacent on the chromosome [187] and these results were consistent with the findings of Krol et al. [188]. These authors analyzed *nif* transcripts by northern hybridization and proposed the existence of a *nifHDK* operon in *A. vinelandii* as demonstrated for *K. pneumoniae* [189] with the same technique.

In the case of *Azospirillum*, two *A. lipoferum* strains and three *A. brasilense* strains were examined and the size of the homologous *EcoRI* and *Hind*III fragments were different from one strain to another [114]. With *A. brasilense* strain Sp 7, the probe hybridized with a single 6.7 kb *EcoRI* fragment which was contained within a 24 kb *Hind*III fragment and a 22 kb *Bam* HI fragment. With *A. lipoferum* strain Br17 homology was found with two *EcoRI* fragments of 16 and 1.8 kb and with a single 15 kb *Hind*III fragment. The 6.7 kb *EcoRI* fragment from strain Sp 7 was cloned in the λ gt7-ara6 vector, then subcloned in plasmid vectors pACYC184 and pRK290 to yield respectively plasmids pAB1 [114] and pAB35 [46]. The physical map of the plasmids was established and derivatives containing *in vitro* generated deletions were constructed [46]. Heteroduplex analysis of phages containing the *K. pneumoniae* or *Azospirillum nif* DNA was performed

[114]. Homology on a sequence approximately 5 kb long was detected. This was consistent with the existence of a complete *nifHDK* gene cluster in *Azospirillum*. Moreover the direction of transcription could be determined from the heteroduplex analysis.

As already reported (see section on Endogenous Plasmid on p. 321) there is no clear demonstration that the *nif* genes which share homology with the *nif* DNA of *K. pneumoniae* are located on plasmids in *Azospirillum* or *Azotobacter*. Using pAB1 as the *nif* probe no hybridization was detected with plasmid DNA of strain Sp7 (author's unpublished results).

Characterization of Nif⁻ mutants

Nif⁻ mutants isolated from *A. vinelandii* strain UW [8, 173] were analyzed biochemically for (i) the activity of nitrogenase components by complementation with pure MoFe protein or Fe protein from *A. vinelandii* [8, 173]; (ii) antigen production with antisera prepared against each nitrogenase component [173]; (iii) EPR signal [173] (the wild type gives a signal characteristic of a functional MoFe protein with a g value of 3.65 [158] and thus the relative amount of functional MoFe protein can be estimated [173]); (iv) complementation with FeMoCo [135]; and (v) iron-staining material detected after migration of cell extracts in non-denaturating gels [190].

Genetic mapping of the Nif⁻ mutants was performed by transformation [69] using the technique of Page and Sadoff [52]. This technique, however, gave imprecise results [69]. Thus, genetically speaking, no *nif* genes were defined. According to the biochemical properties of the mutants and the preliminary genetic map, the Nif⁻ mutants can be distributed into six groups:

i) regulatory mutants; for example, strains UW1 [8] and UW2 [173] which were devoid of activity [173], iron staining material [190] and cross reacting material (CRM) [173] for both nitrogenase components. From UW2 Nif⁺ revertants that fix nitrogen in the presence of ammonia were isolated [191];

ii) mutants possibly impaired in a structural gene for the MoFe protein (subunit α or β): for example, strain UW10 [8], which produced CRM [173], active Fe protein [173] and inactive MoFe protein [173] and which could not be activated by the FeMoCo [134];

iii) mutants impaired in the FeMoCo: for example strain UW45 which has features similar to UW10 except that UW45 MoFe protein can be reactivated by the FeMoCo [134–135, 170];

iv) mutants possibly impaired in the structural gene for the Fe protein: for example, strain UW91 [173], which produced CRM, active MoFe protein with a g value similar to the wild type [173], iron staining material [190] but which also produced an inactive Fe protein;

v) mutants that hyperproduce the MoFe protein: for example, strain UW38 [173, 192] which was devoid of CRM for the MoFe protein but contained an active Fe protein. This mutation mapped very close to the mutation of strain UW6 [69] which had the same phenotype but did not overproduce the Fe protein [173]. The same mutation was responsible for the hyperproduction of Fe protein and the absence of the MoFe protein synthesis as suggested by spontaneous Nif$^+$ revertants isolated from strain UW38 [192]. For strain UW6, though no CRM was detected for the FeMo protein, the corresponding polypeptides were present on two dimension gel electrophoresis [172]; and

vi) mutant which produced no CRM for the Fe protein and low activity for the MoFe protein: for example, strain UW3 [173].

Marker rescue tests, by transformation of plasmids pBL1 and pBL3 DNA into the Nif$^-$ mutants showed that the *nif6*, *nif38* and *nif10* mutations are clustered within a region of 4 kb on the chromosome of *A. vinelandii* [187]. No mutants impaired in electron transport were described, though possible physiological electron carriers were identified [123, 193]. It is also likely that mutants of the oxygen stabilizing factor would appear as Nif$^-$ [145-146] but no such mutant has yet been described.

In *Azospirillum*, only a single *nif* mutant has been characterized to date. This mutant, designated 7571, is a derivative of *A. brasilense* Sp7 and possibly is impaired in a structural gene for the MoFe protein [46]. Evidence was as follows: (i) the mutant did not reduce acetylene under conditions of derepression of nitrogen fixation; (ii) nitrogen fixation was restored after introduction of plasmid pAB35 which carries the wild type *nifHDK* cluster but nitrogen fixation was not restored after the introduction of plasmid pAB36 which possibly shows deletion for *nifD* and *K* but which still carries *nifH*; and (iii) biochemical complementation of crude extracts was observed after the addition of pure *K. pneumoniae* MoFe protein but not by Fe protein.

Regulation of nif genes expression

Regulatory mutants have been described for *Azotobacter* since 1968 [172, 191-192, 194]. Mutants that fix nitrogen in the presence of ammonia were isolated [54, 191, 194-195], particularly among those resistant to methylalanine [194-195]. It was first proposed that methylalanine was a corepressor of nitrogenase biosynthesis [194], but this hypothesis was later rejected [159, 192]. Gordon and Brill [192] suggested the existence of a common regulatory gene that controlled the expression of the nitrogenase structural genes, since among revertants of strain UW2 (a mutant unable to synthesize neither of the two nitrogenase components) some were found to be partially constitutive. To explain the properties of strain UW38 (that hyperproduced the Fe protein but did not contain CRM for the MoFe protein) Shah et al. [192] proposed that the common regulatory gene coded for an activator that could exist in two forms A′ and A″. Each form would activate respectively Fe protein and MoFe protein polypeptides synthesis and the two forms would be inactivated in medium containing ammonia. Thus a single mutation that rendered impossible the conversion of A′ in A″ would cause the hyperproduction of the Fe protein [192]. Genetic mapping, however, showed that mutations in UW2 and UW38 were not closely linked [69] suggesting they could not be in the same regulatory gene.

In *Azospirillum*, mutants impaired in glutamine synthetase activity [29] and in glutamate synthase activity [61] were isolated. It is likely that both enzymes are responsible for NH_4^+ assimilation under nitrogen limitation conditions as previously demonstrated for other diazotrophs [196]. Glutamine synthetase and glutamate synthase deficient mutants were found impaired in nitrogen fixation [29, 61]. In particular Nif^- and Nif^c (that fix nitrogen in the presence of NH_4^+) phenotypes were observed among the glutamine auxotrophs [29]. These phenotypes were also described in *Klebsiella pneumoniae* mutants [197-199]. The glutamate synthase deficient mutants, which displayed an Asm^- phenotype since they could not utilize a series of nitrogen sources [196], were unable to grow on nitrogen-free medium [61]. None of the *Azospirillum* mutants was genetically or biochemically characterized as impaired in the structural gene for the corresponding enzyme. Thus the involvement of glutamine synthetase or glutamate synthase in *nif* genes regulation should be examined with caution, especially in light of what has been recently found in *K. pneumoniae.*

In *K. pneumoniae,* where the regulation of *nif* gene expression is now well documented, two regulatory mechanisms have been identified: a *nif* specific mechanism that involves the products of the *nifLA* transcription unit (the *nifA* gene product being an activator and the *nifL* gene product a repressor) [121, 152, 177, 200], and a non *nif* specific mechanism involving *gln* genes that control the utilization of various nitrogen sources [201]. Therefore, the possible involvement and the identification of genes analogous to *glnF, glnG* and *glnL* (also designated *ntrA, ntrB, ntrC*) [200-204] in the regulation of *nif* gene expression in *Azospirillum* and *Azotobacter* should be considered.

The possibility of intergeneric complementation using plasmids carrying *K. pneumoniae nif* genes introduced by conjugation into *Azotobacter* or *Azospirillum* was investigated. Plasmid pRD1 [176] was found to complement some Nif^- mutants of *A. vinelandii* [205]. It was also introduced into *Azospirillum* [58] where it was stable but no complementation was observed since no Nif^- mutant were available. Kennedy and Robson [55] constructed the plasmid pCK1, that carried the *nifA* gene of *K. pneumoniae* (see cloning system section on p. 322). When introduced in *A. vinelandii* or *A. chroococcum* wild type strain plasmid pCKl conferred a Nif constitutive phenotype to their hosts which became capable of fixing nitrogen in the presence of ammonia. Thus the positive effector role of the *nifA* gene product demonstrated in *K. pneumoniae* [121, 177, 200] could be extended to *Azotobacter.* Moreover, Nif^- mutants of *A. vinelandii* or *A. chroococcum* that had a regulatory phenotype, such as strain UW1, were also constitutive in the presence of ammonia when they contained pCK1 [55]. Kennedy and Robson [55] proposed the existence of a regulatory gene analogous to the *K. pneumoniae nifA* in *Azotobacter.* As *nifA* was also found to activate a *nifH-lac* fusion of *R. meliloti* [206] in *E. coli*, Kennedy and Robson [55] suggested that activation by *nifA* is ubiquitous among diazotrophs.

CONCLUSION

Though *Azotobacter* has been studied for a long time and *Azospirillum* has generated much interest these past few years, the genetics of these bacteria is still at its beginning. The lack of large collection of mutants can partly account for this situation. Recent development of *in vitro* DNA recombinant techniques should allow rapid development of the genetics and the regulation of nitrogen fixation in the near future.

ACKNOWLEDGEMENTS

The author wishes to thank Drs. P.E. Bishop, J.K. Gordon, A. Hartmann and C. Kennedy for communications of results prior to publication and Drs. J.-P. Aubert and C. Kennedy for improving the manuscript. The secretarial work of Mrs. M. Ferrand is greatly appreciated. Work in the author's laboratory was supported by a grant from Elf Bio-Recherches and Entreprise Minière et Chimique.

REFERENCES

1. Becking, J.H. and Johnstone, D.B. The family Azotobacteraceae, pp. 253-261, in *Bergey's Manual of Determinative Bacteriology* (Editors, R.E. Buchanan and N.E. Gibbons), 8th edition, The Williams and Wilkins Co., Baltimore (1974).
2. Tarrand, J.J., Krieg, N.R. and Döbereiner, J. A taxonomic study of the *Spirillum lipoferum* group, with description of a new genus, *Azospirillum* gen. nov. and two species, *Azospirillum lipoferum* (Beijerinck) comb. nov. and *Azospirillum brasilense* sp. nov., *Canadian Journal of Microbiology 24*, 967-980 (1978).
3. Krieg, N.R. Taxonomic studies of *Spirillum lipoferum*, pp. 463-472, in *Genetic engineering for nitrogen fixation*, Volume 9 (Editors, A. Hollander), Plenum Press, New York, London (1977).
4. Becking, J.H. The family Azotobacteraceae, pp. 795-817, in *The Prokaryotes*, Volume 2 (Editors, M.P. Starr, H. Stolp, H.G. Trüper, A. Balows, H.G. Schlegel), Springer-Verlag, Berlin, Heidelberg, New York (1981).
5. Wyss, O. and Bedell Wyss, M. Mutants of *Azotobacter* that do not fix nitrogen, *Journal of Bacteriology 59*, 287-291 (1950).
6. Bulen, W.A., Burns, R.C. and Le Comte, J.R. Nitrogen fixation: cell-free system with extracts of *Azotobacter*, *Biochemical and Bio-*

physical Research Communications 17, 265–271 (1964).

7. Strandberg, G.W. and Wilson, P.W. Formation of the nitrogen fixing enzyme system in *Azotobacter vinelandii, Canadian Journal of Microbiology 14*, 25–31 (1968).
8. Fisher, R. and Brill, W.J. Mutant of *Azotobacter vinelandii* unable to fix nitrogen, *Biochimica et Biophysica Acta 184*, 99–105 (1969).
9. Krieg, N.R. Biology of the chemoheterotrophic spirilla, *Bacteriological Reviews 40*, 55–115 (1976).
10. Becking, J.H. Fixation of molecular nitrogen by an anaerobic *Vibrio* or *Spirillum*, *Antonie van Leeuwenhoek 29*, 326 (1963).
11. Döbereiner, J. and Day, J.M. Associative symbiosis in tropical grasses: characterization of microorganisms and dinitrogen-fixing sites, pp. 518–536, in *Proceedings of the 1st International Symposium on Nitrogen Fixation* (Editors, W.E. Newton and C.J. Nyman), Washington State University Press, Pullman (1976).
12. Döbereiner, J. and DePolli, H. Diazotrophic rhizocoenoses, pp. 301–333, in *Nitrogen Fixation*, Proceedings of the Phytochemical Society of Europe Symposium (Editors, W.D.P. Stewart and J.R. Gallon), Academic Press, London, New York, Toronto, Sydney, San Francisco (1980).
13. Van Berkum, P. and Bohlool, B.B. Evaluation of nitrogen fixation by bacteria in association with roots of tropical grasses, *Microbiological Reviews 44*, 491–517 (1980).
14. Patriquin, D.G. New developments in grass-bacteria associations, pp. 139–190, in *Advances in Agricultural Microbiology* (Editor, N.S. Subba Rao), Oxford & IBH Publishing Co., New Delhi (1982).
15. Bush, J.A. and Wilson, P.W. A non-gummy chromogenic strain of *Azotobacter vinelandii, Nature* (London) *184*, 381 (1959).
16. Dalton, H. and Postgate, J.R. Effect of oxygen on growth of *Azotobacter chroococcum* in batch and continuous cultures, *Journal of General Microbiology 54*, 463–473 (1969).
17. Claus, D. and Hempel, W. Specific substrates for isolation and differentiation of *Azotobacter vinelandii, Archiv für Mikrobiology 73*, 90–96 (1970).
18. Jensen, V. Rhamnose for detection and isolation of *Azotobacter vinelandii* Lipman, *Nature* (London) *190*, 823–833 (1961).
19. Bulen, W.A. and LeComte, J.R. Isolation and properties of a yellow green fluorescent peptide from *Azotobacter* medium, *Biochemical and Biophysical Research Communications 9*, 523–528 (1962).
20. Corbin, L. and Bulen, W.A. The isolation and identification of 2,3-dihydroxybenzoic acid and 2-N, 6-N-di-(2,3-dihydroxy-benzoyl)-L-lysine formed by iron-deficient *Azotobacter vinelandii, Biochemistry 8*, 757–762 (1969).
21. Socolofsky, M.D. and Wyss, O. Resistance of the *Azotobacter* cyst,

Journal of Bacteriology 84, 119-124 (1962).

22. Socolofsky, M.D. and Wyss, O. Cysts of *Azotobacter*, *Journal of Bacteriology 81*, 946-954 (1961).
23. Sadoff, H.L., Berke, E. and Loperfido, B. Physiological studies of encystment in *Azotobacter vinelandii*, *Journal of Bacteriology 105*, 185-189 (1971).
24. Döbereiner, J., Marriel, I.E. and Nery, M. Ecological distribution of *Spirillum lipoferum* Beijerinck, *Canadian Journal of Microbiology 22*, 1464-1473 (1976).
25. Lakshmi Kumari, M., Kavimandan, S.K. and Subba Rao, N.S. Occurrence of nitrogen fixing *Spirillum* in roots of rice, sorghum, maize and other plants, *Indian Journal of Experimental Biology 14*, 638-639 (1976).
26. Haahtela, K., Wartiovaara, T., Sundman, V. and Skujins, J. Root-associated N_2 fixation (acetylene reduction) by Enterobacteraceae and *Azospirillum* in cold climate spodosols, *Applied Environmental Biology 41*, 203-206 (1981).
27. Nur, I. Okon, Y. and Henis, Y. Comparative studies of nitrogen fixing bacteria associated with grasses in Israel with *Azospirillum brasilense*, *Canadian Journal of Microbiology 26*, 174-178 (1980).
28. Von Bülow, J.F.M. and Döbereiner, J. Potential for nitrogen fixation in maize genotypes in Brazil, *Proceedings of the National Academy of Sciences* (USA) *72*, 2389-2393 (1975).
29. Gauthier, D. and Elmerich, C. Relationship between glutamine synthetase and nitrogenase in *Spirillum lipoferum*, *FEMS Microbiology Letters 2*, 101-104 (1977).
30. Eskew, D.L., Focht, D.D. and Ting, I.P. Nitrogen fixation, denitrification, and pleomorphic growth in a highly pigmented *Spirillum lipoferum*, *Applied Environmental Microbiology 34*, 582-585 (1977).
31. Nur, I., Steinitz, Y.L., Okon, Y. and Henis, Y. Carotenoid composition and function in nitrogen-fixing bacteria of the genus *Azospirillum*, *Journal of General Microbiology 122*, 27-32 (1981).
32. Okon, Y., Albrecht, S.L. and Burris, R.H. Factors affecting growth nitrogen fixation of *Spirillum lipoferum*, *Journal of Bacteriology 127*, 1248-1254 (1976).
33. Biggins, D.R. and Postgate, J.R. Nitrogen fixation by cultures and cell free extracts of *Mycobacterium flavum* 301, *Journal of General Microbiology 56*, 181-193 (1969).
34. Franche, C. and Elmerich, C. Physiological properties and plasmid content of several strains of *Azospirillum brasilense* and *A. lipoferum*, *Annales de Microbiologie, Institut Pasteur 132A*, 3-17 (1981).
35. Okon, Y., Albrecht, S.L. and Burris, R.H. Carbon and ammonia metabolism of *Spirillum lipoferum*, *Journal of Bacteriology 128*, 592-597 (1976).

36. Okon, Y., Houchins, J.P., Albrecht, S.L. and Burris, R.H. Growth of *Spirillum lipoferum* at constant partial pressures of oxygen, and the properties of its nitrogenase in cell free extracts, *Journal of General Microbiology 98*, 87–93 (1977).
37. Barak, R., Nur, I., Okon, Y. and Henis, Y. Aerobic response of *Azospirillum brasilense*, *Journal of Bacteriology 152*, 643–649 (1981).
38. Berg, R.H., Tyler, M.E., Novick, N.J., Vasil, V. and Vasil, I.K. Biology of *Azospirillum*-sugarcane association: Enhancement of nitrogenase activity, *Applied Environmental Microbiology 39*, 642–649 (1980).
39. Lamm, R.B. and Neyra, C.A. Characterization and cyst production of azospirilla isolated from selected grasses growing in New Jersey and New York, *Canadian Journal of Microbiology 27*, 1320–1325 (1981).
40. Papen, H. and Werner, D. Organic acid utilization, succinate excretion, encystation and oscillating nitrogenase activity in *Azospirillum brasilense* under microaerobic conditions, *Archive of Microbiology 132*, 57–61 (1982).
41. Neyra, C.A., Dobereiner, J., Lalande, R. and Knowles, R. Denitrification by nitrogen-fixing *Spirillum lipoferum*, *Canadian Journal of Microbiology 23*, 300–305 (1977).
42. Neyra, C.A. and Van Berkum, P. Nitrate reduction and nitrogenase activity in *Spirillum lipoferum*, *Canadian Journal of Microbiology 23*, 306–310 (1977).
43. Nelson, L. and Knowles, R. Effect of oxygen and nitrate on nitrogen fixation and denitrification by *Azospirillum brasilense* grown in continuous culture, *Canadian Journal of Microbiology 24*, 1395–1403 (1978).
44. Sadoff, H.L., Shimei, B. and Ellis, S. Characterization of *Azotobacter vinelandii* deoxyribonucleic acid and folded chromosomes, *Journal of Bacteriology 138*, 871–877 (1979).
45. Franche, C., Canelo, E., Gauthier, D. and Elmerich, C. Mobilization of the chromosome of *Azospirillum brasilense* by plasmid R68-45, *FEMS Microbiology Letters 10*, 199–202 (1981).
46. Jara, P., Quiviger, B., Laurent, P. and Elmerich, C. Isolation and genetic analysis of *Azospirillum brasilense* Nif^- mutants. *Canadian Journal of Microbiology* (in press) (1983).
47. Terzaghi, B.E. Ultraviolet sensitivity and mutagenesis of *Azotobacter*, *Journal of General Microbiology 118*, 271–273 (1980).
48. Green, M., Alexander, M. and Wilson, P.W. Mutants of the *Azotobacter* unable to use N_2 *Journal of Bacteriology 66*, 623–624 (1953).
49. Karlsson, J.L. and Barker, H.A. Induced biochemical mutants of *Azotobacter agilis*, *Journal of Bacteriology 56*, 671–677 (1948).
50. Sorger, G.J. and Trofimenkoff, D. Nitrogenaseless mutants of *Azotobacter vinelandii*, *Proceedings of the National Academy of Sciences* (USA) *65*, 74–80 (1970).

51. Page, W.J. Transformation of *Azotobacter vinelandii* strains unable to fix nitrogen with *Rhizobium* spp. DNA, *Canadian Journal of Microbiology 24*, 209–214 (1977).
52. Page, W.J. and Sadoff, H.L. Physiological factors affecting transformation of *Azotobacter vinelandii*, *Journal of Bacteriology 125*, 1080–1087 (1976).
53. Mishra, A.K. and Wyss, O. An adenine-requiring mutant of *Azotobacter vinelandii* blocked in inosinic acid synthesis, *Experentia 21*, 85 (1969).
54. Terzaghi, B.E. A method for the isolation of *Azotobacter* mutants derepressed for Nif, *Journal of General Microbiology 118*, 275–278 (1980).
55. Kennedy, C. and Robson, R.L. Activation of *nif* gene expression in *Azotobacter* by the *nifA* gene product of *Klebsiella pneumoniae*, *Nature* (London) *301*, 626–628 (1983).
56. Miller, J.H. Experiments in molecular genetics. 2nd edition, Cold Spring Harbor Laboratory, Cold Spring Harbor, New York (1972).
57. Wood, A.C., Menezes, E.M., Dykstra, C. and Duggan, D.E. Methods to demonstrate the megaplasmids (or minichromosome) in *Azospirillum*, pp. 18–34, in *Azospirillum Genetics, Physiology, Ecology* (Editor, Klingmüller), Experentia supplementum 42, Birkhaüser, Basel, Boston, Stuttgart (1982).
58. Polsinelli, M., Baldanzi, E., Bazzicalupo, M. and Gallori, E. Transfer of plasmid pRD1 from *Escherichia coli* to *Azospirillum brasilense*, *Molecular and General Genetics 178*, 709–711 (1980).
59. Mishra, A.K., Roy, P. and Bhattacharya, S. Deoxyribonucleic acid-mediated transformation of *Spirillum lipoferum*, *Journal of Bacteriology 137*, 1425–1426 (1979).
60. Magalhaes, I.M.S., Neyra, C.A. and Döbereiner, J. Nitrate and nitrite reductase negative mutants of N_2-fixing *Azospirillum* spp., *Archives of Microbiology 117*, 247–252 (1978).
61. Bani, D., Barberio, C., Bazzicalupo, M., Favilli, F., Gallori, E. and Polsinelli, M. Isolation and characterization of glutamate synthase mutants of *Azospirillum brasilense*, *Journal of General Microbiology 119*, 239–244 (1980).
62. Hartmann, A. Antimetabolite effects on nitrogen metabolism of *Azospirillum* and properties of resistant mutants, pp. 59–68, in *Azospirillum Genetics, Physiology, Ecology* (Editor, W. Klingmüller), Experentia supplementum 42, Birkhaüser, Basel, Boston, Stuttgart (1982).
63. Barberio, C., Bazzicalupo, M., Gallori, E. and Polsinelli, M. Regulation of ammonium assimilation and N_2 fixation in *Azospirillum brasilense*, pp. 47–53, in *Azospirillum Genetics, Physiology, Ecology* (Editor, W. Klingmüller), Experentia supplementum 42, Birkhaüser, Basel, Boston, Stuttgart (1982).

64. Hartmann, A., Singh, M., Klingmüller, W. Isolation and characterization of *Azospirillum* mutants excreting high amounts of indole acetic acid, *Canadian Journal of Microbiology* (in press) (1983).
65. Elmerich, C. and Franche, C. *Azospirillum* genetics: Plasmids, bacteriophages and chromosome mobilization, pp. 9-17, in *Azospirillum Genetics, Physiology, Ecology* (Editor, Klingmüller), Experentia supplementum 42, Birkhaüser, Basel, Boston, Stuttgart (1982).
66. Beringer, J., Beynon, J.L., Buchanan-Vollaston, A.V. and Johnston, A.W.B. Transfer of the drug-resistance transposon Tn*5* to *Rhizobium*, *Nature* (London) *275*, 633-634 (1978).
67. Simon, R., Priefer, U. and Pühler, A. Vector plasmids for *in vivo* and *in vitro* manipulations of Gram negative bacteria, in *Molecular Genetics of the Bacteria-Plant Interactions* (Editor, A. Pühler), Springer Verlag, Berlin, Heidelberg, New York (in press) (1983).
68. Sen, M. and Sen, S.P. Interspecific transformation in *Azotobacter*, *Journal of General Microbiology 41*, 1-6 (1965).
69. Bishop, P.E. and Brill, W.J. Genetic analysis of *Azotobacter vinelandii* mutant strains unable to fix nitrogen, *Journal of Bacteriology 130*, 954-956 (1977).
70. Sen, M., Pal, T.K. and Sen, S.P. Intergeneric transformation between *Rhizobium* and *Azotobacter*, *Antonie van Leeuwenhoek 35*, 533-540 (1969).
71. Bishop, P.E., Dazzo, F.B., Appelbaum, E.R., Maier, R.J. and Brill, W.J. Intergeneric transfer of genes involved in the *Rhizobium* legume symbiosis, *Science 198*, 938-940 (1977).
72. Maier, R.J., Bishop, P.E. and Brill, W.J. Transfer from *Rhizobium japonicum* to *Azotobacter vinelandii* of genes required for nodulation, *Journal of Bacteriology 134*, 1199-1201 (1978).
73. Page, W.J. and Sadoff, H.L. Control of transformation competence in *Azotobacter vinelandii* by nitrogen catabolite derepression, *Journal of Bacteriology 125*, 1088-1095 (1976).
74. Page, W.J. and von Tigerstrom, M. Induction of transformation competence in *Azotobacter vinelandii* iron-limited cultures, *Canadian Journal of Microbiology 24*, 1590-1594 (1978).
75. Page, W.J. and Doran, J.L. Recovery of competence in calcium limited *Azotobacter vinelandii*, *Journal of Bacteriology 146*, 33-40 (1981).
76. Page, W.J. Optimal conditions for induction of competence in nitrogen-fixing *Azotobacter vinelandii*, *Canadian Journal of Microbiology 28*, 389-397 (1982).
77. Page, W.J. and von Tigerstrom, M. Optimal conditions for transformation of *Azotobacter vinelandii*, *Journal of Bacteriology 139*, 1058-1061 (1979).
78. Schenk, S.P., Earhart, C.F. and Wyss, O. A unique envelope protein

in *Azotobacter vinelandii*, *Biochemical and Biophysical Research Communications* 77, 1452–1458 (1977).

79. David, M., Tronchet, M. and Dénarié, J. Transformation of *Azotobacter vinelandii* with plasmids RP4 (IncP-1 group) and RSF1010 (IncQ group), *Journal of Bacteriology 146*, 1154–1157 (1981).
80. Cohen, S.N., Chang, A.C.Y. and Hue, L. Nonchromosomal antibiotic resistance in bacteria: Genetic transformation of *Escherichia coli* by R-factor DNA, *Proceedings of the National Academy of Science* (USA) *69*, 2110–2114 (1972).
81. Datta, N. and Hedges, R.W. Compatibility groups among fi- R factors, *Nature* (London) *234*, 222 (1971).
82. Guerry, P., Van Embden, J. and Falkow, S. Molecular nature of two nonconjugative plasmids carrying drug resistance genes, *Journal of Bacteriology 117*, 619–630 (1974).
83. Duff, J.T. and Wyss, O. Isolation and characterization of a new series of *Azotobacter* bacteriophages, *Journal of General Microbiology 24*, 273–289 (1961).
84. Chuml, V.A., Thompson, B.J., Smiley, B.L. and Warner, R.C. Properties of *Azotobacter* phage PAV-1 and its DNA, *Virology 102*, 262–266 (1960).
85. Monsour, V., Wyss, O. and Kellog, D.S. Junior. A bacteriophage for *Azotobacter*, *Journal of Bacteriology 70*, 486–487 (1955).
86. Bishop, P.E., Supiano, M.A. and Brill, W.J. Technique for isolating phage for *Azotobacter vinelandii*, *Applied Environmental Microbiology 23*, 1007–1008 (1977).
87. Thompson, B.J., Domingo, E. and Warner, R. Properties of *Azotobacter* phage A14 and its DNA, *Virology 56*, 523–531 (1973).
88. Domingo, E., Gordon, N. and Warner, R. *Azotobacter* phages: Properties of phage A12, A21, A31, A41 and their constituent DNAs, *Virology 49*, 439–452 (1972).
89. Bradley, D.E. Ultrastructure of bacteriophages and bacteriocins, *Bacteriological Reviews 31*, 230–314 (1967).
90. Knovicka, J., Pope, L. and Wyss, O. Morphology and nucleic acid composition of *Azotobacter* bacteriophages, *Journal of Virology 10*, 150–152 (1972).
91. Thompson, B.J., Domingo, E. and Warner, R.C. Pseudolysogeny of *Azotobacter* phages, *Virology 102*, 267–277 (1980).
92. Thompson, B.J., Wagner, M.S., Domingo, E. and Warner, R.C. Pseudolysogenic conversion of *Azotobacter vinelandii* by phage A21 and the formation of a stably converted form, *Virology 102*, 278–285 (1980).
93. Wyss, O. and Nimeck, M.W. Interspecific transduction in *Azotobacter*, *Federation Proceedings 21*, 348 (1962).
94. Franco-Lemos, M.W. and Maringolo, V.L. Isolation and characteri-

zation of a bacteriophage for the genus *Azospirillum*, in *Abstracts of the IV International Symposium on Nitrogen Fixation*, Canberra (1980).

95. Elmerich, C., Quiviger, B., Rosenberg, C., Franche, C., Laurent, P. and Döbereiner, J. Characterization of a temperate bacteriophage for *Azospirillum*, *Virology 122*, 29-37 (1982).
96. Ikeda, H. and Tomizawa, J. Prophage Pl an extrachromosomal replication unit, *Cold Spring Harbor Symposium of Quantitative Biology 33*, 791-798 (1968).
97. Heulin, T., Bally, R. and Balandreau, J. Isolation of a very eficient N_2-fixing bacteria from the rhizosphere of rice, pp. 92-99, in *Azospirillum Genetics, Physiology, Ecology* (Editor, W. Klingmüller), Experentia supplementum 42, Birkhaüser, Basel, Boston, Stuttgart (1982).
98. Singh, M. and Wenzel, W. Detection and characterization of plasmids in *Azospirillum*, pp. 44-51, in *Azospirillum Genetics, Physiology, Ecology* (Editor, W. Klingmüller), Experentia supplementum 42, Birkhaüser, Basel, Boston, Stuttgart (1982).
99. Meyers, J.A., Sanchez, D., Elwell, I.P. and Falkow, S. Simple agarose gel electrophoresis method for the identification and characterization of plasmid deoxyribonucleic acid, *Journal of Bacteriology 127*, 1529-1537 (1976).
100. Eckhardt, T. A rapid method for the identification of plasmid DNA in bacteria, *Plasmid 1*, 584-588 (1978).
101. Casse, F., Boucher, C., Julliot, J.S., Michel, M. and Dénarié, J. Identification and characterization of large plasmids in *Rhizobium meliloti* using agarose gel electrophoresis, *Journal of General Microbiology 113*, 229-242 (1979).
102. Banfalvi, Z., Sakanyan, V., Koncz, C., Kiss, A., Dusha, I. and Kondorosi, A. Location of nodulation and nitrogen fixation genes on a high molecular weight plasmid of *R. meliloti, Molecular and General Genetics 184*, 318-325 (1981).
103. Hombrecher, G., Brewin, N.J. and Johnston, A.W.B. Linkage of genes for nitrogenase and nodulation ability on plasmids in *Rhizobium leguminosarum* and *R. phaseoli*, *Molecular and General Genetics 182*, 133-136 (1981).
104. Prakash, R.K., Schilperoort, R.A. and Nuti, M.P. Large plasmids of fast-growing rhizobia: Homology studies and location of structural nitrogen fixation (*nif*) genes, *Journal of Bacteriology 145*, 1129-1136 (1981).
105. Rosenberg, C., Boistard, P., Dénarié, J. and Casse-Delbard, F. Genes controlling early and late functions in symbiosis are located on a megaplasmid in *Rhizobium meliloti*, *Molecular and General Genetics 184*, 326-333 (1981).

106. Uozumi, T., Barraquio, W.L., Wang, P.L., Murai, F., Chung, K.S. and Beppu, T. Plasmids and *nif* genes in *Rhizobia* and nitrogen-fixing bacteria in the rhizosphere of rice, *Proceedings of the Fourth International Symposium on Genetics of Industrial Microorganisms*, Kyoto (1982).

107. Zöphel, G., Steinbauer, J. and Hess, D. *Azospirillum*: Plasmid content and location of the *nif* genes, *First International Symposium on Bacteria-Plant Interactions*, Bielefeld (1980).

108. Robson, R.L. Detection and functions of indigenous plasmids of *Azotobacter*, 91st Ordinary Meeting of the Society for General Microbiology, Cambridge (1981).

109. Yano, K., Anazawa, M., Murai, F., Fukuda, M. and Kuyohara, H. Indigenous plasmids of *Azotobacteraceae* and their functions, in *Proceedings of the Fourth International Symposium on Genetics of Industrial Microorganisms*, Kyoto (1982).

110. Haas, D. and Holloway, B.W. R factor variant with enhanced sex factor activity in *Pseudomonas aeruginosa*, *Molecular and General Genetics 144*, 243-251 (1976).

111. Haas, D. and Holloway, B.W. Chromosome mobilization by the plasmid R68-45: A tool in *Pseudomonas* genetics, *Molecular and General Genetics 158*, 229-237 (1978).

112. Kondorosi, A., Vincze, E., Johnston, A.W.B. and Beringer, J.E. A comparison of three *Rhizobium* linkage maps, *Molecular and General Genetics 178*, 403-408 (1980).

113. Barberio, C., Bazzicalupo, M., Gallori, E. and Polsinelli, M. Studies on genetic recombination in *Azospirillum brasilense*, *Proceedings of the Associazione Italiana de Genetica*, Italie (1982)

114. Quiviger, B., Franche, C., Lutfalla, G., Rice, D., Haselkorn, R. and Elmerich, C. Cloning of a nitrogen fixation (*nif*) gene cluster of *Azospirillum brasilense*, *Biochimie 64*, 495-502 (1982).

115. Ditta, G., Stanfield, S., Corbin, D. and Helsinki, D. Broad host range DNA cloning system for Gram negative bacteria: Construction of a gene bank of *Rhizobium meliloti*, *Proceedings of the National Academy of Science* (USA) 77, 7347-7351 (1980).

116. Bagdasarian, R.L., Lurz, R., Rückert, B., Franklin, F.C.H., Bagdasarian, M.M., Frey, J. and Timmis, K.N. Specific-purpose plasmid cloning vectors: Broad host range, high copy number, RSF1010-derived vectors, and a host vector system for gene cloning in *Pseudomonas*, *Gene 16*, 237-247 (1981).

117. Friedman, A.M., Long, S.R., Brown, S.E., Buikema, W.J. and Ausubel, F.M. Construction of a broad host range cosmid cloning vector and its use in the genetic analysis of *Rhizobium* mutants, *Gene 18*, 289-296 (1982).

118. Singh, M. Transfer of bacteriophage Mu and transposon Tn5 into

Azospirillum, pp. 35–43, in *Azospirillum Genetics, Physiology, Ecology* (Editor, W. Klingmüller), Experentia Supplementum 42, Birkhaüser, Basel, Boston, Stuttgart (1982).

119. Panopoulos, N.J., Staskawicz, B.J. and Sandlin, P. Search for plasmids associated traits and for a cloning vector in *Pseudomonas phaseolicola*, pp. 365–372, in *Plasmid of Medical, Environmental and Commercial Importance* (Editors, K.N. Timmis and A. Pühler), Elsevier North Holland Biochemical Press, Amsterdam (1979).
120. David, M., Vielma, M. and Julliot, J.S. Introduction of IncQ plasmids into *Rhizobium meliloti*. Isolation of a host range mutant of RSF1010 plasmid, *FEMS Microbiology Letters 16*, 335–341 (1983).
121. Buchanan-Wollaston, V., Cannon, M.C., Beynon, J.L. and Cannon, F.C. Role of *nifA* gene product in the regulation of *nif* expression in *Klebsiella pneumoniae*, *Nature* (London) *294*, 776–778 (1981).
122. Hohn, B. and Collins, J. A small cosmid for efficient cloning of large DNA fragments, *Gene 10*, 291–298 (1980).
123. Yates, M.G. and Jones, C.W. Respiration and nitrogen fixation in *Azotobacter*, pp. 97–137, in *Advances in Microbial Physiology*, volume 11 (Editors, A. Rose and D.W. Tempest), Academic Press (1974).
124. Eady, R.R. and Postgate, J.R. Nitrogenase, *Nature* (London) *249*, 805–810 (1974).
125. Mortenson, L.E. and Thorneley, R.N.F. Structure and function of nitrogenase, *Annual Reviews of Biochemistry 48*, 387–418 (1979).
126. Brill, W.J. Biochemical genetics of nitrogen fixation, *Microbiological Reviews 44*, 449–467 (1980).
127. Robson, R.L. and Postgate, J.R. Oxygen and hydrogen in biological nitrogen fixation, *Annual Reviews of Microbiology 34*, 183–207 (1980).
128. Eady, R. regulation of nitrogenase activity, pp. 172–182, in *Current Perspectives in Nitrogen Fixation* (Editors, A.H. Gibson and W.E. Newton), Australian Academy of Science Press, Canberra City (1981).
129. Emerich, D.W. and Burris, R.H. Complementary functioning of the component proteins of nitrogenase from several bacteria, *Journal of Bacteriology 134*, 936–943 (1978).
130. Bulen, W.A. and LeComte, J.R. The nitrogenase system from *Azotobacter*: two enzyme requirement for N_2 reduction, ATP dependent H_2 evolution, and ATP hydrolysis, *Proceedings of the National Academy of Science 56*, 979–986 (1966).
131. Burns, R.C., Holsten, R.D. and Hardy, R.W.F. Isolation and cristallization of the MoFe protein of *Azotobacter* nitrogenase, *Biochemical and Biophysical Research Communications 39*, 90–99 (1970).
132. Shah, V.K. and Brill, W.J. Nitrogenase IV. Simple method of purification to homogeneity of nitrogenase components from *Azotobacter*

vinelandii, Biochimica et Biophysica Acta 305, 445–454 (1973).

133. Swisher, R.H., Landt, M.L. and Reithel, F.J. The molecular weight of, and evidence for two types of subunits in, the molybdenum-iron-protein of *Azotobacter vinelandii* nitrogenase, *Biochemical Journal 163*, 427–432 (1977).
134. Shah, V.K. and Brill, W.J. Isolation of an iron-molybdenum cofactor from nitrogenase, *Proceedings of the National Academy of Sciences* (USA) *74*, 3249–3253 (1977).
135. Nagatani, H.H., Shah, V.K. and Brill, W.J. Activation of inactive nitrogenase by acid treated component I, *Journal of Bacteriology 120*, 697–701 (1974).
136. Nelson, M.J., Levy, M.A. and Orme-Johnson, W.H. Metal and sulfur composition of iron-molybdenum cofactor of nitrogenase, *Proceedings of the National Academy of Sciences* (USA) *80*, 147–150 (1983).
137. Hausinger, R.P. and Howard, J.B. Comparison of the iron proteins from the nitrogen fixation complexes of *Azotobacter vinelandii, Clostridium pasteurianum*, and *Klebsiella pneumoniae, Proceedings of the National Academy of Sciences* (USA) 77, 3826–3830 (1980).
138. Kelly, M. Some properties of purified nitrogenase of *Azotobacter chroococcum, Biochemica et Biophysica Acta 171*, 9–22 (1969).
139. Yates, M.G. and Planqué, K. Nitrogenase from *Azotobacter chrooc occum*: purification and properties of the component proteins, *European Journal of Biochemistry 60*, 467–476 (1970).
140. Ludden, P.W., Okon, Y. and Burris, R.H. The nitrogenase system of *Spirillum lipoferum, Biochemical Journal 173*, 1001–1003 (1978).
141. Ludden, P.W. and Burris, R.H. An activating factor for the iron protein of nitrogenase from *Rhodospirillum rubrum, Science 194*, 424–426 (1976).
142. Oppenheim, J., Fisher, R.J., Wilson, P.W. and Marcus, L. Properties of a soluble nitrogenase in *Azotobacter, Journal of Bacteriology 101*, 292–296 (1970).
143. Haaker, H. and Veeger, C. Involvement of the cytoplasmic membrane in nitrogen fixation by *Azotobacter vinelandii, European Journal of Biochemistry* 77, 1–10 (1977).
144. Oppenheim, J. and Marcus, L. Correlation of ultrastructure in *Azotobacter vinelandii* with nitrogen source of growth, *Journal of Bacteriology 101*, 286–291 (1970).
145. Sherings, G., Haaker, H. and Veeger, L. Regulation of nitrogen fixation by Fe-S protein II in *Azotobacter vinelandii, European Journal of Biochemistry* 77, 621–630 (1977).
146. Robson, R.L. Characterization of an oxygen stable nitrogenase complex isolated from *Azotobacter chroococcum, Biochemical Journal 181*, 569–575 (1979).
147. Lees, H. and Postgate, J.R. The behavior of *Azotobacter chroococ-*

cum in oxygen- and phosphate-limited chemostat culture, *Journal of General Microbiology 75*, 161–166 (1973).

148. Drozd, J. and Postgate, J.R. Effect of oxygen on acetylene reduction, cytochrome content and respiratory activity of *Azotobacter chroococcum, Journal of General Microbiology 63*, 63–73 (1970).

149. Yates, M.G. Control of respiration and nitrogen fixation by oxygen and adenine nucleotides in N_2 grown *Azotobacter chroococcum, Journal of General Microbiology 60*, 393–401 (1970).

150. Gadkari, D. and Stolp, H. Influence of nitrogen source on growth and nitrogenase activity in *Azotobacter vinelandii, Archive of Microbiology 96*, 135–144 (1974).

151. Robson, R.L. O_2 repression of nitrogenase synthesis in *Azotobacter chroococcum, FEMS Microbiology Letters 5*, 259–262 (1979).

152. Hill, S., Kennedy, C., Kavanagh, E., Goldberg, R.B. and Hanau, R. Nitrogen fixation gene (*nifL*) involved in oxygen regulation of nitrogenase synthesis in *K. pneumoniae, Nature* (London) *290*, 424–426 (1981).

153. Ahmad, M.H. Influence of nitrogen on growth, free amino acids and nitrogenase activity in *Spirillum lipoferum, Journal of General Applied Microbiology 24*, 1271–1278 (1978).

154. Volpon, A.G.T., De-Polli, H. and Döbereiner, J. Physiology of nitrogen fixation in *Azospirillum lipoferum* Br17 (ATCC29709), *Archive of Microbiology 128*, 371–375 (1981).

155. Dalton, H. and Postgate, J.R. Growth and physiology of *Azotobacter chroococcum* in continuous culture, *Journal of General Microbiology 56*, 307–319 (1969).

156. Gordon, J.K. and Brill, W.J. Derepression of nitrogenase synthesis in the presence of excess NH_4^+, *Biochemical Biophysical Research Communications 59*, 967–971 (1974).

157. Shah, V.K., Davis, L.C. and Brill, W.J. Nitrogenase I. Repression and derepression of the iron molybdenum and iron proteins of nitrogenase in *Azotobacter vinelandii, Biochimica et Biophysica Acta 256*, 498–511 (1972).

158. Davis, L.C., Shah, V.K., Brill, W.J. and Orme-Johnson, W.H. Nitrogenase II. Changes in the EPR signal of component I (Iron-molybdenum protein) of *Azotobacter vinelandii* during repression and derepression, *Biochimica et Biophysica Acta 256*, 512–523 (1972).

159. St. John, R.T. and Brill, W.J. Inhibitory effect of methylalanine on glucose grown *Azotobacter vinelandii, Biochimica et Biophysica Acta 261*, 63–69 (1972).

160. Drozd, J.W., Tubb, R.S. and Postgate, J.R. A chemostat study of the effect of fixed nitrogen sources on nitrogen fixation, membranes and free amino acids in *Azotobacter chroococcum, Journal of General Microbiology 73*, 221–232 (1972).

161. Scott, D.B., Scott, C.A. and Döbereiner, J. Nitrogenase activity and nitrate respiration in *Azospirillum* spp., *Archive of Microbiology 121*, 141-145 (1979).
162. Bothe, H., Klein, B., Stephan, M.P. and Döbereiner, J. Transformations of inorganic nitrogen by *Azospirillum* spp., *Archive of Microbiology 130*, 96-100 (1981).
163. Elmerich, C., Gauthier, D. and Houmard, J. Regulation of nitrogenase biosynthesis in *Azospirillum brasiliensis*, in *Proceedings of the Steenbock-Kettering International Symposium on Nitrogen Fixation*, Madison (1978).
164. Gordon, J.K., Shah, V.K. and Brill, W.J. Feedback inhibition of nitrogenase, *Journal of Bacteriology 148*, 884-888 (1981).
165. Laane, C., Krone, W., Konings, W., Haaker, H. and Veeger, C. Short term effect of ammonium chloride on nitrogen fixation by *Azotobacter vinelandii* and by bacteroids of *Rhizobium leguminosarum*, *European Journal of Biochemistry 103*, 39-46 (1980).
166. Zumft, W.G., Alef, K. and Mümmler, S. Regulation of nitrogenase activity in Rhodospirillaceae, pp. 190-193, in *Current Perspective in Nitrogen Fixation* (Editors, A.H. Gibson and W.E. Newton), Australian Academy of Science Press, Canberra City (1981).
167. Hartmann, A. and Kleiner, D. Ammonium (methylammonium) transport by *Azospirillum* spp. *FEMS Microbiology Letters 15*, 65-67 (1982).
168. Gordon, J.K. and More, R.A. Ammonium and methylammonium transport by the nitrogen fixing bacterium *Azotobacter vinelandii*, *Journal of Bacteriology 148*, 435-442 (1981).
169. Nagatani, H.H. and Brill, W.J. The effect of Mo, W and V on the synthesis of nitrogenase components in *Azotobacter vinelandii*, *Biochimica et Biophysica Acta 362*, 160-166 (1974).
170. Pienkos, P.T., Klevickis, S. and Brill, W.J. *In vitro* activation of inactive nitrogenase component I with molybdate, *Journal of Bacteriology 145*, 248-256 (1981).
171. Pienkos, P.T. and Brill, W.J. Molybdenum accumulation and storage in *Klebsiella pneumoniae* and *Azotobacter vinelandii*, *Journal of Bacteriology 145*, 743-751 (1981).
172. Bishop, P.E., Jarlenski, D.M.L. and Hetherington, D.R. Evidence for an alternative nitrogen fixation system in *Azotobacter vinelandii*, *Proceedings of the National Academy of Sciences* 77, 7342-7346 (1980).
173. Shah, V.K., Davis, L.C., Gordon, J.K., Orme-Johnson, W.H. and Brill, W.J. Nitrogenase III. Nitrogenaseless mutants of *Azotobacter vinelandii*: Activities, cross-reactions and EPR spectra, *Biochimica et Biophysica Acta 292*, 246-255 (1973).
174. Bishop, P.E., Jarlenski, D.M.L. and Hetherington, D.R. Expression

of an alternative nitrogen fixation system in *Azotobacter vinelandii*, *Journal of Bacteriology 150*, 1244–1251 (1982).

175. Kennedy, C. Linkage map of the nitrogen fixation (*nif*) genes in *Klebsiella pneumoniae*, *Molecular and General Genetics 157*, 199–204 (1977).
176. Dixon, R., Cannon, F.C. and Kondorosi, A. Construction of a P plasmid carrying nitrogen fixation genes from *Klebsiella pneumoniae*, *Nature* (London) *260*, 268–271 (1976).
177. Kennedy, C., Cannon, F., Cannon, M., Dixon, R., Hill, S., Jensen, J., Kumar, S., McLean, P., Merrick, M., Robson, R. and Postgate, J. Recent advances in the genetics and regulation of nitrogen fixation, pp. 146–156, in *Current Perspectives in Nitrogen Fixation* (Editors, A.H. Gibson and W.E. Newton), Australian Academy of Sciences Press, Canberra City (1981).
178. MacNeil, T., MacNeil, D., Roberts, G.P., Supiano, M.A. and Brill, W.J. Fine structure mapping and complementation analysis of *nif* (nitrogen fixation) genes in *Klebsiella pneumoniae*, *Journal of Bacteriology 136*, 253–266 (1978).
179. Merrick, M., Filser, M., Dixon, R., Elmerich, C., Sibold, L. and Houmard, J. The use of translocatable elements to construct a fine structure map of the *Klebsiella pneumoniae* nitrogen fixation (*nif*) cluster, *Journal of General Microbiology 117*, 509–520 (1980).
180. Pühler, A. and Klipp, W. Fine structure analysis of the gene region for N_2-fixation (*nif*) of *Klebsiella pneumoniae*, pp. 276–286, in *Biology of inorganic nitrogen and sulfur* (Editors, H. Bothe and A. Trebst), Springer Verlag, Berlin, Heidelberg, New York (1982).
181. Sibold, L. The polar effect on *nifM* of mutations in the *nifU*, *-S*, *-V* genes of *Klebsiella pneumoniae* depends on their plasmid or chromosomal location, *Molecular and General Genetics 186*, 569–571 (1982).
182. Cannon, F.C., Riedel, G.E. and Ausubel, F.M. Overlapping sequences of *K. pneumoniae nif* DNA cloned and characterized, *Molecular and General Genetics 174*, 59–66 (1979).
183. Ruvkun, G.B. and Ausubel, F.M. Interspecies homology of nitrogenase genes, *Proceedings of the National Academy of Sciences* (USA) 77, 191–195 (1980).
184. Mazur, B., Rice, D. and Haselkorn, R. Identification of blue green algal nitrogen fixation genes by using heterologous DNA hybridization probes, *Proceedings of the National Academy of Sciences* (USA) 77, 186–190 (1980).
185. Rice, D.G., Mazur, B.J. and Haselkorn, R. Isolation and physical mapping of nitrogen fixation genes from the cyanobacterium *Anabaena* 7120, *Journal of Biological Chemistry 257*, 13157–13163 (1982).
186. Quinto, C., de la Vega, H., Flores, M., Fernandez, L., Ballado, T., Soberon, G. and Palacios, R. Nitrogen fixation genes are reiterated in

Rhizobium phaseoli, *Nature* (London) *299*, 724-726 (1982).

187. Bishop, P.E. and Bott, K.F. Molecular cloning of *nif* DNA from *Azotobacter vinelandii*, *Journal of Bacteriology* (in press) (1983).

188. Krol, A.J.M., Hontelez, J.G.J., Roozendaal, B. and van Kamen, A. On the operon structure of the nitrogenase genes of *Rhizobium leguminosarum* and *Azotobacter vinelandii*, *Nucleic Acids Research 10*, 4147-4157 (1981).

189. Kaluza, K. and Hennecke, H. The nitrogenase genes of *Klebsiella pneumoniae* are transcribed into a single polycistronic RNA subsequent to *nifLA* expression, *FEMS Microbiology Letters 15*, 57-60 (1980).

190. Brill, W.J., Westphal, J., Stieghorst, M., Davis, L.C. and Shah, V.K. Detection of nitrogenase components and other nonheme iron proteins in polyacrylamide gels, *Analytical Biochemistry 60*, 237-241 (1974).

191. Gordon, J.K. and Brill, W.J. Mutants that produce nitrogenase in the presence of ammonia, *Proceedings of the National Academy of Sciences* (USA) *69*, 3501-3503 (1972).

192. Shah, V.K., Davis, L.C., Stieghorst, M. and Brill, W.J. Mutant of *Azotobacter vinelandii* that hyperproduces nitrogenase component II, *Journal of Bacteriology 117*, 917-919 (1974).

193. Benemann, J.R. and Valentine, R.C. Pathways of dinitrogen fixation, pp. 59-104, in *Advances in Microbial Physiology*, Volume 8 (Editors, A. Rose and D.W. Tempest), Academic Press (1972).

194. Sorger, G.J. Regulation of nitrogen fixation in *Azotobacter vinelandii* OP and in an apparently partially constitutive mutant, *Journal of Bacteriology 95*, 1721-1726 (1968).

195. Gordon, J.K. and Jacobson, M.R. Isolation and characterization of *Azotobacter vinelandii* mutant strains with potential as bacterial fertilizer, *Canadian Journal of Microbiology* (in press) (1983).

196. Nagatani, H., Shimizu, M. and Valentine, R.C. The mechanism of ammonia assimilation in nitrogen fixing bacteria, *Archive of Microbiology 79*, 164-175 (1971).

197. Streicher, S.L., Shanmugam, K.T., Ausubel, F., Morandi, C. and Goldberg, R.B. Regulation of nitrogen fixation in *Klebsiella pneumoniae*: Evidence for a role of glutamine synthetase as regulator of nitrogenase synthesis, *Journal of Bacteriology 120*, 815-821 (1979).

198. Leonardo, J.M. and Goldberg, R.B. Regulation of nitrogen metabolism in glutamine auxotrophs of *Klebsiella pneumoniae*, *Journal of Bacteriology 142*, 99-110 (1980).

199. Espin, G., Alvarez-Morales, A. and Merrick, M. Complementation analysis of *glnA* linked mutations which affect nitrogen fixation in *Klebsiella pneumoniae*, *Molecular and General Genetics 184*, 213-217 (1981).

200. Sibold, L. and Elmerich, C. Constitutive expression of nitrogen fixation (*nif*) genes of *Klebsiella pneumoniae* due to a DNA duplication, *EMBO Journal 1*, 1551-1558 (1982).
201. Merrick, M.J. A new model for nitrogen control, *Nature* (London) *267*, 362-363 (1982).
202. Garcia, E., Bancroft, S., Rhee, S.G. and Kustu, S. The product of a newly identified gene, *glnF*, is required for the synthesis of glutamine synthetase in *Salmonella*, *Proceedings of the National Academy of Sciences* (USA) *74*, 1662-1666 (1977).
203. McFarland, N., McCarter, L., Artz, S. and Kustu, S. Nitrogen regulatory locus *glnR* of enteric bacteria is composed of cistrons *ntrB* and *ntrC*: Identification of their protein products, *Proceedings of the National Academy of Sciences* (USA) *78*, 2135-2139 (1981).
204. Espin, G., Alvarez-Morales, A., Cannon, F., Dixon, R. and Merrick, M. Cloning of the *glnA*, *ntrB*, *ntrC* genes of *Klebsiella pneumoniae* and studies of their role in regulation of the nitrogen fixation (*nif*) gene cluster, *Molecular and General Genetics 186*, 518-524 (1982).
205. Cannon, F.C. and Postgate, J.R. Expression of *Klebsiella* nitrogen fixation genes (*nif*) in *Azotobacter*, *Nature* (London) *260*, 271-272 (1976).
206. Sundaresan, V., Jones, J.D.G., Ow, D.W. and Ausubel, F.M. *Klebsiella pneumoniae nifA* product activates the *Rhizobium meliloti* nitrogenase promoter, *Nature* (London) *301*, 728-732 (1983).

Index